Water and Wastewater Management

Global Problems and Measures

Series Editors

Müfit Bahadir, Institut für Ökologische und Nachhaltige Chemie, Technische Universität Braunschweig, Braunschweig, Germany

Andreas Haarstrick, Leichtweiss-Institut für Wasserbau, Exceed, Technische Universität Braunschweig, Braunschweig, Germany

Water and wastewater management are among the greatest challenges of our century and the challenges posed by climate change will become even greater. Unfortunately, however, most efforts, especially in developing countries but also in the so-called developed countries, have been less than optimal or not optimal at all. In particular, there are still too many people who have to live without clean water and decent sanitation. Today, 2.2 billion people lack access to safely managed drinking water and wastewater, and 4.2 billion people lack safely managed sanitation services. The question, why this is the case - especially in developing countries - as well as other urgent water and wastewater management issues, are discussed in this book-series. Contributions therein present in more detail critical reviews, discussions, and analysis of the water and wastewater situation and management aspects in different regions and countries worldwide.

More information about this series at https://link.springer.com/bookseries/16756

Müfit Bahadir • Andreas Haarstrick
Editors

Water and Wastewater Management

Global Problems and Measures

Springer

Editors
Müfit Bahadir
Institut für Ökologische und
Nachhaltige Chemie
Technische Universität Braunschweig
Braunschweig, Germany

Andreas Haarstrick
Leichtweiss-Institut für
Wasserbau, Exceed
Technische Universität Braunschweig
Braunschweig, Germany

ISSN 2731-3166 ISSN 2731-3174 (electronic)
Water and Wastewater Management
Global Problems and Measures
ISBN 978-3-030-95287-7 ISBN 978-3-030-95288-4 (eBook)
https://doi.org/10.1007/978-3-030-95288-4

This Springer imprint is published by the registered company Springer Nature Switzerland AG
The registered company address is: Gewerbestrasse 11, 6330 Cham, Switzerland

Preface

Without doubt, water and wastewater management are among the greatest challenges of our century, and there is also no doubt that the challenges posed by climate change will become even greater.

Unfortunately, however, most efforts, especially not only in developing countries but also in the so-called developed countries, have been less than optimal or not optimal at all. In particular, there are still too many people who have to live without clean water and decent sanitation. According to UN-Water (www.unwater.org/water-facts/), today, 2.2 billion people lack access to safely managed drinking water and wastewater and 4.2 billion people lack safely managed sanitation services. The question of why this is so, and why in many cases in developing countries, can be answered in large part by the fact that in these countries political systems prevail that are composed of extractive political institutions and extractive economic action. However, the scarcely practiced conservation of resources, which is to be considered in this context, can also be traced back to industrialised countries. The currently prevailing economic action does not prioritise sustainability and resource conservation in the agenda. There is definitely a need to put one's own house in order.

Governments have the responsibility for many governance functions, such as formulating policy, developing legal frameworks, planning, coordination, funding and financing, capacity development, data acquisition and monitoring, and regulation. In this context, good water governance comprises many elements, but it principally includes effective, responsive, and accountable state institutions that respond to change, openness, and transparency providing stakeholders with information, and giving citizens and communities a say and role in decision-making; this is the framework of an inclusive political and economic system. At this point, the participation and multi-stakeholder engagement are important parts of policy processes, although measuring their effectiveness is still in its infancy. The importance of having a transparent, universal, and neutral platform for government and citizen groups in place to mobilise available resources and seek alternative means of ensuring improved water services and sanitation has proven to be essential and complementary to local government support.

Unfortunately, an acute lack of capacity is constraining water resource development and management in all its facets across most developing countries, particularly not only in Sub-Saharan Africa and South and South-Eastern Asia but also in Latin America and Middle East North Africa.

Human resource shortages are reported in all key areas, including agriculture and irrigated farming, water-related risk management, water and sanitation services, wastewater treatment, recycling and reuse technologies, and desalination.

Another aspect that plays a decisive role besides the problems mentioned so far is the question of education and capacity building. An aspect, by the way, that will also be taken up in a book series, which will follow after this book. For the time being, it should be emphasised at this point that if not only the respective developing countries but also the development policy strategies of the rich industrial nations do not change their direction towards immense education and capacity building measures, then all efforts and invested money will continue to be wasted and many of the SDGs will fail. It is not for no reason that SDG 4.c supports this by calling for *Substantially increasing the supply of qualified teachers, including through international cooperation for teacher training in developing countries, especially least-developed countries and small-island developing states*.

The presented book addresses the situation of water and wastewater management from a global angle, underpinned by selected case studies. As mentioned, the publication of this book will also be the start of a book series that in more detail critically reviews, discusses, and analyses the water and wastewater situation and management in different regions and countries worldwide.

Braunschweig, Germany

Andreas Haarstrick
Müfit Bahadir

About This Book

The book addresses the situation of water and wastewater management from a global angle, underpinned by selected case studies. The publication of this book will also be the start of a book series that in more detail critically reviews, discusses, and analyses the water and wastewater situation and management in different regions and countries worldwide. Further, the book provides a useful resource for scientists, researchers, and practitioners dealing with water and wastewater management.

Contents

About the Editors

Müfit Bahadir studied chemistry at the Free University Berlin and Bonn University in Germany, received his Ph.D. in 1975 from Bonn University and his Assoc. Prof. 1988 from Munich Technical University. He became Full Professor of Environmental and Sustainable Chemistry at the Technische Universität Braunschweig in 1989. Since 1997, he is an honorary doctor and visiting professor at the Selcuk University Konya in Turkey. His research fields cover environmental chemistry and analyses, environmental pollution through industrial processes and products, pesticide chemistry and metabolism in soil and water, ecotoxicology, sustainable chemistry, renewable feed stocks, and biodiesel and bio-lubricants. During the last 20 years, his research focused on R&D of sustainable water and wastewater management in developing countries. He retired in 2016.

Andreas Haarstrick has studied Chemistry at the Technische Universität Braunschweig (1983–1989) and received his doctorate in 1992 in the field of biotechnological up- and downstreaming processes of biopolymers. Since 2006, he is Professor for Bioprocess Engineering at the TU Braunschweig. His teaching and research cover modelling biological and chemical processes in heterogeneous systems, development of models predicting pollutant reduction in and emission behaviour of landfills, growth kinetics at low substrate concentrations under changing environmental conditions, Advanced Oxidation Processes (AOP), and groundwater management. Since 2012, he is the Managing Director of the DAAD exceed-Swindon Project dealing with sustainable water management in developing countries.

Part I
Introduction

Water and its Global Meaning

1

Andreas Haarstrick and Müfit Bahadir

Abstract

In times of climate change, the question of safe and sustainable use of freshwater resources is becoming increasingly urgent, and in some regions around the equator belt, even more urgent. It is imperative that water management is reformed towards sustainability and that the awareness of both the responsible governments and the population is raised even more than is the case at present. A "business as usual" attitude will even more jeopardise human lives and lead to environmental destruction than is already the case. We have the choice! It is evident that increasing water demand follows population growth, economic development and changing consumption patterns. Global water demand has increased by 600% over the past 100 years and will grow significantly over the next two decades in all the three components, industry, domestic, and agriculture. Industrial and domestic demand will grow faster than agricultural demand but demand for agriculture will remain the largest—Global water demand for all uses, presently about 4,600 km^3 per year, will increase by 20–30% by 2050, up to almost 5,500–6,000 km^3 per year. By 2040/50 the global population will increase to between 9.4 and 10.2 billion people and most of the population growth will occur in Africa. This chapter addresses the most pressing issues of water resource sustainability and puts a finger on the wound that, if left untreated, will become an uncontrollable inflammatory problem.

Keywords

Water crisis · Water management · Water scarcity

1.1 Water on Earth

About 4.5 billion years ago, our planet Earth was formed. Since then, natural forces have shaped our planet. Water became the most important factor of life. Not only the origin of life, but also the human civilization depends on water. The earliest ancient civilizations developed at river basins. Bricks were produced from water and soil for building homes. Distribution and consumption of water were the triggers of formation tribes and states. Settlements at oceans were able to

A. Haarstrick (✉)
Leichtweiss-Institut für Wasserbau, Exceed, Technische Universität Braunschweig, Braunschweig, Germany
e-mail: a.haarstrick@tu-bs.de

M. Bahadir
Institut für Ökologische und Nachhaltige Chemie, Technische Universität Braunschweig, Braunschweig, Germany

M. Bahadir and A. Haarstrick (eds.), *Water and Wastewater Management*, Water and Wastewater Management, https://doi.org/10.1007/978-3-030-95288-4_1

develop to world metropoles. But today, these metropoles are severely endangered due to sea level rise. Meanwhile, big cities are built in deserts. They can only exist due to water transported from long distances of hundreds of kilometers. Nowadays, water has become scarce and is going to be much scarcer with the growing world population. Human being shaped the earth that significantly calling the period "*Anthropocene*"—the age of men.

Water is a versatile stuff. The total amount of it on earth remains constant. It takes just different aggregate states like liquid, solid and vapor (gaseous). Due to this versatility, the hydrologic cycle is triggered, continuously generating freshwater. Each drop of freshwater we drink is recycled numerously over the time. To the best of our knowledge, our earth is the sole planet in the universe having liquid water. But it is still controversial, where the water on earth came from. Water is assumed to be imported from comets, consisting of frozen water and dust that hit our planet in earth history in great number.

Through solar irradiation water evaporates from water surfaces (oceans, lakes) and forms clouds in the atmosphere through condensation at deeper temperatures since cold air can store less amounts of water than warm air. It starts raining. The precipitation is taken up by plants and evaporated over the leaves again. The same with soil surfaces. The water cycle runs continuously. The large forest areas of equatorial belt in Latin America, Africa, South Asia, Oceania, etc. are called "rain forests", being mostly responsible for this phenomenon. Plants take up water from the soil through their roots, transport it through the stems to the leaves and evaporate it again—called evapotranspiration. The vapor condenses in the atmosphere forming clouds and raining and precipitating again. This way, the rain forests organize their water demand by them own. Without the plants, water would soak away to the aquifer or run off to the oceans.

The total volume of water on our planet remains always the same—about 1.4 billion km^3. This is a huge amount of water. But just less than 3% of it is non-saline (fresh) water that is needed for most living species. 2% of freshwater is stored in polar and glacier ice, which is not available for human consumption. Just less than 1% of the total water amount on earth can be used as drinking water and water for different human activities like agriculture, industry, and public water supply. Almost the half of this 1% of water is stored partly in deep aquifers and is renewed rather slowly. This groundwater is pumped to surface and available for human consumption. But at the same time, the groundwater reserves deplete. Today, roughly 2/3 of freshwater is used for irrigation in agriculture, followed by the industry and public supply (each approx. 1/6 of freshwater amount). Agriculture is the greatest consumer of freshwater worldwide. It is estimated that almost one billion people worldwide do not have access to clean drinking water, whilst around one third of mankind does not have suitable sanitary facilities or wastewater treatment. This situation becomes worsened through the future scenario of climate change particularly in already drought regions of our planet.

Water is essential for all major socio-economic sectors, contributing to each of them in a different way. For instance, agriculture requires large quantities of water for irrigation and food production. Energy requires water for powering turbines, cooling power plants, and growing biofuels. Access to safe water supply and basic sanitation is necessary for maintaining public health. Water is needed to keep the ecosystems healthy, which in turn provide crucial environmental goods and services. The benefits from each of these sectors are provided through water. Managing water for all is not only a question of availability of resources and money, but equally a matter of public participation and good governance. Water is a local issue and involves numerous stakeholders at basin, municipal, regional, national, and international levels. If effective public governance is missed to manage interdependencies across policy areas and between levels of governmental bodies, policymakers will face obstacles designing and implementing measures for sustainable water management. Even mitigating the impacts of natural disasters through extreme events following climate change like flash floods and inundations, recently occurring almost every year, is not only a problem of Developing Countries, but also of the developed world.

Many water resources have trans-boundary character. Rivers of Nile, Euphrates and Tigris, and Jordan in the Middle East, but also Rivers Elbe and Rhine in the heart of Europe deserve joint solutions of the riparian countries. Basin wide cooperation is essential for sustainable water management for prosperity and peace in the respective regions. Many projects in development cooperation fail for ignoring the importance of socio-cultural aspects, and solely concentrating on technological problems. The end of the active involvement of experts from industrialized countries often leads to failure in the implementation stage of projects. Especially, projects related to water supply and use gain little recognition if the various political, cultural, and social meanings of water in different regions are not considered.

1.2 The Presence and Future of Water

The next time one opens a bottle of water one should think about where the water used comes from and under what conditions it was extracted. Indian beverage cans contain treated rainwater, while the water for soft drinks in the Maldives is obtained from seawater. In most cases, however, drinking water is still obtained from surface and groundwater reservoirs. It should be emphasized, however, that the above examples are no longer exceptions, and more and more water has to come from different sources to meet demand. There is a reason for this: we are at the beginning of a global freshwater crisis.

Given that 70% of the Earth's surface is covered with water, and that volume remains constant (at 1,386,000,000 km^3; Fig. 1.1), how is a water shortage even possible? Well, it must be remembered that 97.5% of all water is seawater that is unsuitable for direct human consumption. And both populations and temperatures are ever rising, meaning that the freshwater we do have is under severe pressure.

Global water demand is expected to increase by 55% by 2050. Much of the demand will be used in agriculture. Demand accounts for more than 70% of global freshwater consumption. To feed the growing population, food production will have to increase by 69% by 2035. The same is true for water withdrawals for energy. Here, a decline of more than 20% is expected due to the likewise increasing demand [1]. With this development, it once again becomes quite clear that a global water crisis will certainly occur, if intelligent measures for the sustainable use of drinking water resources are not initiated as soon as possible.

Mexico City, which was built on old lake beds, is experiencing an average subsidence of 30–40 cm/a in some areas [2, 3]. The reason is that the aquifer beneath the city has been depleted. Once horizontal streets now consist of hills and valleys. Meanwhile, the city must import 40% of its fresh water from 2000 m lower regions (partly also from seawater desalination plants) and has become more extremely vulnerable than ever before [4].

This situation is also comparable to California (USA). The state experienced the most intense drought periods from 2012–2014 compared to those reconstructed over the past 1200 years according to the Palmer Drought Severity Index (PDSI). The large aquifer volumes declined at a rate of almost 20 bill. m^3/a during this time [5]. Even subsequent extreme precipitation was unable to fully recharge the reservoirs. Groundwater reservoirs generally have a regeneration time of up to 300 years.

What else could a global freshwater shortage lead to? One gloomy theory refers to the possibility of armed water conflicts. Particularly vulnerable regions in this respect are the Middle East and the Arabian Peninsula [6]. Another theory deals with climate change and the spread of deserts. Devastating examples of this can be found in north-eastern parts of China, in India and in Sub-Saharan Africa [6].

But it is worth looking at some bright spots. Some nations have found noteworthy solutions—Australia, for example. Australia survived a "millennium drought" that lasted from 1997 to

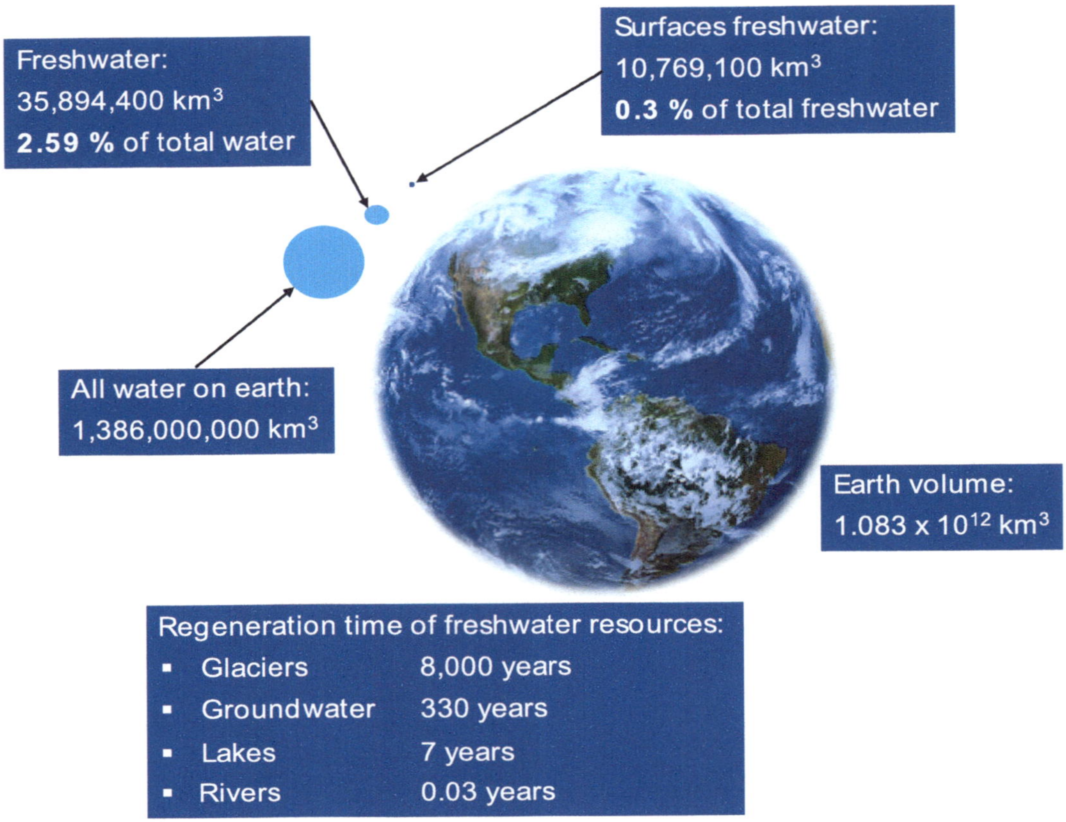

Fig. 1.1 Water on earth

2009. The response was a rapid implementation of measures that halved water consumption by businesses and households. This was achieved, among other things, by introducing a price for water that made it a tradable commodity [7]. This worked well, but one must not forget that water trading involves many dangers, especially on the stock exchange. Despite successful regulation of use through price, it must be ensured that fresh water supplies are subject to equitable distribution. The right to water must have a strong legal and ethical basis that is not negotiable.

Another notable example is Israel, which considers water availability a matter of national security. By recycling wastewater, including domestic sewage, the Shafdan Wastewater Treatment Facility near Tel Aviv, for example, provides about 140,000,000 m^3/a of water for agricultural use. Thus, over 87% of Israel's agricultural water needs are now met by wastewater [8]. Spain as one of the biggest European exporters of vegetables and fruits, for example, manages only 19%. However, it should be noted, that Israel is pursuing a conflict-laden policy in the Middle East regarding freshwater resources, which is one of the most controversial sets of issues in this region [9].

Can desalination of seawater be an effective solution to meet the growing demand for water in industry and agriculture and, moreover, avoid political crises? Currently, desalination via reverse osmosis is highly costly and maybe also for the next decades only a thinkable option for rich countries. The capital costs are presently going to be higher than a treatment plant to treat freshwater. However, the further development of renewable energy and related technologies may contribute to reduce those costs making desalinated water affordable and contribute to ease political tensions.

A simpler, cheaper, and quick-to-implement solution is rainwater harvesting. In Melbourne, Australia, one of the largest rainwater harvesting tanks can hold four million liters [10]. Authorities such as Kerala, Bermuda and the Virgin Islands have mandated that all new buildings be equipped with a rainwater harvesting system. For Malaysia/Singapore, it has been calculated that up to 75% of domestic water demand can be met by rainwater harvesting [11].

As earlier mentioned, around the world, more than 70% of freshwater is used for irrigation. Moreover, in many cases, the agricultural irrigation techniques are inefficient. Further, with respect to excessive freshwater use, thermal power plants (nuclear, coal, natural gas) require vast amounts of water for cooling. Renewables for the most part—solar and wind—do not! The faltering shift to renewables has much to do with heavy-handed, lobby-driven government policies. Bold and consistent action is needed to incentivize society and the economy and to support smart investments.

To achieve water-efficient societies, there are undoubtedly many ways, in which this can be done. In fact, it should be quite simple, either by increasing the efficiency, with which every drop of water is used, or simply by moving away from water-intensive uses and increasing the use of appropriate environmental technologies with sustainability credentials.

Closely linked with the question of the future fate of freshwater resources is the debate on sustainability. In 1987, Gro Harlem Brundtland (former Norwegian Prime Minister) in her report popularized the term "sustainability" by referencing critical environmental and development problems on global level. In this context, she stressed the following items [12]:

- Establishment of a way of acting that considers the needs of current and future generations without any compromises;
- Equal distribution of natural resources amongst users in an area shall not only spatially happen but also be temporally during the time of usage.

This approach sounds like a "wonderful" idea. But is it realistic? In this debate, one has to consider three main issues. The first one is an inherent problem that deals with the implementation of conceptual ideas and acceptance by majority society. The second issue raises the question how it could become successful. The answer may only result from changes, which must be obtained in a way that the total dependence on earth's resources deeply penetrate human consciousness. The third issue relates to basic requirements: An all-over accepted definition and manifestation of what is really needed, and an adaptation to sustainable action in daily routine.

The World Water Council says: "*There is a water crisis today. But the crisis is not about having too little water to satisfy our needs. It is a crisis of managing water so badly that billions of people—and the environment—suffer badly.*" [13]. It is still possible to take corrective measures to prevent the water crisis from worsening. However, what is depressing and sad is that despite the growing awareness that our freshwater resources are limited and need to be protected in terms of both quantity and quality, nothing significant is changing.

Regardless of the use of freshwater (agriculture, industry, household), huge water savings and improved water management are possible. Water is wasted almost everywhere, and if people do not face water scarcity, they take access to water for granted and as a natural thing. With population growth, urbanization and advancing industrialization, water consumption is bound to increase. In order not to fall further into a massive crisis, several measures need to be taken to increase the proportion of people with sustainable access to safe drinking water and sanitation:

- Guarantee the right to water;
- Decentralize the responsibility for water;
- Develop know-how at local level;
- Increase and improve financing;
- Evaluate and monitor water resources.

Urgent action needs to be taken about agriculture, transboundary cooperation, resource conservation, and ecological diversity. The increasing demand for water by humans not only reduces the amount of fresh water, but also has a profound impact on aquatic ecosystems and the species that depend on them. There is a danger that ecological balances will be massively disturbed and thus no longer be able to serve their regulating role.

The UN Water Report 2019 [6] says that global water demand is expected to continue increasing at a similar rate until 2050, accounting for an increase of 20–30% above the current level of water use, mainly due to rising demand in the industrial and domestic sectors. Over 2 billion people live in countries experiencing high water stress, and about 4 billion people experience severe water scarcity during at least one month a year. Stress levels will continue to increase as demand for water grows and the effects of climate change intensify. Moreover, these processes are intensified by changing climate conditions and the increase of extreme weather events (Table 1.1).

Water-related natural hazards such as floods and droughts affect already water supply and sanitation infrastructure, leading to significant economic and social losses and impacts. It can already be observed that such hazards are increasing in frequency and intensity because of climate change. Short- and long-term impacts of water-related extreme events also include the spread of communicable diseases, disruptions to water and food supplies, and the damage to financial assets and social disruption.

1.3 Water and Wastewater Management in the Context of Sustainable Development Goals

The Agenda 2030 of the United Nations established 17 Sustainable Development Goals (SDGs) and 169 global targets, related to development outcomes and means of implementation for the period 2015–2030. The SDGs were designed to be integrated and indivisible to balance the social, economic, and environmental dimensions of sustainable development [14].

The anchoring of SDG 6 in the 2030 Agenda, "*Ensure availability and sustainable management of water and sanitation for all*", reflects the most pressing issues in water and sanitation. The Agenda addresses rising inequalities, depletion of natural resources, environmental degradation and climate change as the greatest challenges of the present. It points out that social development and economic prosperity depend on sustainable management of freshwater resources and ecosystems, and it emphasizes the integrated nature of the SDGs.

Water-related ecosystems and the environment have always provided natural sites for human settlements and civilizations, bringing benefits such as transportation, natural purification, irrigation, flood protection and habitats for

Table 1.1 Average annual impacts of inadequate drinking water and sanitation services, water-related disasters, epidemics, earthquakes, and conflicts

	Water-related			Others	
	Inadequate water and sanitation	Drought	Flooding	Earthquakes and epidemics	Conflicts
People affected during a period of emergency	No data	55 million	106 million	6 million	65 million
People killed	780,000 by infection deseases	1100	6000	56,000	75,000 war deaths
Economic damage	No data	5.4 billion USD	31.4 billion USD	30 billion USD	No data

biodiversity. However, population growth, agricultural intensification, urbanization, industrial production and pollution, and climate change are already overwhelming and undermining nature's ability to provide key functions and services. Consequently, poor, and marginalized populations will be disproportionately affected, further exacerbating rising inequalities.

The use of freshwater in agriculture, industry and households produces highly polluted wastewater that pollutes freshwater resources. In many countries, much of this wastewater is still discharged into natural waters and freshwater areas without any treatment.

Against this background, the accomplishment of SDG 6 is highly challenging. The challenge is seen above all in the governments' decision-making and prioritisation processes. Each government must decide how to incorporate them into national planning processes, policies and strategies based on national realities, capacities, levels of development and priorities. In best case, they should cover the entire water cycle including provision of drinking water (SDG target 6.1), sanitation and hygiene services (6.2), treatment and reuse of wastewater and ambient water quality (6.3), water-use efficiency and scarcity (6.4), Integrated Water Resources Management (IWRM) also through transboundary cooperation (6.5), protecting and restoring water-related ecosystems (6.6), international cooperation and capacity-building (6.a), and participation in water and sanitation management (6.b). The sub-targets mentioned here, and indicators linked to them are far from perfect. However, they provide an unprecedented basis for working systematically on the future of water security worldwide [15].

The implementation of good water governance and sustainable water management depends on the participation of a range of stakeholders that includes local communities. Governments have responsibility for many governance functions, such as formulating policy, developing legal frameworks, planning, coordination, funding and financing, capacity development, data acquisition and monitoring, and regulation. Good water governance comprises many elements, but it principally includes effective, responsive, and accountable state institutions that respond to change, openness and transparency providing stakeholders with information, and giving citizens and communities a say and role in decision-making. At this point, the participation and multi-stakeholder engagement are important parts of policy processes, although measuring their effectiveness is still in its infancy. The importance of having a transparent, universal, and neutral platform for government and citizen groups in place to mobilize available resources and to seek alternative means of ensuring improved water services has proven to be essential and complementary to local government support.

It should also be kept in mind that for reliable water management data acquisition and monitoring plays an important role. Data underpin the governance elements of accountability, transparency, and participation. They enable progress to be monitored and service providers, governments, and development partners to be held accountable. Many developing and emerging countries lack the financial, institutional, and human resources to acquire and to analyse data to support governance. Less than half of developing countries have comparable data available on progress towards meeting each of the global SDG 6 targets. Further, data acquisition and monitoring require political commitment to transparency that includes efforts related to accessibility and sharing of data. Increased utilization of the latest Earth observations, citizen science and private sector data should be incorporated into data-monitoring systems to complement existing data-collection efforts.

All the considerations and reminded measures are ultimately of little use if the capacity development leaves far behind. Unfortunately, an acute lack of capacity is constraining water resources development and management in all its facets across most developing countries, particularly in Sub-Saharan Africa and South and South-Eastern Asia. Human resource shortages are reported in all key areas, including agriculture and irrigated farming, water-related risk management, water and sanitation services, wastewater treatment, recycling and reuse technologies, and desalination. This is not a new phenomenon and has been a leading concern and constraint on water-related development for many decades. If the respective developing countries, but also the development policy strategies of the rich industrial nations do not change their direction towards immense education and capacity building measures, then all efforts and invested money will continue to be wasted, and many of the SDGs will fail. So, this is no news to say that education, further education, and training are essential for achieving SDG 6. SDG 4.c supports this by calling for "*substantially increasing the supply of qualified teachers, including through international cooperation for teacher training in developing countries, especially least-developed countries and small-island developing states*" [16].

Unfortunately, reality shows something else, e.g., Fig. 1.2 provides an indication of the numbers of students enrolled in tertiary education in 2015 in various regions, though it was not disaggregated into sectors. There were notable low numbers in Sub-Saharan Africa, which has a population of about one billion. In 2015, however, only 3.5% of students enrolled in tertiary education [17].

Several developing countries have basic education institutional structures in place, but they are too often in need of strengthening and funding. There may be options for regional training, where concerned countries share facilities and costs. However, this is not easy to implement unless countries and their schools, training centres and universities plan, fund and implement capacity development programmes that meet the expectations of national development plans.

In his speech about research in Africa 2004, the UN Secretary General Kofi Annan highlighted: "*The knowledge required for Sub-Saharan Africa to achieve its own green revolution is not lacking. What is lacking as ever, is the will to turn knowledge into practice.*" [18]. This sentence was not meant to criticise the quality of research in Africa, but to urge the focus of research and the effectiveness of the pathway from research to policy and practice. Thus, in many cases, this kind of effectiveness would contribute to a significant and visible improvement in sustainable water management and people's daily lives.

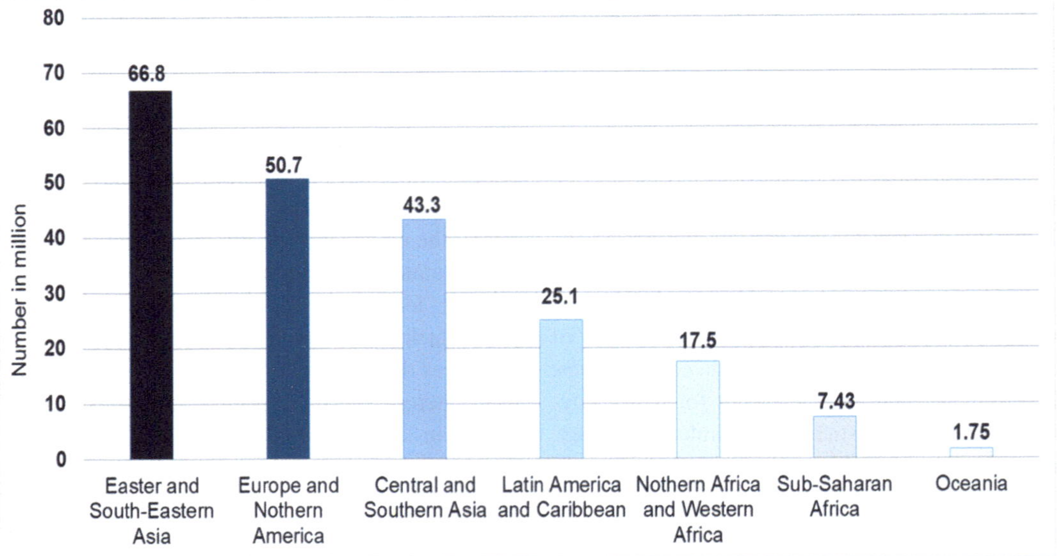

Fig. 1.2 Students enrolled in tertiary education in SDG world regions in 2015 [17]

1.4 Future Perspectives and Challenges of Sustainable Water Management in Water-Scarce Developing Countries

The most severe water problems are found in countries located in the Middle East-North Africa (MENA), South-East Asia (SEA), and Sub-Saharan Africa (SSA) regions. These different regions individually face challenges in providing safe, affordable, and sustainable water and sanitation for all. The challenges could not be more different, just as the regions themselves are different.

The **MENA region** is the most water scarce region in the world. According to FAO the "water poverty/scarcity" threshold of renewable water resources is 1,000 m^3 per capita per year, while there is in average less than 200 m^3 per capita per year in the MENA region [19]. It is already clear that water scarcity on a per capita basis is increasing and will continue to increase due to population growth and climate change. The result will be a rapid increase in groundwater demand. Impacts on agriculture, such as the loss of arable land and consequently decline in agricultural production, also contribute to increasing numbers of people migrating from rural to urban areas or fleeing to other countries. According to WHO, some 51 million people (or 9% of the total population) lacked a basic drinking water service in 2015 in the entirety of the MENA region [6]. Besides climate change and population growth, another problem is that if armed and religious conflicts continue in the region, hopes for consistent sustainable water management, access to water services for all and the establishment of safe water infrastructure are illusory. Already today, natural disasters linked to climate change impacts have resulted in the displacement of over 2,400,000 people across the Arab Region in 2016, many of them being in the Arab Least Developed Countries (LDCs) (98%): 1,230,000 in Sudan, 810,000 in Somalia, and 360,000 in Yemen [6]. The international community would do well to make every effort to end these conflicts. But the parties and governments in the region must also work resolutely to achieve this. Peaceful and free access to water and the intelligent use of resources are indeed a challenge, but only a certain prosperity can grow in the region and inspire hope for a future and secure water supply, infrastructure, and sanitation.

The **SEA region** is mostly affected by flood and drought disasters. The Asian Development Bank (ADB) reported in 2016 that 48 countries in the region are classified as water insecure regions due to low water availability and unsustainable groundwater extraction. For reasons like those in the MENA region, the need for agricultural irrigation has also steadily increased in the region. The consequence here, too, is disproportionate decreases in groundwater levels and thus a significant increase in water stress. This development is particularly evident in the North China Plain and in northwest India. In addition, high levels of water pollution are worsening the situation in terms of drinking water availability, caused by alarming rates of untreated sewage discharged into surface waters. The situation is exacerbated by high levels of chemical pollution in runoff water [20]. In the Ganges and Mekong basins, high concentrations of arsenic compounds are increasingly endangering groundwater quality. Arsenic comes from the erosion of pyrite minerals on high Mount Everest and run off into the Everest originated river systems, being transported over long distances till the Gulf of Bengal and Mekong Delta in South Vietnam. The challenges for sustainable water management are huge. This becomes clear, when one looks at the rapid growth of the region's urban population in most Asian countries, which has more than doubled since 1950, and so the problem cities face with developing adequate infrastructure are a task of the century to keep pace with rising water and sanitation needs. Looking at the rural areas in the Asian Regions with long time unsustainable practices and unequal access to irrigation water, the impact on agricultural productivity and poverty alleviation is devastating and leads more and more to loss of livelihoods, while most rural poor people are dependent on agriculture. Accordingly, the concept of water security is gaining importance not only in Asia. The concept contributes to improving the resilience of water and sanitation services and is the key to optimizing and securing access to clean water in a climatically uncertain future, while at the same time providing governmental authorities with realistic information to take tailored measures.

In the **Sub-Saharan Africa** region, periodic and chronic water scarcity represents a major challenge to Africa's path to development and prosperity. The lack of water management infrastructure both in terms of storage and supply as well as of improving drinking water provision and sewage disposal are decisive factors that are responsible for the hardly changing poverty [6].

Climate change processes that negatively affect rainfall and temperature trends threaten water availability, agricultural productivity, and ecosystem balance in almost all regions of Sub-Saharan Africa. Additional challenges for sustainable water management arise from the growing population, which is expected to reach 1.3 billion out of 2.2 billion people worldwide by 2050. An additional aspect, which is linked to the mentioned challenge, relates the education sector. While it is estimated that 85% of primary school teachers worldwide were trained in 2016, the proportion for Sub-Saharan Africa was only 61% [17]. If equal opportunities, adequate education and training of young women and men are ensured, the intellectual contribution from these segments of the population could help Africa on its way to achieving SDG 6. Besides the academic elite, a broad and good education is a basic prerequisite for economic and political stabilization and for overcoming mass poverty. If governments continue to ignore this and do not bring about any concrete changes, the disastrous situation of many states in Sub-Saharan Africa will not change.

And yet, contrary to all the current negative signs, governments and humanity still have it in their hands to realize a positive turnaround. Only the honest and firm will must reign to bring about the positive turnaround. In all the regions discussed, there were positive initiatives and promising solutions, especially in the water sector. Countries like Cost Rica, Rwanda, Kenya, South Africa, Thailand, and Vietnam are just few examples. Perhaps the changes in one or another developing country were also based on participatory processes that brought in new and diverse voices and ideas, so that people could influence decisions as rights holders and deeply rooted and unconscious prejudices and discrimination could be overridden by changing attitudes and norms within water institutions at all levels. This is the only way to achieve sustainable water management and thus sustainable development.

1.5 The Way Forward

Water availability can be seen as a function of two distinct but inseparable characteristics. The first relates to water supply, which is the amount of water that can be sustainably drawn from surface, underground and unconventional sources. These include desalination of seawater, reuse of treated wastewater, and collection of rainwater and fog. Increasing water use efficiency in all major water use sectors (agriculture, energy, industry, and municipalities/households) can also go a long way towards reducing overall demand, freeing up water supplies for other users, including ecosystems. The second characteristic relates to access to freshwater, which means that water must be transported from the source and made available to the various users in sufficient quantity and suitable quality for the intended uses.

The need to improve and to secure water resource management is particularly critical for areas suffering from chronic or recurrent water scarcity, where demand exceeds sustainable supply, or where supply is affected by pollution, land degradation or other phenomena. The need to technically improve and sustainably optimize access to freshwater exists in all types of hydrological regimes, even in places with relative water abundance. Barriers to improved accessibility and sustainable use are often social, political, and economic. Against this background, sustainable governance structures are indispensable. Such structures must guarantee fair and equitable and sustainable use of water resources for all. It is a high priority to ensure that sufficient water of suitable quality is available to meet the basic needs of all people, both for domestic use and subsistence. However, very often, links between water and broader decisions regarding food and energy security, humanitarian crises, economic development, and environmental sustainability often remain unrecognized or poorly understood. Worsening extreme events, environmental degradation, declining water availability and quality, population growth, rapid urbanization, unsustainable and inequitable patterns of production and consumption within and between countries, current and potential conflicts, and unprecedented migration flows are among the interconnected pressures facing mankind, which through their impact on water often hit those in vulnerable situations the hardest.

Moving forward while making progress requires a renegotiation of power relations at all levels, equitable participation and representation of all groups, and new partnerships to transform the economic, social, and political processes that guide water resource management and drive the provision of safe and affordable water and sanitation. In this context, the right priorities of the government and its active support, as well as shared awareness in the majority society are the basis for real change and improvement of living conditions.

However, given the complexity of this issue, other aspects need to be added, such as the role of society and the state. It is not enough to look at the tools available to water managers to solve the problems, but the commitment of society and the state is needed.

Water governance thus steps out of the traditional context, which was primarily about the question of supply and demand. Here, formal, and informal structures, procedures and processes operate in an integrated way at the national level. However, good water governance must also be considered globally. Global cooperation is essential in this regard. Without this cooperation, many strategies and national solutions will fail sooner or later. Neglecting the global dimension of water management carries the risk that too much national development outside the field of water management will overshadow and possibly even nullify good intentions.

After all, it is evident that according to UN-Water 2 billion people today do not have access to safely managed drinking water and wastewater, and 1.7 billion people do not have access to safely managed sanitation services [21]. The question of why this is so, and in many cases in developing countries, can be answered in large part by the prevalence in these countries of political systems composed of extractive political institutions and extractive economic action. However, the scarcely practiced resource conservation to be considered in this context can also be traced back to industrialized countries. The currently prevailing economic and political action does not prioritize sustainability and resources conservation on the agenda.

Good water governance comprises many elements, but it essentially includes effective, responsive, and accountable government institutions that respond to change, openness and transparency that provide information to stakeholders and give citizens and communities a voice and role in decision-making. This is the framework of an inclusive political and economic system. The importance of having a transparent, universal, and neutral platform for government and citizen groups to mobilize available resources and to seek alternative ways to ensure improved water and sanitation management has been demonstrated as well as the importance of complementing local government support is proven.

Let's not fool ourselves, the genuine will alone to do this is ultimately the driving force. Basically, we have no choice but to want positive change.

References

1. Burek P, Satoh Y, Fischer G, Kahil MT, Scherzer A, Tramberend S, Nava LF, Wada Y, Eisner S, Flörke M, Hanasaki N, Magnuszewski P, Cosgrove B, Wiberg D. Water futures and solution: fast track initiative (Final Report). In: IIASA working paper. Laxenburg, Austria, International Institute for Applied Systems Analysis (IIASA); 2016. pure.iiasa.ac.at/13008/.
2. Avilés J, Pérez Rocha L. Regional subsidence of Mexico City and its effects on seismic response. Soil Dyn Earthq Eng. 2010;30:981–9.
3. Gutierrez J. Water scarcity and supply challenges in Mexico City's informal settlements. Penn Institute for Urban Research 2; 2019.
4. Kimmelman M. Mexico City, parched and sinking, faces a water crisis. New York Times—Climate; 2017. https://www.nytimes.com/interactive/2017/02/17/world/americas/mexico-city-sinking.html.
5. Griffin D, Anchukaitis KJ. How unusual is the 2012–2014 California drought? Geogr Res Lett. 2014;41(24):9017.
6. WWAP (UNESCO World Water Assessment Programme). The United Nations world water development report: leaving no one behind. Paris: UNESCO; 2019.
7. Australian Government, Bureau of Meteorology. Australian water markets report 2017–18: national overview section; 2019. ISSN 2207-1733.
8. Marin P, Tal S, Yeres J, Ringskog K. Water management in Israel key innovations and lessons learned for water-scarce countries: World Bank—Water global practice. 2017;18–21
9. Fröhlich CJ. Security and discourse: the Israeli–Palestinian water conflict. Conflict Secur Dev. 2012;12(2):123–48. https://doi.org/10.1080/14678802.2012.688290.
10. Imteaz MA, Shanableh A, Rahman A, Ahsan A. Optimisation of rainwater tank design from large roofs: A case study in Melbourne, Australia. Resour Conserv Recycl. 2011;55(11):1022–29.
11. Hafizi Md Lani N, Yusop Z, Syafiuddin A. A review of rainwater harvesting in Malaysia: prospects and challenges water. 2018;10(506). https://doi.org/10.3390/w10040506.
12. Keeble BR. The Brundtland report: our common future. Med War. 1988;4(1):17–25.
13. World Water Council. 2020. https://www.worldwatercouncil.org/en/water-crisis.
14. Scharlemann JPW, Brock RC, Balfour N, Brown C, Burgess ND, Guth MK, Ingram DJ, Lane R, Martin JGC, Wicander S, Kapos V. Towards understanding interactions between sustainable development goals: the role of environment–human linkages. Sustain Sci. 2020;15:1573–84.
15. Guppy L, Mehta P, Qadir M. Sustainable development goal 6: two gaps in the race for indicators. Sustain Sci. 2019;14:501–13.
16. Isgida Y. How does the newly added DAC evaluation criterion "coherence" contribute to achieving the SDG target 4c for teachers? J Int Coop Educ. 2020;22(2)/23(2):15–29.
17. Renata A, Ortigara C, Kay M, Uhlenbrook S. A review of the SDG 6 synthesis report 2018 from an education, training, and research perspective. Water. 2018;10:1353. https://doi.org/10.3390/w10101353.
18. Kofi Annan. 2004. https://www.un.org/press/en/2004/sgsm9405.doc.htm.
19. Mualla W. Water demand management is a must in MENA countries... But is it enough? J Geol Resour Eng. 2018;6:59–64.
20. UN ESCAP. Paths to 2015—MDG priorities in Asia and the Pacific. Asia-Pacific MDG Report; 2010/11.
21. UN-Water. Summary Progress Update 2021—SDG 6—water and sanitation for all. Version: July 2021. Geneva, Switzerland; 2021.

Part II
Hydrology and Climate Change Impacts

Climate Change Impacts on Water Resources

2

Veysel Yildiz, Murat Ali Hatipoglu, and S. Yurdagül Kumcu

Abstract

Climate change is global warming resulted by human and natural effects and as a large scale change on weather pattern. As the climate change is affecting water cycle, it reduces the predictability of water availability, worsens water quality, exacerbates water scarcity, and threatens sustainable development and biodiversity worldwide. The increase of concentrations of greenhouse gases in the atmosphere causes climate change by trapping heat, which leads to more severe weather events. This results in the increased frequency of intense climate-related natural disasters. However, there is different views on the magnitude of change of climatic variables. The climate change affects water resources in complex ways. It is changing almost all stages of water cycle diagram; evaporation, evapotranspiration, precipitation, surface runoff, infiltration, etc. This induces the disruption of the hydrological cycle by altering when, where, and how much precipitation falls. Climate change affects demand for water resources, the way we use water, and the amount we need as well as it is altering the water cycle and the availability of water resources. Droughts and floods occur naturally and involve many factors, but climate change could possibly exacerbate these water related events. Recent studies concluded that latest intense precipitation events are the fingerprints of climate change. Rising temperature results in increase in evaporation from the earth's surface, change in the regime and increase intensity of precipitation. This is possibly leading to more often occurrence of severe flood and drought with a longer duration. What is more, as the ocean warms, freshwater glaciers melt at a high rate. The water from the melted glaciers eventually reaches the ocean, which leads to a rise in sea level. Thus, groundwater could easily be contaminated by saltwater, which causes reducing in the quantity and quality of the reserves. Sustainable development and management of water resources is key to cope with climate change especially in vulnerable areas of the world. A holistic approach is a must for countries to cope and to adapt the ability of resilience towards the adverse impacts of climate change on water resources.

V. Yildiz
Department of Civil and Structural Engineering, The University of Sheffield, Sheffield, England

M. A. Hatipoglu
General Directorate of State Hydraulic Works, Ankara, Turkey

S. Y. Kumcu (✉)
Department of Civil Engineering, Necmettin Erbakan University, Konya, Turkey
e-mail: yurdagulkumcu@gmail.com

M. Bahadir and A. Haarstrick (eds.), *Water and Wastewater Management*,
Water and Wastewater Management, https://doi.org/10.1007/978-3-030-95288-4_2

Keywords

Global water cycle · Climate change · Water resources · Uncertainty

2.1 Introduction

Reliable, sustainable and clean water is essential for the continuation of life. Water is also required for crucial sectors, such as drinking and domestic water, agriculture, energy production, navigation, recreation, industry and manufacturing. Increasing water demand for these sectors requires protection of water resources.

While the amount of water resources is adequate globally, its spatial and temporal distribution is not homogeneous across the world. Thus, water resources are not sufficient in most parts of the world. This is now a paramount problem because of the increasing rate of population growth and higher living standards, and thus, water pollution and climate change effects [1]. Nearly half a billion people face severe water scarcity in the world during all year [2]. Moreover, one third of the world's population has water scarcity especially in arid and semiarid regions [3, 4], and within a couple decades, two-thirds of the world's population may face water shortages [5, 6].

Climate change affects all major economic sectors, natural resources, and biodiversity [7]. It is also the cause of unexpected weather conditions, which shows extremes in terms of frequency, intensity, spatial extent and duration [8]. These changes in addition to the amount, effect on their time distribution and in response leads to degradation of water quality as well as stress creation on fragile balance of water resources and water uses [9]. Indeed, it is predicted that that climate change is responsible about 20% of the increased water scarcity around globe [10]. Additionally, it will have a significant influence on the sustainability of water supply, both quantitatively and qualitatively, human health, and the food supply in future decades [11]. Therefore, Water resources development and management is a very important issue [12]. Paris Agreement and Sustainable Development Goals (SDGs) aim to strengthen the global response to the adverse impacts of climate change [13, 14].

Many researchers have studied the relations between climate change and water resources [15–21]. The impacts of climate change are generally evaluated through the application of rainfall-runoff models. These evaluations have found that stream-flow variability is closely associated with climate change. The impacts are foreseen to vary regionally with major impacts on local natural environments and human systems [22], and are likely to include changes in hydro-climate patterns as well as increase in the probability of extreme events.

2.2 Water Cycle and Availability

Water is steadily in motion between the atmosphere, the lithosphere and the hydrosphere. Water Cycle is the chain of events that water follows as it reaches the atmosphere from liquid to gaseous state, and then condenses again and returns to the earth as precipitation. Water cycle as shown in Fig. 2.1 has a balance of precipitation, evaporation, and all other cycle parameters in between.

Climate and water resources are closely related to each other. Indeed, every change in the climatic system induces a change in the water system, and the other way round [23]. The impacts of climate change on water resources can be comprehensive as it is affecting almost all stages of water cycle diagram: evaporation, evapotranspiration, precipitation, surface runoff, infiltration, etc. [24]. This induces the disruption of the hydrological cycle by altering when, where, and how much precipitation falls [25]. In other words, it shifts the precipitation quantity, form and patterns, causes change in the frequency and magnitude of floods, droughts and sea levels [26].

Increasing trend in temperature is predicted to continue in the future [27]. Thus, increasing in the temperatures rise the rate of evaporation of water into the atmosphere, this in turn, enhances

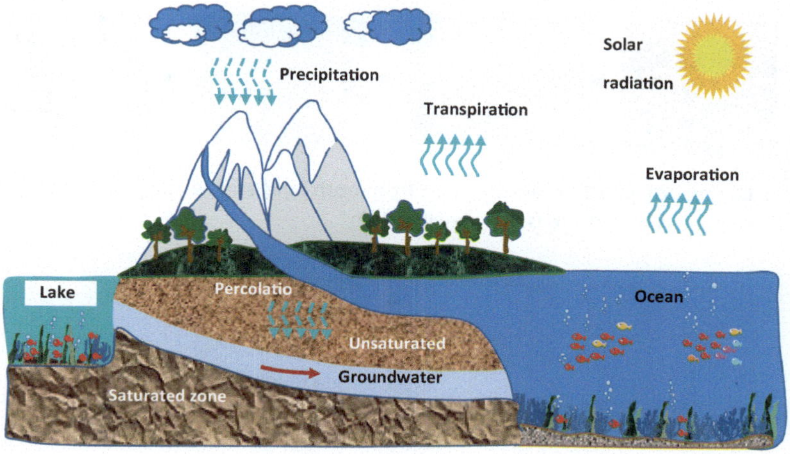

Fig. 2.1 Projected changes in the water cycle

the atmosphere's capacity to hold water [28]. Since the processes involved are highly dependent on temperature, a change in one element of the water cycle has significant effects on others. Climate change has widely differing effects on water resources depending on the region. Water related extremes such as in runoff, flooding, or sea level rise are more critical than water shortages in some regions. While increased evaporation rate might dry out some regions, fall as excess precipitation on other areas. What is more, warmer air can hold more water vapor, which can lead to more intense rainstorms, causing major problems like extreme flooding in coastal communities around the world [29]. Some climate related models predict that while the weather will be wetter in coastal regions, it will become drier in the middle of continents in the future. Also, some models estimate that the increasing amount of evaporation and rainfall over oceans will be higher compared with that over lands [30]. Indeed, annual precipitation and average water availability trends in high latitudes and in wet tropical regions are increasing. In mid-latitudes and dry tropical regions, on the contrary, it has fallen that means those areas face water stress as a result of decreasing of water availability.

That is to say, melting of glaciers and snow cover because of rising temperatures and more precipitation to fall as rain rather than snow leads to a decrease in water supplies, thereby reducing water availability during dry seasons in some areas that depend on melt-water from mountain ranges. For example, in parts of the Northern Hemisphere, an earlier arrival of spring-like conditions is leading to early snow melting resulting river flows [31]. In response, seasons with the highest water demand such as irrigation, are being adversely impacted by a reduced availability of fresh water. Figure 2.2 shows some results of the climate change.

2.3 Water Supply and Demand

Increasing trends in temperature are predicted to continue into the foreseeable future, which will shift the spatial and temporal occurrence of rain fall and the availability of water resources. This, in response, adversely affects water supply systems. Water supply systems are already under stress because of aging infrastructure, population growth, increasing living standards, varying consumption patterns, and increased demand for agriculture. Climate change is another crucial factor in the current challenges, contributing to the vulnerability of water supply systems. Decrease in water availability thanks to climate change and other factors may cause or prolong social conflicts between municipalities, people and authorities over future water use [32].

Climate change is affecting how we use water and how much we need. People and animals demand more water to maintain their health and

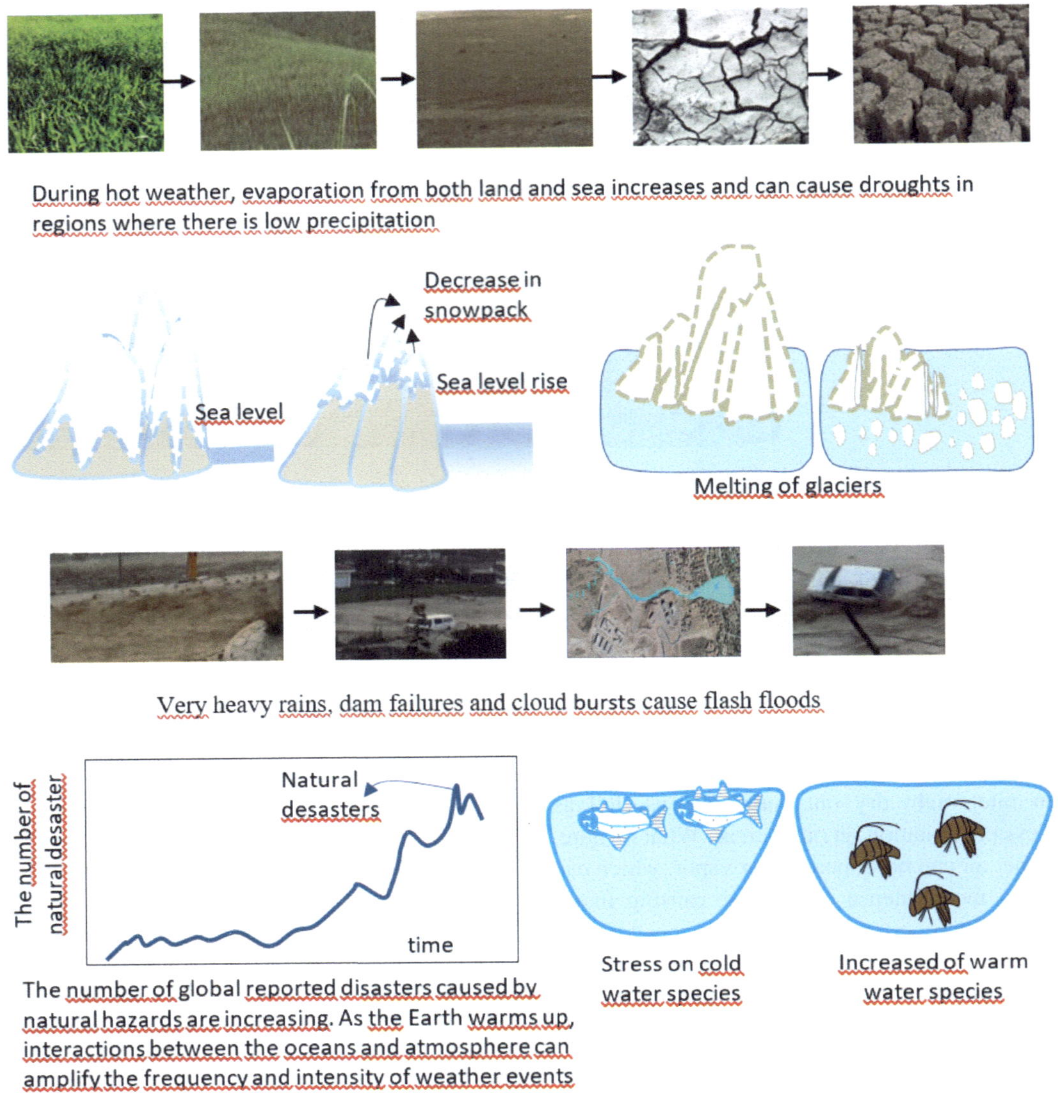

Fig. 2.2 Some results of the climate change

thrive as temperature rises [33]. Moreover, water demand, especially in irrigation, generally increases with temperature rise and decreases with precipitation rise [34]. This leads to an increase demand for water resources while shrinking available water supplies, resulting in higher pressure on the already limited water supplies of many regions. In response, water supplies in some regions are already under pressure by increasing demand as well as decreasing runoff and groundwater recharge. As a result of increased surface temperatures and alteration of hydrologic cycle, the amount of winter precipitation received as rain is expected to be higher with a decreasing rate of the snow form. Snow pack levels are also expected to form later in the winter, accumulates in smaller quantities, and melt earlier in the season, leading to reduced summer flows [35]. As a consequence of altering cycle with warming, snow-fed basins are likely to face more frequent summer droughts. Besides, such shifts in the form and timing of precipitation and runoff will alter timing and availability of water supply, affecting agricultural, municipal, and public uses [36]. The amount of water available for many important economic activities, which depend on water, such as hydropower, agriculture, and livestock will decrease. What is more, shrinking of mountain glaciers through global warming threatens drinking water supplies for millions of people [37]. Increase of evaporation because of increasing temperature consequently adversely affects management of multipurpose reservoir systems [38]. Decrease in inflow into the reservoir and increase of dam downstream demand as a result of climate change coupled with population growth and industrial development may cause conflicts between end users, which may create crisis conditions.

2.4 Water Quality

Declining water quality is another consequence of climate change all around the world. Shifts in the spatial and temporal occurrence of precipitation, increase of flooding due to heavy rainfall events together with higher water temperatures, are likely to exacerbate water quality problems [38]. Water temperature rises in streams, lakes, and reservoirs as air temperature rises. Rising air and water temperatures will also impact water quality by increasing primary production, organic matter decomposition, and nutrient cycling rates in lakes and streams, resulting in lower dissolved oxygen levels hence more stress on the fish, insects, crustaceans and other aquatic animals that rely on oxygen [39].

The increased frequency and intensity of rainfall events result in flooding. These extreme events will likely cause increased runoff and erosion, while there is little or no observed evidence yet that soil erosion and sediment loads have been altered significantly due to changing climate [40]. It is projected that more sediments including mineral matter, chemicals and trash, pollutants, and organic material will be conveyed into water resources such as streams and groundwater systems and will worsen the water quality. Water quality may be further worsened, if water supply shortage causes nutrients and contaminants to become more concentrated [41]. Naturally, the pollution load in streams and rivers will tend to be carried to larger bodies of water downstream like lakes, estuaries, coastal ocean, where one of the more dramatic consequences of heavy runoff can be blooms of harmful algae and bacteria [42].

Besides, sea level rise through thermal expansion of the sea water and melting of glaciers and ice sheets can lead to saltwater intrusion into fresh groundwater bodies, which may contaminate the supply from groundwater, especially in low-lying, gently sloping coastal areas, which causes reducing the quantity and quality of the reserves. At the same time, increases in runoff, flooding, or sea level rise are more critical than water shortages in some areas. These effects can reduce the quality of available water related extremes such as in runoff, flooding could overwhelm and damage the infrastructure of water distribution network like sewer systems, and water treatment plants result in untreated sewage into drinking water supplies.

2.5 Impacts of Changes in Water Resources on Other Sectors

The impacts of climate change on water resources and availability and water quality can affect a number of crucial sectors such as hydropower, infrastructure, recreation, and agriculture. Climate change remains a key driver for hydropower production, since it is ultimately controlled by weather and precipitation trends. In other words, hydropower generation is closely linked to hydrological conditions of a watershed and sensitive to seasonal changes in water quantity [43, 44]. In response, there has been a dramatic increase in research of climate change impacts on water resources and more specifically on hydropower generation [45, 46]. Climate change impacts on hydropower is rather complex, and even more so if analyses include interactions between the technical, physical and economical components of the systems [47]. Climate change would probably increase the intensity and frequency of extreme events by altering the annual mean and seasonality of runoff, which influence the availability and stability of hydropower production, and at the same time increase the value of the storage role of large hydropower plants [48–52]. Indeed, predicted increase of streamflow, caused by more future extremes, does not necessarily translate increased power generation because of plant capacity limitations and spillage of water. Furthermore, the increasing trend of extreme hydrological events adversely affects the financial viability of existing and potential hydro schemes, meaning large uncertainties in the future projections. Moreover, these extremes reduce the utilization efficiency of water. What is more, decreased hydropower potential is likely resulting in increases of the energy price. Therefore, extreme events need to be taken into account during hydropower development to mitigate possible adverse influences on hydropower generation. The studies of climate change impacts before and during implementation of hydropower projects can result in timely responses and adaptation to climate change with a potential of considerable cost savings [48]. Water resources planners and decision makers are rightly concerned about the potential effects of future uncertainties, especially of climate change on hydropower plants because of high investment costs, social costs due to population displacement, and environmental costs [53]. It is of great importance to incorporate uncertainty assessment and risk analysis due to the inherent uncertainty associated with climate change [54].

Climate change related extremes are adding to pressures on global agricultural and food systems. Indeed, they have a complex cause-effect relationship. Significant quantities of gas emissions are produced by the agricultural sector that impacts climate. The increase in the concentration of greenhouse gases result in temperature rise as well as changes in the precipitation regime, which affect quality and stability of the agricultural production, but also on the natural environment, in which agriculture is practiced [55]. What is more, the changing climate is also adding to resource problems, such as water scarcity, pollution and soil degradation. Increases in the frequency and intensity extreme of weather events can also hamper food delivery, resulting in fluctuation of food prices. In addition, climate change can hinder food availability, reduce access to food, and worsen food quality [56].

As climate change impacts vary regionally, certain areas are expected to experience more droughts. Therefore, more hydraulic structures would be built in near future to convey available water to areas facing water shortages. Moreover, increased probability of flooding may necessitate infrastructure changes to minimize its impacts. Both of these essential measures may result in more emissions and higher energy demand. Changes in water availability have already proven to trigger refugee dynamics and political instability [57].

Increased water temperatures are likely to cause the habitat ranges of many fish species to shift, which could disrupt ecosystems. A shift in hydrologic cycle to increased winter flow can also affect the life cycle of fish species, such as salmon, which depend on late spring flows to “flush” young salmon to the ocean, and on summer flows to moderate water temperatures. Indeed, the number of survived salmon smolt will decrease and more frequent fish kills will occur from lethal stream water temperatures [58]. Additionally, reduced snow pack and earlier spring snow melt put traditional winter sports, such as skiing and snowmobiling, at risk.

2.6 Conclusions

Changes in climate significantly influence the hydrological cycle and hence affect the quality, quantity and availability of water resources. In other words, increased temperature, changed precipitation patterns and snow cover, and increased frequency of flooding and droughts are the main consequences of climate change. 90% of all natural disasters are water-related hazards. The duration and intensity of floods and droughts are expected to increase near future due to climate change, as the whole global water cycle is affected by global warming. While available water resources are decreasing in many regions, an increasing demand for water supply is expected. As a result of reduced groundwater recharge, the availability of groundwater for drinking water in some regions will likely decline. In particular, water quality is likely to be worsened by warmer water temperatures, changed precipitation pattern, duration and intensity. These will cause adverse impacts on the ecosystem, human health, and water system reliability and operating costs. Therefore, the management of water resources in terms of quantity and quality are becoming increasingly complex in many places. The precise consequences of climate change on water resources remain uncertain, which makes adaptation challenging. While an increase trend in temperature is projected by Global Circulation Models (GCMs), in general, precipitation projection remains inconsistent. Despite improvements in climate science, the GCMs formed and developed to project climate futures generate a wide range of projections that in general disagree on magnitude of precipitation changes. However, increasing trend in temperature might change precipitation pattern in winter (from snow to rain) and cause to early snow melting in spring and more evaporation, which likely result in increasing trend of extreme hydrological events.

Altering precipitation pattern coupled with rise of temperatures is of concern to water resource managers and decision makers in a number of settings such as hydropower generation, irrigated agriculture, and water supply. Overall, climate change could make it more difficult to manage hydropower production, to grow crops, to raise animals, and to catch fish by altering precipitation pattern. Indeed, management strategies must be adapted to minimize these potential climate change effects. Climate change is an inevitable phenomenon of natural and human origins, to which reduction and adaptation are needed in order to reduce the magnitude of impact and vulnerability.

The adverse effects of climate change on water resources could be minimized through a number of settings. Among these are (1) adapting hydropower operating policies, (2) changing crops pattern and practices that are robust to changing conditions, (3) raising awareness, (4) rehabilitating water supply infrastructure, and (5) supporting water transfer opportunities. Furthermore, future projections should inform management strategies to increase the ability to efficiently prepare for and to adapt to future water resource challenges.

Acknowledgements Authors would like to thank the State Hydraulic Works, Turkey, Necmettin Erbakan University, Konya, Turkey, and EXCEED Swindon project funded by DAAD (German Academic Exchange Service).

References

1. Stocker T. Climate change 2013: the physical science basis: Working Group contribution to the Fifth assessment report of the Intergovernmental Panel on Climate Change. Cambridge University Press; 2014.
2. Mekonnen MM, Hoekstra AY. Four billion people facing severe water scarcity. Sci Adv. 2016;2(2): e1500323.
3. Hofste RW, Reig P, Schleifer L. 17 countries, home to one-quarter of the world's population, face extremely high water stress. 2019. https://www.wri.org/insights/17-countries-home-one-quarter-worlds-population-face-extremely-high-water-stress. Accessed 23 Oct 2021.
4. Ashraf S, AghaKouchak A, Nazemi A, Mirchi A, Sadegh M, Moftakhari HR, Hassanzadeh E, Miao C-Y, Madani K, Baygi MM, et al. Compounding effects of human activities and climatic changes on surface water availability in Iran. Clim Change. 2019;152 (3):379–91.
5. Cosgrove WJ, Loucks DP. Water management: current and future challenges and research directions. Water Resour Res. 2015;51(6):4823–39.
6. Boretti A, Rosa L. Reassessing the projections of the world water development report. NPJ Clean Water. 2019;2(1):1–6.
7. Sabbaghi MA, Nazari M, Araghinejad S, Soufizadeh S. Economic impacts of climate change on water resources and agriculture in Zayandehroud river basin in Iran. Agric Water Manage. 2020;241:106323.
8. Seneviratne S, Nicholls N, Easterling D, Goodess C, Kanae S, Kossin J, Luo Y, Marengo J, McInnes K,

Rahimi M et al. Changes in climate extremes and their impacts on the natural physical environment. 2012. https://www.ipcc.ch/site/assets/uploads/2018/03/SREX-Chap3_FINAL-1.pdf. Accessed 23 Oct 2021.
9. Schewe J, Gosling SN, Reyer C, Zhao F, Ciais P, Elliott J, Francois L, Huber V, Lotze HK, Seneviratne SI, et al. State-of-the-art global models underestimate impacts from climate extremes. Nat Commun. 2019;10(1):1–14.
10. Simonovic SP. Managing water resources: methods and tools for a systems approach. Routledge; 2012. ISBN 9781844075546.
11. Bates B, Kundzewicz Z, Wu S, Palutikof J. Observed and projected changes in climate as they relate to water. Clim Change Water. 2008;20.
12. Sheffield J, Wood EF, Pan M, Beck H, Coccia G, Serrat-Capdevila A, Verbist K. Satellite remote sensing for water resources management: potential for supporting sustainable development in data-poor regions. Water Resour Res. 2018;54(12):9724–58.
13. Tarawneh QY, Chowdhury S. Trends of climate change in Saudi Arabia: implications on water resources. Climate. 2018;6(1):8.
14. Donevska K, Panov A. Climate change impact on water supply demands: case study of the city of Skopje. Water Supply. 2019;19(7):2172–8.
15. Hodgkins G, Dudley R, Huntington T. Changes in the timing of high river flows in New England over the 20th century. J Hydrol. 2003;278(1–4):244–52.
16. Novotny EV, Stefan HG. Stream flow in Minnesota: indicator of climate change. J Hydrol. 2007;334(3–4):319–33.
17. Wang S, Kang S, Zhang L, Li F. Modelling hydrological response to different land-use and climate change scenarios in the Zamu river basin of northwest China. Hydrol Process Int J. 2008;22(14):2502–10.
18. Tu J. Combined impact of climate and land use changes on streamflow and water quality in eastern Massachusetts, USA. J Hydrol. 2009;379(3–4):268–83.
19. Li F, Zhang G, Xu YJ. Assessing climate change impacts on water resources in the Songhua river basin. Water. 2016;8(10):420.
20. Wang W, Shao Q, Yang T, Peng S, Xing W, Sun F, Luo Y. Quantitative assessment of the impact of climate variability and human activities on runoff changes: a case study in four catchments of the Haihe river basin, China. Hydrol Process. 2013;27(8):1158–74.
21. Gebremeskel G, Kebede A. Estimating the effect of climate change on water resources: integrated use of climate and hydrological models in the Werii watershed of the Tekeze river basin, northern Ethiopia. Agric Nat Resour. 2018;52(2):195–207.
22. Dodd C. The water cycle. WorldAtlas. 2021. https://www.worldatlas.com/the-water-cycle.html. Accessed 23 Oct 2021.
23. Kundzewicz ZW. Climate change impacts on the hydrological cycle. Ecohydrol Hydrobiol. 2008;8(2–4):195–203.
24. Pimentel D, Berger B, Filiberto D, Newton M, Wolfe B, Karabinakis E, Clark S, Poon E, Abbett E, Nandagopal S. Water resources: agricultural and environmental issues. Bioscience. 2004;54(10):909–18.
25. Huntington TG. Evidence for intensification of the global water cycle: review and synthesis. J Hydrol. 2006;319(1–4):83–95.
26. Naschen K, Diekkrüger B, Leemhuis C, Seregina LS, van der Linden R. Impact of climate change on water resources in the Kilombero catchment in Tanzania. Water. 2019;11(4):859.
27. Meehl GA, Stocker TF, Collins WD, Friedlingstein P, Gaye AT, Gregory JM, Kitoh A, Knutti R, Murphy JM, Noda A et al. Global climate projections. Chapter 10. 2007. https://www.ipcc.ch/site/assets/uploads/2018/02/ar4-wg1-chapter10-1.pdf. Accessed 23 Oct 2021.
28. Karl TR, Melillo JM, Peterson TC, Hassol SJ. Global climate change impacts in the United States. Cambridge University Press. 2009. https://www.nrc.gov/docs/ML1006/ML100601201.pdf. Accessed 23 Oct 2021.
29. Project TCR. How is climate change impacting the water cycle? 2021. https://www.climaterealityproject.org/blog/climate-change-impacting-water-cycle. Accessed 29 Oct 2021.
30. The water cycle and climate change. 2019. https://www.kildwick.com/the-water-cycle-and-climate-change. Accessed 23 Oct 2021.
31. The water cycle. NASA. 2010. https://earthobservatory.nasa.gov/features/Water. Accessed 23 Oct 2021.
32. Brun J et al. Adapting to impacts of climate change on water supply in Mexico city. Human Development Report Office, UNDP; 2007.
33. Climate impacts on water resources, Dec 2016. https://19january2017snapshot.epa.gov/climate-impacts/climate-impacts-water-resources.html. Accessed 23 Oct 2021.
34. Change OC et al. Intergovernmental panel on climate change. World Meteorological Organization. 2007.
35. Garrido A, Dinar A et al. Managing water resources in a time of global change. Routledge; 2009. ISBN 9780415619776.
36. Luce CH. Effects of climate change on snowpack, glaciers, and water resources in the northern Rockies. In: Climate change and rocky mountain ecosystems. Springer; 2018. pp. 25–36.
37. Asia's glaciers to shrink by a third by 2100, threatening water supply of millions, Sep 2017. https://www.theguardian.com/environment/2017/sep/14/asia-glaciers-shrink-threatening-water-supply. Accessed 23 Oct 2021.
38. Jamali S, Abrishamchi A, Madani K. Climate change and hydropower planning in the Middle East:

implications for Iran's Karkheh hydropower systems. J Energy Eng. 2013;139(3):153–60.
39. Extreme temperature change: water temperatures, Jul 2021. https://climatechange.lta.org/climate-impacts/water-temperatures/. Accessed 23 Oct 2021.
40. Field CB, Barros VR. Climate change 2014–impacts, adaptation and vulnerability: regional aspects. Cambridge University Press; 2014.
41. Majumder M. Impact of urbanization on later shortage in face of climatic aberrations. Springer; 2015. ISBN 978-981-4560-73-3.
42. Ahuja S. Handbook of water purity and quality. Academic Press; 2009. eBook ISBN 9780080921129.
43. Schaeffer R, Szklo AS, de Lucena AFP, Borba BSMC, Nogueira LPP, Fleming FP, Troccoli A, Harrison M, Boulahya MS. Energy sector vulnerability to climate change: a review. Energy. 2012;38(1):1–12.
44. Teot′onio C, Fortes P, Roebeling P, Rodriguez M, Robaina-Alves M. Assessing the impacts of climate change on hydropower generation and the power sector in Portugal: a partial equilibrium approach. Renew Sustain Ener Rev. 2017;74:788–99.
45. Turner SW, Ng JY, Galelli S. Examining global electricity supply vulnerability to climate change using a high-fidelity hydropower dam model. Sci Total Environ. 2017;590:663–75.
46. Hamududu B, Killingtveit A. Assessing climate change impacts on global hydropower. Energies. 2012;5(2):305–22.
47. Gaudard L, Romerio F, Dalla Valle F, Gorret R, Maran S, Ravazzani G, Stoffel M, Volonterio M. Climate change impacts on hydropower in the Swiss and Italian Alps. Sci Total Environ. 2014; 493:1211–21.
48. Vicuna S, Dracup JA, Dale L. Climate change impacts on two high-elevation hydropower systems in California. Clim Change. 2011;109(1):151–69.
49. de Oliveira VA, de Mello CR, Viola MR, Srinivasan R. Assessment of climate change impacts on streamflow and hydropower potential in the headwater region of the Grande river basin, southeastern Brazil. Int J Climatol. 2017;37(15):5005–23.
50. Cherry JE, Knapp C, Trainor S, Ray AJ, Tedesche M, Walker S. Planning for climate change impacts on hydropower in the far north. Hydrol Earth Syst Sci. 2017;21(1):133–51.
51. Chang J, Wang X, Li Y, Wang Y, Zhang H. Hydropower plant operation rules optimization response to climate change. Energy. 2018;160:886–97.
52. Shrestha A, Shrestha S, Tingsanchali T, Budhathoki A, Ninsawat S. Adapting hydropower production to climate change: a case study of Kulekhani hydropower project in Nepal. J Cleaner Prod. 2020;279:123483.
53. Ray PA, Bonzanigo L, Wi S, Yang Y-CE, Karki P, Garcia LE, Rodriguez DJ, Brown CM. Multidimensional stress test for hydropower investments facing climate, geophysical and financial uncertainty. Glob Environ Chang. 2018;48:168–81.
54. Carvajal PE, Li FG, Soria R, Cronin J, Anandarajah G, Mulugetta Y. Large hydropower, decarbonisation and climate change uncertainty: Modelling power sector pathways for Ecuador. Energy Strat Rev. 2019;23:86–99.
55. Agovino M, Casaccia M, Ciommi M, Ferrara M, Marchesano K. Agriculture, climate change and sustainability: the case of EU-28. Ecol Ind. 2019;105:525–43.
56. Brown M, Antle J, Backlund P, Carr E, Easterling B, Walsh M, Ammann C, Attavanich W, Barrett C, Bellemare M et al. Climate change, global food security and the us food system. 2015.
57. UN-Water. Water and climate change. https://www.unwater.org/water-facts/climate-change/. Accessed 23 Oct 2021.
58. Adams RM, Peck DE. Effects of climate change on water resources. Choices. 2008;23(316–2016–6682):12–14.

3 Drought Management

I. Ethem Karadirek

Abstract

Drought is a phenomenon that adversely affects a great variety of human activities in virtually all climatic regions. Drought is defined as a period that a region experiences below normal precipitation. Drought can last for days, weeks, months, and years. The period of drought lasts longer, the affects get greater. Drought is a natural disaster that should be managed by implementing sustainable disaster management strategies. National drought plans are of great importance to reduce impacts of drought phenomena. This chapter aims to provide a brief summary about drought management under changing climate.

Keywords

Climate change · Drought indices · Drought management · Drought risk assessment

3.1 Introduction

Water resources are under pressure due to population growth, industrialization, expansions in energy and agricultural sectors, and climate change. Climate change and water resources have a close relationship. Impacts of climate change on water resources have been comprehensively discussed over the years [1–3]. Climate change does not only result in temperature increase, but it also causes changes in precipitation behaviour and pattern. Climate change affects the hydrological cycle ending up with an increase in drought events, which have significant impacts on agriculture, ecosystem, and societies [4, 5]. Drought episodes start with meteorological drought resulting from lack of precipitation, then affects soil moisture resulting in agricultural drought, afterwards causes hydrological drought [6].

Climate change affects the frequency, severity, and duration of droughts. Drought, as a natural disaster, plays vital role for water resources planning and management, and requires sustainable risk management strategies in order to reduce its impacts. This chapter provides an understanding of droughts. The main objective is to provide a broad review of drought phenomena.

I. E. Karadirek (✉)
Department of Environmental Engineering, Akdeniz University, Antalya, Turkey
e-mail: ethemkaradirek@akdeniz.edu.tr

M. Bahadir and A. Haarstrick (eds.), *Water and Wastewater Management*,
Water and Wastewater Management, https://doi.org/10.1007/978-3-030-95288-4_3

3.2 Drought Phenomena

Drought is a phenomenon that adversely affects a great variety of human activities in virtually all climatic regions [7–9]. There are different definitions for drought. However, drought is defined as "*a temporary lack of water, which is, at least partly, caused by abnormal climate conditions and is damaging to an activity, group, or the environment*" [10]. Drought is a condition originating from precipitation deficiency that causes water shortage for activities, group and/or environment. Water shortage associated with drought should be considered as a relative condition [11]. Drought differs from aridity. Drought is a temporary condition, whereas aridity is a long-term climatic condition [12]. Drought depends on many factors such as effectiveness and intensity of rainfall and number of rainfall events [11], timing and spatial distribution of rain, hydro-environmental factors, usage of water depending on needs in terms of quantity and timing [10]. Drought can last from days to years depending on water resources [13]. Drought is generally classified into four categories as *meteorological*, *hydrological*, *agricultural* and *socio-economic drought* [11].

Meteorological drought, which is often regional and presumably based on a thorough understanding of regional climatology, is defined as lack of precipitation for a given period over a region [14–16]. Meteorological measurements are the leading indicators to express drought. An ongoing meteorological drought event can intensify quickly or end abruptly. Drought periods are generally defined as the number of days with precipitation below specified threshold values [17].

Hydrological drought refers to the decrease and deficiencies in surface and ground waters that occur because of a long-term lack of precipitation. River flow, lake, reservoir, and groundwater level measurements are generally used for hydrologic drought analysis. Hydrological measurements are not solely first indicator for drought, as there is a time gap between lack of rain and the lack of water in streams and reservoirs. Hydrological drought may last longer after the meteorological drought has ended [18–20].

Agricultural drought is a condition, in which there is not enough moisture in the root zone of the plant for growth. Agricultural drought occurs, if there is not enough soil moisture during a certain critical period during the growth period when a particular plant needs water. Agricultural drought is a typical situation that occurs after meteorological drought and before hydrological drought. Agricultural drought can seriously reduce crop yields, even when the soil is deeply saturated. In addition to the lack of precipitation, higher temperatures and relative humidity exacerbate the effects of agricultural drought [21–23].

Socio-economic drought, which is closely related with human life, occurs, while demand of a society cannot be met by water resources due to lack of precipitation [24, 25].

Meteorological, hydrological, and agricultural droughts are physical phenomena, while socio-economic drought is usually associated with insufficient water supply [26].

3.3 Drought Indices

Many indices have been developed to assess and to monitor drought episodes. In this section, the most widely utilized and recognized drought indices such as standardized precipitation index (SPI), Palmer drought severity index (PDSI), percent of normal index (PNI), surface water supply index (SWSI), crop moisture index (CMP), standardized precipitation evapotranspiration index (SPEI) and decile index (DI) are summarized.

3.3.1 Standardized Precipitation Index (SPI)

Standardized precipitation index (SPI) is the foremost broadly utilized index for drought assessment, as it is easy to use. SPI was firstly introduced in 1993 by Mckee et al. [27]. SPI only requires precipitation data, which can be easily acquired from rainfall gauges and/or rainfall satellite data [16]. SPI is easy to use for monitoring drought and determining lack of precipitation at varying time scales such as 3, 6, 12, 24, and 48 months [27]. SPI is computed based on a long-term precipitation data for a certain period. Precipitation data set is fitted to probability distribution that is then altered to normal distribution, meaning mean SPI is zero for a certain region at a certain period [28]. As precipitation data is available and applicable for all climate conditions, SPI that is admirably requires a data set of at least 30 years can be calculated for varying time periods [27].

Drought magnitude (DM) according to SPI is computed based on Eq. (3.1) [27]:

$$DM = -\sum_{j=1}^{x} SPI_{ij} \tag{3.1}$$

where j represents the first month of drought and keeps increasing till the end of drought (x) for any of time scale (i). The unit of DM is months and might be equivalent to duration of drought in case of SPI = −1 for each month [27]. Drought magnitude can have values starting from −2 representing extreme drought to +2 standing for extremely wet. Preliminary precipitation record is important, as SPI is computed based on the precipitation data set. The fundamental of SPI is based on distributions of precipitation data. Therefore, numerical differences should be taken into account during computing SPI values when different lengths of precipitation data are used [16]. Probability distribution of precipitation data is important for computing SPI values, and the most widely utilized approach is gamma distribution [29].

3.3.2 Palmer Drought Severity Index (PDSI)

Palmer drought severity index was developed in 1965 by Palmer [30]. PDSI, which is generally used for monitoring drought, areal extent and severity of drought periods at regional scale, requires available precipitation and temperature data for estimation of relative dryness [30]. The PDSI is most widely utilized regional drought index and based on supply and demand for soil moisture. The PDSI is an outstanding index for long term analysis of drought [31, 32]. Effects of global warming on drought episodes can be demonstrated as PDSI uses temperature and physical water balance data. PDSI is usually calculated on monthly basis. Precipitation, temperature, and soil moister capacity are used as inputs, which are helpful for evaluation of evapotranspiration, surface flow and moisture loss from the surface. On the one hand, PDSI is delicate to precipitation and temperature data sets, as these data sets are the main input parameters. On the other hand, PDSI does not account snow/ice and human practices such as irrigation that affect the water balance [33]. Positive value of PDSI represents the wet, whereas negative values stand for drought periods.

3.3.3 Percent of Normal Index (PNI)

The percent of normal index (PNI) is the simplest drought index and is calculated as a percentage, mainly by dividing the amount of precipitation over the average precipitation of a specified period. Periods of precipitation of 12 months or less can also be used to calculate the PNI [34]. The PNI is calculated based on Eq. (3.2) [34]:

$$PNI = \left(\frac{P_i}{P}\right) * 100 \tag{3.2}$$

where P_i stands for precipitation in time increment i (mm), and P is the average precipitation for the studied period (mm).

The PNI, which is easy to understand and to calculate, can be calculated for varying time periods ranging from a month to a couple of months representing a certain season or annual [34, 35]. The PNI was firstly developed to assess meteorological drought. However, it can also be utilized for calculation of streamflow drought severity using streamflow data [35, 36].

3.3.4 Surface Water Supply Index (SWSI)

Shafer and Dezman developed the surface water supply index (SWSI) in 1982 in order to assess the hydrological drought [37]. The priority of SWSI is to monitor fluctuations in surface water resources, which makes SWSI is a helpful tool for monitoring the impacts of hydrological drought on different sectors such as urban and industrial water supply [37, 38]. Snowpack, streamflow, precipitation and reservoir storage are the required parameters for calculation of the SWSI [37]. Snowpack data may be replaced with streamflow data during summer seasons [16]. There are also studies that replacing snowpack component with groundwater data to calculate the SWSI in some regions, where groundwater is more important than snowpack [39].

3.3.5 Crop Moisture Index (CMI)

Crop moisture index (CMI) was developed for assessment of short-term moisture. Calculation of the CMI requires weekly temperature and precipitation data [40]. The CMI has limitations for long-term assessment of drought [16], whereas the CMI has been reported as an effective tool for assessment of agricultural drought in warmer seasons [41].

3.3.6 Standardised Precipitation Evapotranspiration Index (SPEI)

Standardized precipitation evapotranspiration index (SPEI) is a multi-scalar drought index based on climatic data. The SPEI, which can be calculated ranging from 1–48 months, is calculated based on precipitation and potential evapotranspiration (PET) [42]. The SPEI, which can be implemented for determination of onset, duration and magnitude of drought episodes, combines a great variety of timescales of the SPI with evapotranspiration data, which makes it effective for climate change studies [42, 43]. The SPEI that is sensitive to the calculation of PET requires more data than the SPI.

3.3.7 Decile Index (DI)

Decile index (DI) is one of the simplest indices for drought monitoring. The DI is defined as the ranking of precipitation in certain time intervals over the historical periods [44]. Rainfall is the only required data for calculation of DI that provides an effective and accurate statistical measurement of precipitation. A cumulative distribution function is created by ranking long-term (generally 30 years and more) precipitation data from the highest to the lowest and then distribution is split into ten groups of each is called a decile [45]. The deciles are classified into five groups, two deciles per each class. Precipitation lower than 20% is classified as much below normal, whereas precipitation higher than 80% is classified as much above normal [46].

3.4 Drought Management Strategies

Drought can be managed based on two basic approaches namely reactive (crisis) and proactive (risk) management. Reactive management approach is based on implementing measures and actions after beginning of a drought episode [47]. Reactive (crisis) management approach for a drought is generally inefficient due to limited time to take actions and is mostly based on relief actions for recovery from drought disaster [48]. On the other hand, proactive (risk) management approaches require implementing all components of disaster management, as given in Fig. 3.1.

Steps to develop a national drought policy are recommended by World Meteorological Organization (WMO) and Global Water Partnership (GWP) [49]. These steps are "*i. appointing a drought management policy, ii. defining goals*

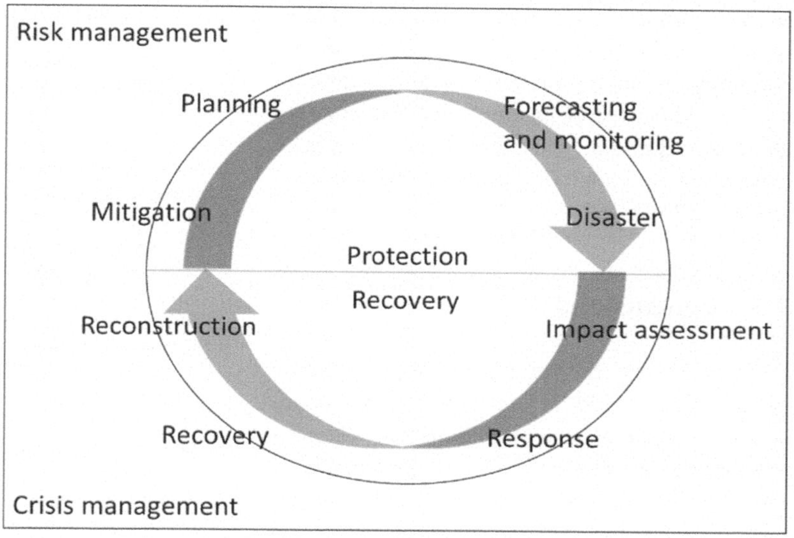

Fig. 3.1 Disaster management cycle [47]

and objectives of a proactive based drought management policy, iii. ensuring stakeholder participation, solving conflicts between water use sectors, iv. specification of risk groups and inventory data and financial resources, v. preparation of drought policy and plans including monitoring, early warning and forecasting, assessment of risks and impacts, mitigation and response, vi. identification of research needs and filling institutional gaps, vii. Integrating science and policy aspects, viii. Promoting drought management policy and preparedness and building public awareness and consensus, ix. Developing educational programs, and x. evaluating and revising drought management policy and supporting preparedness plans" [49, 50].

Drought management should consist of drought assessment, considering system scale outlook, changes in risk over time (climate change, socio-economic development etc.), and sectors and people under risk. Effective drought management strategies should provide an increase in resilience and enhancement in preparedness [51].

3.4.1 Drought Risk Assessment

Drought is the result of the lack of precipitation that causes water shortage for activities such as farming and/or for a group of people such as farmers [11]. Frequency and severity of droughts are affected by climate change [52, 53]. Drought is a natural hazard, which should be managed by implementing sustainable risk management strategies [54]. Developing national drought plans play a key role for drought management [55], and projections on impacts of climate change should be considered for developing drought plans [54]. Forecasting and monitoring of drought, as an early warning system, is crucial for developing drought management plans. Drought monitoring is implemented by applying drought indices that reflect the deviations from normal regional conditions [56].

Mapping drought, which helps determining the regions being most at risk of drought, is helpful for assessing drought risks and developing drought management plans [57]. Drought risk is calculated based on the likelihood of adverse impacts resulted from interactions between hazard, exposure, and vulnerability. Although, an appropriate way to assess drought risk should consist of incorporating vulnerability and hazard [58–60], only few studies have been carried out by integrating vulnerability and hazard so far [59].

3.5 Conclusions

Drought is one of the most complex natural hazards that should be managed by risk management approaches. Relationship between meteorological and hydrological parameters is vital for drought risk assessment and management. Developing national drought plans and implementing sustainable measures are vital to reduce impacts of drought episodes. This chapter provided an understanding of drought, drought indices, drought management strategies and risk assessment.

Acknowledgements The author would like to thank Akdeniz University, Antalya, Turkey, and EXCEED Swindon project funded by DAAD (German Academic Exchange Service).

References

1. Zhang X, Chen YF, Xia J, Yang QC. Impacts of climate change on the availability of water resources and water resources planning. Conference Impacts of climate change on the availability of water resources and water resources planning, Hohai Univ, Nanjing, Peoples R China vol. 350. Int Assoc Hydrological Sciences; 2011. p. 324.
2. Yang N, Men BH, Lin CK. Impact analysis of climate change on water resources. Conference Impact analysis of climate change on water resources, Nanjing, Peoples R China vol. 24. Elsevier Science Bv; 2011. p. 643–8.
3. Kundzewicz ZW, Krysanova V, Benestad RE, Hov O, Piniewski M, Otto IM. Uncertainty in climate change impacts on water resources. Environ Sci Policy. 2018;79:1–8.
4. Breshears DD, Cobb NS, Rich PM, Price KP, Allen CD, Balice RG, et al. Regional vegetation die-off in response to global-change-type drought. Proc Natl Acad Sci USA. 2005;102(42):15144.
5. Madadgar S, AghaKouchak A, Farahmand A, Davis SJ. Probabilistic estimates of drought impacts on agricultural production. Geophys Res Lett. 2017;44(15):7799–807.
6. Tabari H. Climate change impact on flood and extreme precipitation increases with water availability. Sci Rep. 2020;10(1):13768.
7. Van Loon AF, Gleeson T, Clark J, Van Dijk A, Stahl K, Hannaford J, et al. Drought in the anthropocene. Nat Geosci. 2016;9(2):89–91.
8. Bachmair S, Stahl K, Collins K, Hannaford J, Acreman M, Svoboda M, et al. Drought indicators revisited: the need for a wider consideration of environment and society. Wiley Interdiscip Rev-Water. 2016;3(4):516–36.
9. Kchouk S, Melsen LA, Walker DW, van Oel PR. A review of drought indices: predominance of drivers over impacts and the importance of local context. Nat Hazards Earth Syst Sci Discuss. 2021;2021:1–28.
10. Droughts KG. Annu Rev Environ Resour. 2008;33 (1):85–118.
11. Wilhite DA, Glantz MH. Understanding: the drought phenomenon: the role of definitions. Water Int. 1985;10(3):111–20.
12. Maliva R, Missimer T. Aridity and drought. Environmental science and engineering (Subseries: Environmental science); 2012. p. 21–39.
13. Wood EF, Schubert SD, Wood AW, Peters-Lidard CD, Mo KC, Mariotti A, et al. Prospects for advancing drought understanding, monitoring, and prediction. J Hydrometeorol. 2015;16(4):1636–57.
14. Santos MA. Regional droughts—a stochastic characterization. J Hydrol. 1983;66(1–4):183–211.
15. Chang TJ, Kleopa XA. A proposed method for drought monitoring. Water Resour Bull. 1991;27 (2):275–81.
16. Mishra AK, Singh VP. A review of drought concepts. J Hydrol. 2010;391(1):202–16.
17. Estrela MJ, Penarrocha D, Millan M. Multi-annual drought episodes in the Mediterranean (Valencia region) from 1950–1996. A spatio-temporal analysis. Int J Climatol. 2000;20(13):1599–618.
18. Dracup JA, Lee KS, Paulson EG. On the statistical characteristics of drought events. Water Resour Res. 1980;16(2):289–96.
19. Mohan S, Rangacharya NCV. A modified method for drought identification. Hydrol Sci J-J Sci Hydrol. 1991;36(1):11–21.
20. Clausen B, Pearson CP. Regional frequency-analysis of annual maximum streamflow drought. J Hydrol. 1995;173(1–4):111–30.
21. Rhee J, Im J, Carbone GJ. Monitoring agricultural drought for arid and humid regions using multi-sensor remote sensing data. Remote Sens Environ. 2010;114(12):2875–87.
22. Zargar A, Sadiq R, Naser B, Khan FI. A review of drought indices. Environ Rev. 2011;19:333–49.
23. Mannocchi F, Todisco F, Vergni L. Agricultural drought: indices, definition and analysis. Conference Agricultural drought: indices, definition and analysis, Rome Headquarters Italian National Res Council,

Rome, ITALY. Int Assoc Hydrological Sciences; 2004. p. 246–54.
24. Liu SN, Shi HY, Sivakumar B. Socioeconomic drought under growing population and changing climate: a new index considering the resilience of a regional water resources system. J Geophys Res-Atmos. 2020;125(15):21.
25. Shi HY, Chen J. Characteristics of climate change and its relationship with land use/cover change in Yunnan Province, China. Int J Climatol. 2018;38 (5):2520–37.
26. Tu XJ, Wu HO, Singh VP, Chen XH, Lin KR, Xie YT. Multivariate design of socioeconomic drought and impact of water reservoirs. J Hydrol. 2018;566:192–204.
27. McKee TB, Doesken NJ, Kleist J. The relationship of drought frequency and duration to time scales. In: Eighth Conference on applied climatology; 1993. pp. 179–84.
28. Edwards DC. Characteristics of 20th century drought in the United States at multiple time scales. Conference characteristics of 20th century drought in the United States at multiple time scales.
29. Mishra AK, Singh VP. Analysis of drought severity-area-frequency curves using a general circulation model and scenario uncertainty. J Geophys Res-Atmos. 2009;114:18.
30. Palmer WC. Meteorologic drought. US Department of Commerce, Weather Bureau. Research Paper No: 45; 1965. p. 58.
31. Kim TW, Valdes JB. Nonlinear model for drought forecasting based on a conjunction of wavelet transforms and neural networks. J Hydrol Eng. 2003;8(6):319–28.
32. Dai A, Trenberth KE, Qian TT. A global dataset of palmer drought severity index for 1870–2002: relationship with soil moisture and effects of surface warming. J Hydrometeorol. 2004;5(6):1117–30.
33. Alley WM. The palmer drought severity index - limitations and assumptions. J Climate Appl Meteorol. 1984;23(7):1100–9.
34. Willeke G, Hosking JRM, Wallis JR, Guttman NB. The national drought atlas. Institute for Water Resources Report 94-NDS-4, US Army Corps of Engineers; 1994.
35. Nikbakht J, Tabari H, Talaee PH. Streamflow drought severity analysis by percent of normal index (PNI) in northwest Iran. Theor Appl Climatol. 2013;112(3–4):565–73.
36. Boughton WC. Multi-year streamflow drought in eastern Australia. Aust J Water Resour. 2009;13 (1):31–42.
37. Shafer BA, Dezman LE. Development of a surface water supply index (SWSI) to assess the severity of drought conditions in snowpack runoff areas. In: 50th annual western snow conference. Reno, Nevada; 1982.
38. Hoekema DJ, Sridhar V. Relating climatic attributes and water resources allocation: a study using surface water supply and soil moisture indices in the Snake River basin, Idaho. Water Resour Res. 2011;47:17.
39. Jang SH, Lee J-K, Oh JH, Jo JW, Cho Y. The probabilistic drought prediction using the improved surface water supply index in the Korean peninsula. Hydrol Res. 2018;50(1):393–415.
40. Palmer WC. keeping track of crop moisture conditions, nationwide: the new crop moisture index. Weatherwise. 1968;21(4):156–61.
41. Heim RR. A review of twentieth-century drought indices used in the United States. Bull Am Meteorol Soc. 2002;83(8):1149–65.
42. Vicente-Serrano SM, Begueria S, Lopez-Moreno JI. A Multiscalar drought index sensitive to global warming: the standardized precipitation evapotranspiration index. J Clim. 2010;23(7):1696–718.
43. Begueria S, Vicente-Serrano SM, Reig F, Latorre B. Standardized precipitation evapotranspiration index (SPEI) revisited: parameter fitting, evapotranspiration models, tools, datasets and drought monitoring. Int J Climatol. 2014;34(10):3001–23.
44. Gibbs WJ, Maher JV. Rainfall deciles as drought indicators: bureau of meteorology; 1967.
45. Smakhtin VU, Hughes DA. Automated estimation and analyses of meteorological drought characteristics from monthly rainfall data. Environ Model Softw. 2007;22(6):880–90.
46. Morid S, Smakhtin V, Moghaddasi M. Comparison of seven meteorological indices for drought monitoring in Iran. Int J Climatol. 2006;26(7):971–85.
47. Wilhite DA, Hayes MJ, Knutson C, Smith KH. Planning for drought: moving from crisis to risk management. JAWRA J Am Water Resour Assoc. 2000;36(4):697–710.
48. Wilhite DA, Sivakumar MVK, Pulwarty R. Managing drought risk in a changing climate: the role of national drought policy. Weather Clim Extremes. 2014;3:4–13.
49. World Meteorological Organization (WMO) and Global Water Partnership (GWP): National drought management policy guidelines: a template for action (D.A. Wilhite). Integrated drought management programme (IDMP) tools and guidelines series 1 WMO, Geneva, Switzerland and GWP, Stockholm, Sweden; 2014.
50. Wilhite DA. Integrated drought management: moving from managing disasters to managing risk in the Mediterranean region. Euro-Mediterr J Environ Integr. 2019;4(1):42.
51. WorldBank. Assessing drought hazard and risk: principles and implementation guidance. Washington, DC: World Bank. 2019.
52. Cockfield C, Dovers S. The science and policy of climate variability and climate change: intersections and possibilities. In: Botterill LC, Cockfield C, editors. Drought, risk management, and policy. 1st ed. Boca Raton: CRC Press; 2013. p. 29–44.
53. Kim TW, Jehanzaib M. Drought risk analysis, forecasting and assessment under climate change. Water. 2020;12(7):7.

54. Dikici M. Drought analysis with different indices for the Asi Basin (Turkey). Sci Rep. 2020;10(1):20739.
55. Al-Safi HIJ, Sarukkalige PR. Assessment of future climate change impacts on hydrological behavior of Richmond River Catchment. Water Sci Eng. 2017;10 (3):197–208.
56. Dai AG. Drought under global warming: a review. Wiley Interdiscip Rev-Clim Chang. 2011;2(1):45–65.
57. Nunez JH, Verbist K, Wallis JR, Schaefer MG, Morales L, Cornelis WM. Regional frequency analysis for mapping drought events in north-central Chile. J Hydrol. 2011;405(3–4):352–66.
58. Jia H, Pan D, Wang J-a, Zhang W-c. Risk mapping of integrated natural disasters in China. Nat Hazards. 2016;80(3):2023–35.
59. Rajsekhar D, Singh VP, Mishra AK. Integrated drought causality, hazard, and vulnerability assessment for future socioeconomic scenarios: An information theory perspective. J Geophys Res-Atmos. 2015;120(13):6346–78.
60. Dabanli I. Drought hazard, vulnerability, and risk assessment in Turkey. Arab J Geosci. 2018;11 (18):12.

Flood Management Under Changing Climate

4

S. Yurdagül Kumcu

Abstract

Many researchers have worked on global warming during the last decades, because its consequences are affecting all areas of our life. Increasing of planet's temperature is main cause of global warming. According to NASA, the Earth average temperature has increased about 1 degree Fahrenheit during the twentieth century. This increase in the temperature may be seen small change, but its impacts are inevitable problems leading lasting scar on the planet. Global warming has many negative effects on the atmosphere, of which floods are the most devastating ones. Natural processes that cause rain, snowfall, hailstorms, and rise in sea levels are reliant on many diverse factors. If the weather gets warmer, evaporation processes from both land and sea increase. If evaporation increases and cannot be compensated by precipitation, this leads to drought. It means scarcity of water resources and crop famine particularly in the regions where the temperature is already high. If evaporated water falls on unexpected time and unexpected place, it causes flood. Increasing temperature is the reason of the melting of ice and glaciers rapidly, which leads to raising the sea level, which is another form of floods. Heavy rainfalls and more often thunderstorms, which cause flood, are among the global warming effect of climate change. Floods have already caused significant loss of life and properties in human life. It is predicted that the amount of precipitation per unit area will increase with the effect of climate change, and also foreseen the rising of the number and intensity of floods. In this case, it is expected that flood damages will increase, too. It may not be possible to prevent floods, but it is possible to reduce the damages resulting from floods. Structural and non-structural measures should be taken to protect against flooding. An integrated river basin management approach and a logical coordinative plan are essential for a sustainable flood management. However, integrated basin development is complex and implies the application of a holistic and multi-disciplinary approach (IPCC, Climate change 2007: Synthesis report. 2007. https://www.ipcc.ch/site/assets/uploads/2018/02/ar4_syr_full_report.pdf. Accessed 24 Oct 2021). Flood management plan studies include (1) inundated areas under possible floods, (2) preparing flood hazard maps for various return periods, (3) preparing flood risk maps, which have the analysis of the potential negative impacts of flood on people, buildings, agricultural areas and infrastructures, and

S. Y. Kumcu (✉)
Department of Civil Engineering, Necmettin Erbakan University, Konya, Turkey
e-mail: yurdagulkumcu@gmail.com

M. Bahadir and A. Haarstrick (eds.), *Water and Wastewater Management*,
Water and Wastewater Management, https://doi.org/10.1007/978-3-030-95288-4_4

(4) objectives and measures before, during and after the floods. These approaches are presented in this chapter.

Keywords

Climate change · Flood management · Flood hazard maps · Flood risk maps · Hydrology

4.1 Introduction

Climate change and water are closely related to each other. According to scientists, the most important effects of climate change are disruption of the water cycle and water change in quantity and quality. Although water amount in the world is constant because of water cycle, management of water resources in quality and quantity will be difficult in the near future, as the water resources are variable with time and place.

Climate change means in general terms involving changes in many climatological factors such as temperature and precipitation. These changes occur due to global warming. For example, if one looks at on a large scale, Sahel Zone located in the Central Africa became more arid, while USA became rainier in the twentieth century. In California, these changes lead to a reduction of snow mass in Sierra and this has caused water shortage. Rapid glacial melt in Antarctica and Greenland has been influencing ocean currents, causing an increase of sea levels, destructive storms and hurricanes around the planet [1]. As a result, climate change encompasses not only rising average temperatures, but also extreme weather events and additionally increases in the frequency and severity of these events. Climate change is leading to increasing temperatures on land and sea as well as changing of precipitation behavior and patterns. While the weather pattern is varying regionally as changing from arid to rainy weather in arid regions, it is also changing seasonally from snowy to wetter precipitations in winters and from dry to drier weather in summers.

4.2 Floods

Natural hazards vary in magnitude and intensity in time and space. Under certain conditions and influenced by triggering factors, they may cause loss of life, destroy infrastructures and properties, impede economic and social activities, and cause destruction of cultural heritage monuments and the environment. It should be stressed that, during the last few decades, natural hazards have been the cause of the loss of hundreds of thousands of human lives and for damage and losses of billions of euros around the world. Just in the period 1974–2003 more than two million people lost their lives due to natural hazards [2].

Among the most destructive natural hazards are floods caused by river overflows, flash floods in the cities, and coastal floods in the coastal areas. A warmer climate with its increased climate variability will increase the risk of both floods and droughts [3, 4]. As there are a number of climatic and non-climatic drivers influencing flood and drought impacts, the realization of risks depends on several factors. Flooding is an overflowing of water onto land that is normally dry. Flooding may occur as an overflow of water from water bodies, such as a river, lake, or ocean, in which the water overtops or breaks levees, resulting in some of that water escaping its usual boundaries. Floods include river floods, flash floods, urban floods and sewer floods, and can be caused by intense and/or long-lasting precipitation, snowmelt, dam break, or reduced conveyance due to ice jams or landslides. Floods depend on precipitation intensity, volume, timing, antecedent conditions of rivers and their drainage basins (e.g., presence of snow and ice, soil character, wetness, urbanization, and existence of dikes, dams, or reservoirs). Human encroachment into flood plains and lack of flood response plans increase the damage potential.

The severe floods led the researchers to set in force the defense against floods [5]. It is the purpose of this overview to review the advances in flood risk assessment and management both from the scientific and professional point of view. In this context, it highlights some important critical aspects, which should be addressed based on the latest scientific findings, resulting in a detailed modelling of floods and giving reliable flood risk maps and plans.

4.2.1 Steps to Be Taken Before, During and After the Flood

(1) *Steps to be taken before the flood*

(1.1) Definition of hazard and risk
- Investigation of flood return periods
- Determination of the area under potential flood risk
- Preparation of flood hazard and risk maps

(1.2) Reducing the amount of the hazard
- Combining the flood risk management and land use
- Application of structural and non-structural flood measurement techniques
- Estimation of the flood risk and application of the early warning systems
- Preparation of the action plans in order to reduce loss of life and property

(1.3) Planning and preparedness
- Establishing the structural and non-structural measurements for reducing the flood hazard
- Preparing the action plans

(2) *Steps to be taken during and after the flood*

(2.1) Interventions
- Lifesaving activities
- Public health
- Enabling to open access road
- Repairing the critical facilities
- Warning systems
- Providing the health and security of flood intervening group
- Management of social media and VIP
- Operation control and coordination
- Ensuring transportation, accommodation of evacuated people

(2.2) Recovering actions
- Transition from rescue works to recovering actions
- Providing to return normal life
- Conducting damage assessment
- Recovering financial costs of the hazard
- Receiving information
- Recovering services

4.2.2 Flood Risk Assessment

4.2.2.1 Preliminary Flood Risk Assessment

Preliminary flood risk assessment is based on availability or easily accessibility of information, which will be used in the study. Within the scope of the study, the floods that occurred in the past are examined, important historical floods are determined, and by using various methods possible future floods and inundated areas estimated.

Following the preliminary flood risk assessment, areas with a serious potential flood risk are determined, flood hazard maps and flood and flood risk maps are prepared, respectively, in these identified areas [6]. Finally, a flood risk management plan is prepared based on these studies. For this reason, flood risk assessment is very important in terms of carrying out it primarily in those areas having flood risk. Since flood management plans will be reviewed periodically, preliminary flood risk assessment, flood hazard and flood risk maps should also be updated. The preliminary studies include the following studies in the order:

- Data acquisition and evaluation;
- Assessment of serious historical floods;
- Determination of serious historical floods;
- Determination of possible floods;
- Identification of areas under flood risk.

4.2.2.2 Flood Hydrology

The most fundamental part of any flood hydrology analysis is the compilation and analysis of hydrologic and meteorological data accumulated during and after severe flood events. These data are required in the development of criteria by hydro-meteorologists for making PMP (probable maximum precipitation) estimations, for development of unit hydrograph and infiltration parameters necessary to determine the rainfall-runoff relationships for both gauged and ungauged basins, and for preparing discharge-probability relationships. Hydrologic data include records of flood runoff measured at continuous recording streamflow gauges, crest stage streamflow gauges, indirect peak discharge measurements based on flood marks at locations, where there are no stream gauges, and reservoir operation records, from which inflow hydrographs may be determined based on outflow and change of storage relationships. Meteorological data include precipitation, temperature and wind records collected at climatological stations. Other vitally important data include data related to watershed characteristics such as topography, amount and type of vegetation, geologic setting, drainage network development, and degree of development. Return periods and flow rates of floods are computed by various methods. These methods are determined up to the size and characteristics of the basin, details of the work, and availability of data [7].

The main methods that can be used to compute the floods in small and medium scaled basins are:

- Rational methods: Maximum discharge resulted from rainfall, which is distributed uniformly to the peak flow rate to be obtained directly by flow observation stations;
- Statistical methods;
- Flood recurrence curves of the river basin;
- Hydrological models developed for the basin.

4.2.2.3 Flood Hazard Maps

Flood hazard assessment and mapping is useful in order to identify inundated areas at risk under flooding, and consequently to develop flood risk maps and flood risk management including disaster preparedness. Flood hazard assessments and maps typically bring out the depth of flooding in a given location, based on various scenarios (e.g., 100-year events, 50-year events, etc.). These maps are fundamentals of flood risk maps. If the depth of flooding overflows the riverbed and riverbanks, the detailed studies are done for the flood risk maps [8].

Flood hazard maps are important for land use planning in flood-prone areas. It offers opportunities for easily read and understandable, rapidly-accessible figures and maps, which show the identification of inundated areas at risk of flooding [9]. Flood hazard maps enable to increase awareness of flooding among the public, local authorities and other organizations. They also encourage people living and working in flood-prone areas to find out more about the local flood risk and to take appropriate precautions and actions [10].

Flood hazard maps are obtained for various flood return periods. Discharges belonging to return periods tend to change by climate change. Therefore, climate change should be considered, and flood hazard maps should be refreshed as flood hazard mapping typically obtained as a 'snapshot' of flood risk at a given point in time for a given flood return period. Geographic Information Systems (GIS) are generally used to visualize flood hazard maps. These maps compute the water levels and show inundated areas under the flood.

4.2.2.4 Flood Risk Maps

Flood risk assessment is the analysis of the worst possible consequences of flooding. The main purpose of the assessment of flood risk and damage is human safety, flood prevention in floodplains and preventing damage to public and private infrastructure, commercial and other economic activities. Flood risk maps allow deciding, which areas are under the highest risk, and what measures have to be taken [11, 12]. Flood risk map includes:

- Population affected by flood;
- Damage, which is occurred in buildings and goods because of flood;
- Flooded strategic structures and infrastructures.

Flood risk maps show inundated areas for various return periods like Q50, Q100 and Q500. Figure 4.1 shows flood risk maps of Konya and the economic losses and affected population of the related return periods.

4.2.2.5 Flood Management Plans

Flood management plans aim to struggle flood risk by determining, what needs to be done in the management of flood risks, when and by whom the actions should be made. These are items pointed at to be reached in the management of flood risks.

While the flood hazard definition includes the height of the flow, the flow velocity and inundated areas during the flood, and the duration of the flooding, the risk includes both flood hazard and the person and goods that will be affected by the flood.

The flood management plan aims to minimize or even to eliminate the consequences on the population, economic activities, socio-cultural events and environmental factors that may be affected by the flood. In order to work out, this

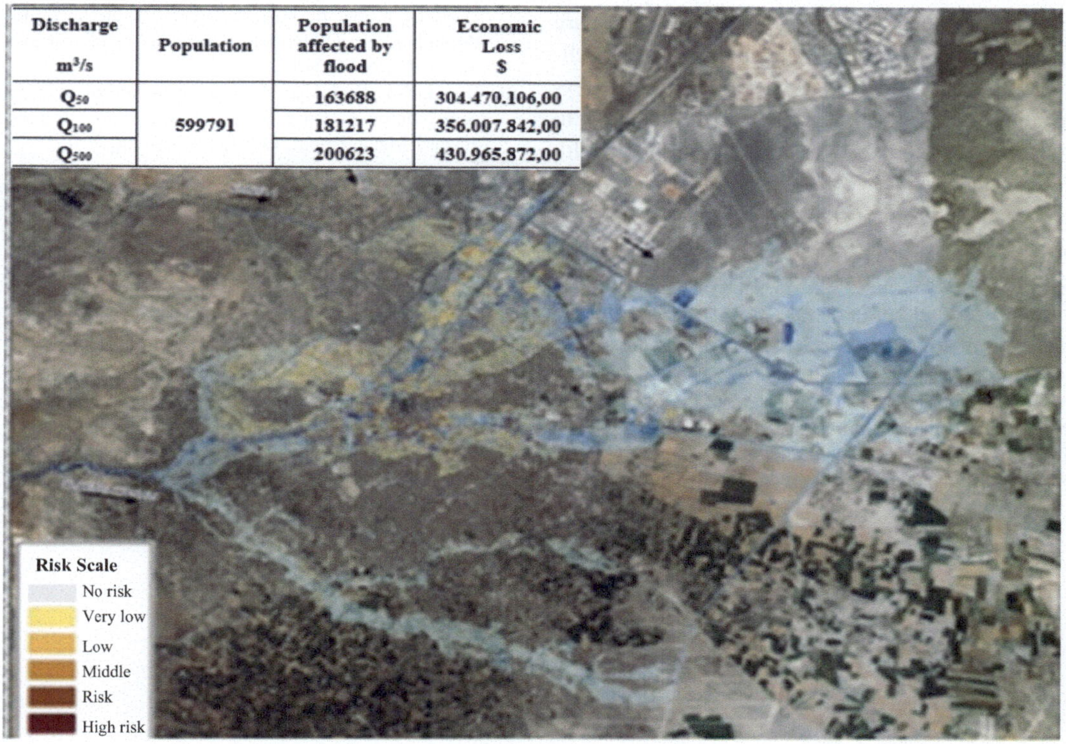

Fig. 4.1 Flood risk map for Konya [12]

plan needs the cooperatively work of governmental organizations, local authorities, industrialists, farmers, tourism sectors and the public.

4.2.2.6 Measures

After all studies, measurements are defined and taken for each location. Since the wounds are difficult to heal during or after flooding, predicting the flood damages and effects beforehand, foreseeing the inundated areas will ensure that the flood-induced damages are minimized. Taking both structural and non-structural measures protect the areas in the likelihood of floods and reduce the impacts of floods in the areas under flood risk.

4.3 Conclusions

Flood is one of the most devastating problems resulting from climate change. Floods are increasing in frequency and intensity due to climate change. Floods can cause widespread devastation, resulting in loss of life and damages to personal property and critical public health infrastructure. Between 1998 and 2021, floods have affected more than 2 million people worldwide. As floods cause loss of life and property, and damage critical infrastructure, flood risk management and flood measurements have vital importance in flood prone zones.

Acknowledgements The author would like to thank Necmettin Erbakan University; Konya, Turkey, and EXCEED Swindon project funded by DAAD (German Academic Exchange Service).

References

1. WWF, Restoring rivers can build resilience against floods: Report. 2021. https://www.wwf.eu/?4332941/Restoring-rivers-can-build-resilience-against-floods-New-Deltares-report. Accessed 24 Oct 2021.
2. Guha-Sapir D, Hargitt D, Hoyois P. Thirty years of natural disasters 1974–2003: The numbers. 2004. https://www.unisdr.org/preventionweb/files/1078_8761.pdf. Accessed 24 Oct 2021.
3. Wetherald RJ, Manabe S. Simulation of hydrologic changes associated with global warming. J Geophys Res. 2003;107(D19–4379):1–15.
4. IPCC, Climate change 2007: Synthesis report. 2007. https://www.ipcc.ch/site/assets/uploads/2018/02/ar4_syr_full_report.pdf. Accessed 24 Oct 2021.
5. Moel HD, Alphen JV, Aerts JCJH. Flood maps in Europe—methods, availability and use. Nat Hazards Earth Syst Sci. 2009;2009(9):289–301.
6. Nagy L. Flood risk of protected floodplain basins. In: Geotechnical hazards. CRC; 1998.
7. Arthur G, Cudworth J. Flood hydrology manual. Denver; 1992.
8. Pappenberger F, Dutra E, Wetterhall F, Cloke HL. Deriving global flood hazard maps of fluvial floods through a physical model cascade. Hydrol Earth Syst Sci. 2012;16:4143–56.
9. Bapalu GV, Sinha R. GIS in flood hazard mapping: a case study of Kosi River Basin, India. Natural Hazard Management, GIS Development; 2005.
10. Environment Agency. Flood and water management act. 29, UK Public General Acts. 2010. https://www.legislation.gov.uk/ukpga/2010/29. Accessed 24 Oct 2021.
11. Merz B, Thieken AH, Gocht M. Flood risk mapping at the local scale: concepts and challenges. In: Flood risk management in Europe. Advances in natural and technological hazards research, Springer; 2007.
12. Ministry of Agriculture and Forestry. Konya closed basin flood management plan. 2020.

Water Resources Allocation and Priorities

5

Burcu Tezcan, Yakup Karaaslan, and Mehmet Emin Aydin

Abstract

As climate variability, population growth, and lifestyle changes amplify the stress on water resources, water management become more pronounced. Spatio-temporal distribution of water resources, political complexities, social and cultural norms add extra complexity on this issue. It is clear that an effective water allocation system is required to handle the impacts of these factors in advance. Therefore, developing a framework for comprehensive water allocation is a critical issue in all countries. This paper discusses water allocation mechanisms in Turkey and the United States. Each country has its own water allocation mechanism and challenges based on its political, hydrological, and cultural structures. Setting a successful water allocation mechanism mostly depends on the country's political circumstances, regional hydrology, capacity of institutions, priorities, and water demands.

Keywords

Water allocation · Water allocation plans · Water allocation priorities · Water scarcity · Water protection

5.1 Introduction

The simplest definition of water allocation is sharing water among users [1]. It is a necessary process, when the natural distribution and availability of water cannot reach to all of water users. In other words, it is the mechanism that determines, who can take water, how much and which purposes they take, from which location and when [2].

Water allocation measures would not be necessary, if water resources were unlimited and always available in terms of water quality, quantity, or reliability, etc. However, with population growth, socio-economic development, higher standards of living, droughts and many other reasons, the demand for water resources are increasing and competition on limited resources is becoming fierce [3].

When water scarcity has increased all over the world, the importance of water allocation plans and agreements become an essential subject to solve the international and local conflicts related with access to water, because reductions in water availability affect some sectors such as agriculture, municipal, energy production, industrial demands, and so on. Therefore, allocation

B. Tezcan · Y. Karaaslan
General Directorate of Water Management, Ankara, Turkey

M. E. Aydin (✉)
Civil Engineering Department, Necmettin Erbakan University, Konya, Turkey
e-mail: meaydin@erbakan.edu.tr

M. Bahadir and A. Haarstrick (eds.), *Water and Wastewater Management*,
Water and Wastewater Management, https://doi.org/10.1007/978-3-030-95288-4_5

objectives have taken on an increasing degree of significance over time, and different ways of approaches related to water resources management strategies have been emerged [3].

The objective of water allocation is to optimize the sufficient use of water in human society, and during this process it aims to protect and to preserve the water resources and the environment. It is important to use water in sustainable limits in order to protect its quality and quantity, and to assure that the quality of basin is maintained [1]. Water allocation process is undertaken within the national policy framework. However, the approach to allocation is different based on the nature of catchment and the political system. For example, while making water allocation decisions, states have more autonomy in federal systems; however, central government has more power to dictate the plans in unitary system [1]. Therefore, each country in the world has set its own system based on its nature of basin conditions, political situation, socio-economic structure, and so on.

Even though there is no certain formulation worldwide to determine how to allocate water, there are number of considerations, which are guided for allocation plans and agreements, such as the fair division of water, physical characteristics and the existing or current use of the basin, population numbers and growth projections for the basin, etc. Nowadays, due to limited water resources for future basin use and growing economic costs from poor allocation methods, many parts of the world have put an emphasis on economic assessments and the water use efficiency. Therefore, the assessment and incorporation of current and future use scenarios in water allocation planning is becoming more important [4].

5.2 Example of Countries in Water Allocation

5.2.1 Water Resources Protection in Turkey

The area of Turkey is 783.577 km^2 and the total annual precipitation amounts to 450 billion m^3. Annual available surface water amount is 94 billion m^3; the safe reserve amount of groundwater is 18 billion m^3, and the total annual usable water amount is 112 billion m^3. Annual average surface flow (natural flow) is 185 billion m^3 [5].

Storage facilities in Turkey contain 186.37 billion m^3 water, of which 92.4 billion m^3 is stored as total active volume. The water in the storage facilities is used for irrigation, domestic and drinking, industrial and hydroelectric purposes during dry periods, when rainfall and thus stream flow is insufficient [5]. Despite increasing water demand, the scarcity of water resources of desired quantity and quality necessitates the protection and improvement of current water resources. In line with the sustainable development goals, it is essential to manage water resources on a basin basis with a holistic approach to ensure the sustainable use of water resources in terms of quantity, quality and ecosystems' health, considering the balance of protection and use.

Marshes, reeds, peatlands and all water that is fresh or brackish, natural or artificial, continuous or temporary, stagnant or running, where its depth does not exceed six meters during the withdrawal of the tidal movements of sea are defined as "wetlands". There are 3 types of wetlands in Turkey: terrestrial, marine/coastal and artificial wetlands. Wetlands have various functions such as water filtering, flood protection, coastal stabilization, water regime regulation, and sediment retention.

In Turkey, "Wetland Management Plans" are prepared in order to define all activities and measures such as protection, use, research, monitoring and control in a holistic approach in order to ensure the rational use of wetlands. Wetland Management Plans are prepared by the General Directorate of Nature Conservation and National Parks. Activities related to the protection of wetlands are carried out under the coordination of the General Directorate of Nature Conservation and National Parks.

Environmental flow rate is the minimum amount of water required for a river and its environment to have a sustainable and healthy ecosystem. In other words, it is the link between management models that is developed for rivers to function properly, and targets of water or land use. There is no need having a negative impact on streams to calculate environmental flow requirement. It can be calculated for streams that does not have a negative impact. In Turkey, the General Directorate of Nature Conservation and National Parks is responsible to determine the amount of environmental flow requirement.

Recent years, pressures on the quantity and quality of water resources in Turkey have increased significantly with the effect of climate change. With the increasing pressure on limited water resources, the importance of water management grows vastly. One of the two important components that make it possible to use water is quantity and quality of water. The first and fundamental step to be taken at the point of water management is to monitor water in terms of quantity and quality in order to determine the current situation. It is possible to determine, whether the water is suitable for use or not by monitoring its quality.

There are 25 water basins in Turkey. Monitoring studies to determine the water quality in the basins are carried out by the General Directorate of State Hydraulic Works within the scope of the Basin Monitoring Programs prepared by the General Directorate of Water Management. Water quality monitoring is carried out for various purposes such as revealing the pollutant concentrations, determining the pressures on surface and ground waters and the measures to be taken, examining and classifying the compliance of waters with the standards, revealing the effectiveness of the measures determined within the scope of the River Basin Management Plans on water resources. In addition, project-based studies and monitoring activities are carried out within the scope of River Basin Management Plans, which are prepared by General Directorate of Water Management, and the results of all these studies are evaluated according to the limits in the regulation on Surface Water Quality. The results of the evaluations made are shared with the relevant institutions. However, measures are determined according to the water quality status within the scope of River Basin Management Plans.

The regulatory Institution in wastewater management in Turkey is the Ministry of Environment and Urbanization. The main financial provider is İlbank Comp. In addition, State Hydraulic Works builds wastewater treatment facilities with loans to municipalities. Furthermore, European Union funds (IPA) and foreign financial resources are also used in wastewater investments. Local responsibilities of wastewater management are municipalities, organized industrial zones, industrial facilities and sub-municipal settlements.

The basic legislation on wastewater management are the "Water Pollution Control Regulation" and "Urban Wastewater Treatment Regulation" together with the Environmental Law. Discharge principles, criteria and principles of discharge permit of wastewater, principles related to wastewater infrastructure facilities, monitoring and inspection procedures, and principles to prevent water pollution are determined by the regulation on Water Pollution Control. Technical and administrative principles regarding the collection, treatment and discharge of urban and certain industrial wastewaters discharged into sewerage systems, monitoring, reporting and inspection of wastewater discharge are regulated by the "Regulation on Urban Wastewater Treatment".

Turkey has been prepared its water allocation plans since 2012 under the General Directorate of Water Management. The plans consider the basin as a whole system and categorize the sectors in basins. Thus, it is called "Sectorial Water Allocation Plans". The aim of the plans is to optimize social justice, satisfaction of each water use sector (domestic, agricultural, environmental flow, energy, industry, etc.) and to increase the net national income per capita by assessing future drought conditions under a changing climate, urbanization and population growth. In the plans, current and future water demands for the sectors and environmental flow objectives are determined under different drought conditions as well as a socio-economic analysis is carried out to understand, how the changes of water management likely impact people, who live in the basin. Decision support models are used to allocate water under various drought conditions. In case, water potential cannot meet the demand, the alternatives are proposed in the plans, such as desalination, wastewater reuse, inter-basin water transfers, optimum crop patterns under drought conditions, etc.

5.2.2 Methodology of Sectorial Water Allocation Plans in Turkey

The first step of the plans is to identify the current water use among sectors in basins. Different water use sectors (i.e., domestic, environmental flow, agriculture, industry, energy and other sectors) are determined, and their water consumptions are assessed. To reduce the likelihood of conflicts among sectors, not only allocating water itself, but also fairly allocation of benefits obtained from water is provided in the Sectorial Water Allocation Plans in Turkey.

With rapid urbanization and the effect of climate change, cities of Turkey become thirstier, especially the largest cities by population, such as Istanbul, Izmir, Ankara, etc. Thus, securing spatial and temporal variability of water becomes a challenge. To reduce the negative impacts of water variability, the sectorial water allocation plans set adaptive strategies by long-term planning. Future water demands for domestic water use sector are analyzed in urban and rural areas by considering population growth, climate change, land use, seasonal water demand changes due to migration and tourism for specific regions.

Another important step of the plans is to protect ecosystems in basins. In the plans, environmental flow objectives are specified for most of the reservoirs' releases under normal and drought conditions. The monthly average environmental flow objectives are calculated for specified water infrastructures in all basins. While allocating water in a way that is fair among sectors, the ecosystems are also protected by releasing their water requirements. Different methodologies are used to calculate environmental flow requirement in basins: Tennat, Global Environmental Flow Calculator (GEFC), wetted-perimeter method, etc.

Water is a critical input for agricultural sector in Turkey. Approximately 77% of the water withdrawals is due to agricultural activities in basins. In addition, agriculture, food, and related industries play an important role in the economic development of the country. According to the sectorial Water Allocation Plan for Konya Basin, approximately 4,800 hm^3 (=million m^3) water are allocated for irrigation purposes in 2016, which resulted in 0.6 billion US Dollar income. Based on the projections in the plan, implementation of the plan has significant impacts on both water consumption and net national income in the basin. In 2040, approximately 2,444 hm^3 of water is projected to use in the basin, while net national income is increasing to 0.8 billion US Dollar [5]. To achieve this, optimum crop patterns are identified for agricultural fields during normal and drought conditions. The identification process of the crop patterns has vital importance for the people in the basin, thereby consideration of the socio-economic structure of the basin is very important. For example, industries supplied by agricultural products, such as sugar beet, etc. has to sustain their business due to development of the region even though crops need intensive volume of water. Furthermore, when the cultivation areas of some crops are changed, the efficiency of the crop decreased. Sectorial Water Allocation Plans consider all of these challenges to secure current and future socio-economic situation of basins while improving water resources in Turkey.

Irrigation water demand projections are determined based on crop pattern, irrigation efficiency and irrigation areas. Crop water requirements are calculated depending on normal and deficit irrigation (irrigation water is reduced by 80, 60, 40, 20%) to respond future drought conditions. Water loss rate due to leakage in irrigation, irrigation efficiency and rate are also considered in calculations [5].

All industrial sectors are categorized based on their activities in the plans, such as textile, food, chemistry, leather, etc. Meanwhile, water use for power plants, livestock, bottled drinking water, mining, recreation, forestry, fisheries and aquaculture activities are assessed both for current and future conditions under normal and drought conditions.

To support decision making and to improve policy evaluation, decision support system models are used to model water resources and their withdrawals by assessing the balance equilibrium between them. In Turkey, sectorial Water Allocation Plans include multiple drought scenarios for long-term planning. Drought conditions are divided into 5 categories: "normal", "mild drought", "moderate drought", "severe drought" and "emergency drought". Two different drought analyses are studied in the plans: meteorological and hydrological droughts.

In planning operation systems, water allocation priorities are categorized based on sectors such as drinking and domestic use, environmental flow requirements, irrigation, industrial and energy. Water allocation based on priority use can be determined at national or basin levels. Furthermore, after identifying the priorities for allocation, assessing the water ability of the basin in different times and under different climate conditions are an essential issue to satisfy its current and future water demands in order to make wise decisions [4].

In sectorial water allocation models, water allocation priorities are categorized based on sectors such as domestic, environmental flow requirements, irrigation, water use for industrial and production of energy, etc. The order of the priorities for basin water allocation in Turkey is defined in the regulation, called "Preparation, Implementation and Tracking of Basin Management Plans". According to the regulation, the order of the water allocation priorities among sectors is:

1. Drinking and Domestic
2. Ecosystem maintenance
3. Irrigation
4. Industrial and energy production
5. Tourism, mining, etc.

5.3 Water Allocation in the United States

With changes in the nation's economic, demographic, and political landscape, the availability and demands for water resources have become very important. To respond these changes, water resource managers are regulating the water management practices, addressing changed priorities, consumption rates, physical conditions, and allocations. Based on the World Bank data, the population of the U.S. increased from around 238 million in 1985 to 309 million in 2010 (Fig. 5.1). While there is an increasing trend in population, total water consumption in the U.S. decreased. When total water consumption in 1985 was approximately 1.502 billion m^3 per day, it was about 1.343 billion m^3 per day in 2010. There was around 10% decreases in total water use from 1985 to 2010 [7].

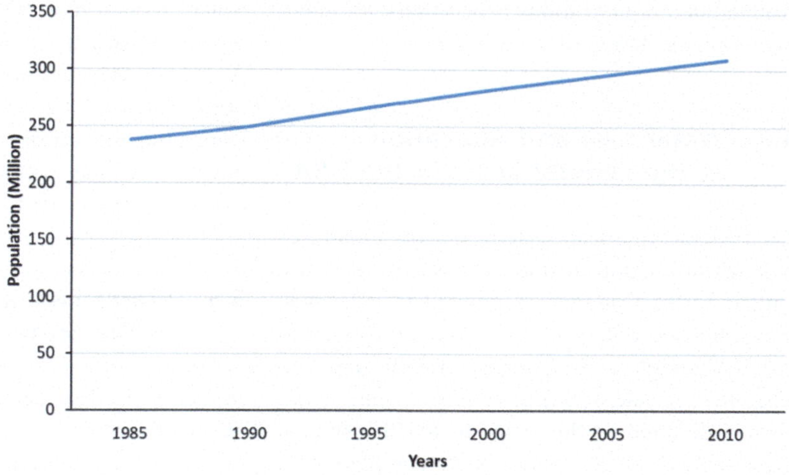

Fig. 5.1 Changes in population in the United States between 1985 and 2010 [6]

In 2010, freshwater withdrawals were 86% of total withdrawals [7]. While fresh surface water withdrawals for public water supplies, irrigation, aquaculture, thermoelectric, and industrial use are mostly in California, Colorado, Idaho, Illinois and Texas, fresh ground water withdrawals for irrigation and public water supply are mostly in Arkansas, California, Florida, Idaho, Nebraska, and Texas [8].

Total water withdrawals in the United States were estimated for eight categories of use in 2010: Thermoelectric power (45%), irrigation (33%), public supply (12%), self-supplied domestic (1%), self-supplied industrial (4%), livestock (1%), mining (1%), and aquaculture (3%) [7].

During the 1970s, 21 major water resources regions and 222 sub-regions have been designed by the U.S. Water Recourses Council to establish a comprehensive planning. Furthermore, 352 hydrological accounting units and 2149 cataloging units based on surface features are defined by the U.S. Geological Survey to manage the national water data. Along with the collection of social and economic data of cities, counties, and states, all of these applications implement a multiscale approach to water management [9].

Water in the U.S. is managed at different levels such as federal, state, and local governments (i.e., municipalities, counties), and adaptive management practices via agency structure, which can change direction [10]. At each level, the decision for water allocation and management is made by a unique set of policies, laws, and customs [11]. Therefore, coordination among all levels of government and organizations is necessary for effective implementation of water policies.

5.4 Water Law and Allocation at the Federal Level in the USA

The Federal American system controls and makes policies in relation to interstate water resources. United States Congress is granted to determine how interstate waters are to be used and allocated [12]. Although U.S. Congress might select to allocate interstate waters directly by statute, it has always deferred to the states to reach a mutual agreement about the issue. After completion of successful water negotiations between states, they must apply to the U.S. Congress for the formation of an Interstate Compact before the agreement becomes binding [11].

Furthermore, the federal government is engaged with local water resources via many national environmental statutes, such as National Environmental Policy Act, Clean Water Act, and Endangered Species Act. These statutes and other federal statues are essential for water resources decision process to establish minimum standards for water quality, wetland protection, protection of endangered species, and other criteria [11]. Arguably, the strategy of managing water in the nation does not have good fragmentation. For example, water allocation from lakes, streams, and ground water is controlled mostly by states. However, the Federal U.S. Environmental Protection Agency (USEPA) is responsible for water pollution. As a result, there is a bifurcated water policy system between the federal government, which dominates water quality issues, and states, which control quantity issues [9].

Perhaps, the U.S. has never passed a national water policy because of the diversity in water governance across the country. However, there are many national water laws, regulations, and policies that form the basis of water management in the U.S. Also, many executive and legislative commissions and committees have worked on policies to facilitate more effective and efficient water coordination [11].

In addition to common methods of allocating water, federal law also comes into play with regards to water on federal lands, and Native American Reservations. These are predefined rights and called federally reserved rights [13]. Depending on certain federal laws, the federal government may secure the use of water in federal and tribal lands for specific reasons [11].

5.5 Water Law and Allocation at the State Level

In the U.S., English common law evolved to govern the water allocation system across much of the country prior to westward migration and began with the riparianism system. The colonists, who began the new society in eastern United States, realized that the physical geography was similar to England, such as humidity, many streams and rivers, etc. Therefore, they started to apply the same type of law, which they used in England [14]. Contrary to the eastern part, the western half of the country suffered from water scarcity and arid climate. Hence, prior appropriation legal system was developed to govern the western part's water resources. These two systems are separated by the 98th meridian, which roughly breaks the country in two parts. A hybrid system of these two doctrines is used around one fifth of U.S. states, thereby, each state has a different water allocation law than another one. In addition, for surface and ground waters, all states have different allocation laws [11].

Riparian rights are basically depending on ownership of land bordering a water body. First, it allowed owners adjoining a watercourse to use water under natural flow theory. Based on this theory, before and after use of water, every user is obliged to maintain the quantity and quality of the water. Then, this doctrine was revised depending on the theory of reasonable use. For example, the riparian users could not hurt the stream (natural flow theory) to the user that should not interfere with the reasonable needs of other riparian (theory of reasonable use). Based on riparian doctrine, when new users are added to the system, other users adjust their rights, which means the riparian doctrine does not protect the established users from the users, who are added to the system later [9]. In case of being injured by another riparian, the landowner must prove this injury to court or state water agency. If it is proven that the use of that amount of water is hurting the others, the landowner might lose the rights to the stream. In addition, the government might seize water rights to a stream for municipal purposes by using the power of eminent domain [11].

Today, the eastern U.S. mostly uses the riparian-based water systems. However, in semiarid western territories, prior appropriation doctrine was developed in response to dry conditions. Based on this system, whoever is the first use the water, s/he has the right to its future use as against its later users. In fact, western water law initially evolved from customs, experiences, and usage within the early mining camps [9].

Based on a priority date, the prior appropriation doctrine allows non-riparian to divert and to use the water for beneficial purposes such as irrigation, mining, industrial application, stock watering, municipal and domestic use, and ecological purposes. This right to use water is acquired by filling a claim for the diverted amount of water with the local court or state agency. The filing date is important, because it establishes the priority of the users' water right in the system. Based on the filing date, the first person has the senior water right, and the others have a junior water right, and these rights persist as long as the right holder consumes the water [11]. However, in times of low flow, water deliveries to most of the junior water right holders are cut off sequentially to make sure that water is delivered to senior water right holders [15].

5.6 Comparison of Water Allocation Mechanism in Countries

It is clear that the success of water allocation plan in achieving its broader social, economic, and environmental goals will depend on the level of compliance. This extends to compliance by water abstractors, different levels of government and government agencies.

Turkey set a successful methodology to prepare water allocation plans and defined the sectorial priorities by regulation. However, the country still encounters institutional problems, such as misalignment, which institute manages what portion, and insufficient enforcement of regulation due to delays of approval of the draft national water law. Sectorial water allocation plans in the country have been successfully prepared for some basins. However, to apply the plans in the field is a big challenge due to insufficient monitoring structure and the aforementioned problems. Particularly, most of water abstractors for irrigation fields and the amount of surface water they consume cannot be efficiently monitored. This is even worse, when it comes to groundwater abstractions. Thus, Turkey needs a technology-based solution to manage water resources, such as a smart national water management system. The current system is mostly managed manually, which increases the chances of errors and time lags.

In the U.S., after establishing a federal system, national and state governments as well as entities within the individual states have taken the unique responsibility to manage the natural resources of the U.S. Even though state governments have the authority and responsibility to manage their water resources, there is a federal involvement in local water resources, especially in cases resolving interstate water conflicts, managing interstate waterways, and promoting common environmental standards [11].

The states have a primary authority to manage water resources within their boundary [11]. There are two different ways to govern water allocation in states: Riparian and prior appropriation system. Each state has its own water codes that outline the details of each system. Application of these water codes is based on custom, culture, geography, legislation, and case law in each state [9].

The types of water allocation systems are evolved in all countries based on their political, hydrological and cultural differences. Ultimately, deciding the right point to settle a water allocation system will mostly depend on the country's particular political circumstances, the capacity of the management agencies, overall policy objectives, local hydrology, priorities and water demands.

5.7 Conclusions

Water allocation has been an important economic, social and environmental subject since ancient times. People have tried to set up laws through the centuries about water allocation to improve economic development, environmental protection and public health [4]. Since 2012, Turkey has been prepared its water allocation plans for basins. The country successfully set up the framework for the plans. However, there are still some obstacles to efficiently execute these plans, such as insufficient water monitoring structure, delays of approval for the draft national water law, etc.

In the U.S., water is managed at different levels (federal, state, and local government), but the primary authority is given to states. The states have their own water code to manage water resources within their boundary. Each state applies its own water code depending on culture, custom, geography, legislation, and case law [9]. There is also federal involvement to manage water resources in the country in case of resolving interstate water conflicts, managing interstate waterways, and promoting common environmental standards [11]. The decision for water allocation and management at each level is made based on unique set of policies, laws, and customs [11].

Each country has its own formulation to determine water allocation system. The political situation of the country, hydrologic characteristics of the basins, institutional arrangements, priorities and water demands play an important role to set a water allocation system.

References

1. Taylor PV. Principles and practices of water allocation among water-use sectors; 2000.
2. Dinar A, Rosegrant MW, Meinzen-Dick RS. Water allocation mechanisms: principles and examples (No. 1779). World Bank Publications; 1997.
3. UN-ESCWA. Sectoral water allocation policies in selected UNESCWA member countries: an evaluation of the economic, social and drought-related impact. E/UNESCWA/SDPD/2003/13. United Nations Economic and Social Commission for West Asia, New York; 2003.
4. Speed R, Yuanyuan L, Zhiwei Z, Le Quesne T, Pegram G. Basin water allocation planning: principles, procedures and approaches for basin allocation planning; 2013.
5. Ministry of Forestry and Water Affairs. Sectoral water allocation plan for Konya closed basin; 2018.
6. The World Bank Data. 2011. Accessed http://data.worldbank.org/indicator/SP.POP.TOTL?end=2010&locations=US&start=1985&view=chart.
7. Maupin MA, Kenny JF, Hutson SS, Lovelace JK, Barber NL, Linsey KS. Estimated use of water in the United States in 2010 (No. 1405). US Geological Survey; 2014.

8. Kenny JF, Barber NL, Hutson SS, Linsey KS, Lovelace JK, Maupin MA. Estimated use of water in the United States in 2005 (No. 1344). US Geological Survey; 2009.
9. Rodrigues DB, Gupta HV, Serrat-Capdevila A, Oliveira PT, Mario Mendiondo E, Maddock T III, Mahmoud M. Contrasting American and Brazilian systems for water allocation and transfers. J Water Resour Plan Manag. 2014;141(7):04014087.
10. Doyle MW. America's rivers and the American experiment. J Am Water Resour Assoc. 2013;49 (4):975–6.
11. Institute for Water Resources (IWR). Aspects of governing water allocation in the U.S.; 2015.
12. Abrams RH. Interstate water allocation: a contemporary primer for Eastern States. UALR L Rev. 2002;25:155.
13. Christian-Smith J, Allen L. Legal and institutional framework of water management. New York: Oxford University Press; 2012. p. 23–51.
14. Thompson SA. Water use, management, and planning in the United States. Academic Press; 1998.
15. Bark RH, Jacobs KL. Indian water rights settlements and water management innovations: the role of the Arizona Water Settlements Act. Water Resour Res. 2009;45(5).

Part III
Urban Water Supply

Water Losses Management in Urban Water Distribution Systems

6

I. Ethem Karadirek
and Mehmet Emin Aydin

Abstract

Management of water losses in water distribution systems is crucial for sustainable management of water resources. A significant amount of water is annually lost in water supply systems. Water losses in a water distribution system consist of apparent and real losses. Real losses are mainly due to leakage from pipes, service connections and overflows at storage tanks, while apparent losses result from unauthorized consumption, metering inaccuracies, and data handling errors. The key step to control water losses is to understand the concept of water losses. Evaluating performance of water supply services by implementing suitable performance indicators (PIs) is also vital for sustainable water management. Afterwards, assessing components of water losses and required intervention tools to control water losses can be implemented. This chapter aims to provide an understanding of water losses management strategies.

I. E. Karadirek (✉)
Department of Environmental Engineering, Akdeniz University, Antalya, Turkey
e-mail: ethemkaradirek@akdeniz.edu.tr

M. E. Aydin
Department of Civil Engineering, Necmettin Erbakan University, Konya, Turkey

Keywords

Apparent losses · Leakage · Real losses · Non-revenue water · Water losses

6.1 Introduction

Water resources, which are unevenly distributed across the Earth's surface, are currently under pressure due to climate change and increasing demand, which can affect the practices of water resources management [1]. Urban water systems have a key role in sustainable management of water resources. According to a World Bank study, each year, around 45 million m^3 of water is being lost in water supply systems [2]. Water utilities are due to supply water in adequate quality and quantity to the end-users. Managing water losses is a challenging task for water utilities. Water losses in water distribution systems are classified as real and apparent losses. Real losses stand for the volume of water that is physically lost, while apparent losses represent the volume of water that is physically used but not invoiced to the end-users. The problem of excessive water losses is not only a revenue problem, but it also results in waste of resources. Controlling water losses reduces demand on water, costs, and energy needs of water from abstraction to the supply. The key step for water losses management is to understand the concept

M. Bahadir and A. Haarstrick (eds.), *Water and Wastewater Management*,
Water and Wastewater Management, https://doi.org/10.1007/978-3-030-95288-4_6

of water losses. This chapter aims to provide a comprehensive review for water losses concept, performance indicators, assessment of water losses components and intervention tools to control water losses.

6.2 Defining Water Losses

6.2.1 Components of Water Losses and Standard Water Balance

Defining components of water losses is the key step to control water losses. International Water Association (IWA) developed a standard water balance (Table 6.1) to standardize the definitions, which are required for quantification and control of water losses [3, 4].

System input volume (SIV) represents the volume of water supplied to the system. Authorized consumption is the sum of metered and unmetered consumption by registered customers, whereas water losses, which consist of real losses and apparent losses, is the difference between SIV and authorized consumption. Non-revenue water consists of water losses and unbilled authorized consumption, whereas revenue water refers to the volume of water that water utilities get paid for it [5]. Real losses are the volume of water that are physically lost in water distribution systems and consist of leakage on mains and pipes resulting from joints, cracks, bursts, fittings, leakage and overflows at water storage tanks, and leakage on service pipes up to metering point. Apparent losses are usually associated with the volume of water that is physically used but not paid for it and result from the metering inaccuracies and data handling errors, and unauthorized (illegal) consumption [3–6].

The volume of real losses represents how water utility manage its assets. Real losses occur in each water distribution systems and cannot be completely avoided. Each water distribution system, even in new systems, has a bottom level of real losses, which is called unavoidable real losses. Real losses, which are mostly leakage and occur on mains, pipes, and service connections, can be the result of many conditions such as poor installation and materials, corrosion, pressure fluctuations and transients, traffic loading, environmental conditions, and lack of proper maintenance [5, 7–10]. Leakage is generally classified as background leakage, reported leakage and unreported leakage, as depicted in Fig. 6.1 [5, 6].

Reported leakage, which has relatively high flow rates, is generally reported by the public and/or water utilities, whereas unreported leakage can be detected by traditional leakage detection programs. Background leakage, which has relatively small flow rates, cannot be detected by traditional methods. Time matters, as the

Table 6.1 IWA standard water balance [3, 4]

System input volume	Authorized consumption	Billed authorized consumption	Billed metered consumption	Revenue water
			Billed unmetered consumption	
		Unbilled authorized consumption	Unbilled metered consumption	Non-revenue water
			Unbilled unmetered consumption	
	Water losses	Apparent losses	Unauthorized consumption	
			Metering inaccuracies and data handling errors	
		Real losses	Leakage on transmissions and distribution mains	
			Leakage and overflows at storage tanks	
			Leakage on service connections up to metering point	

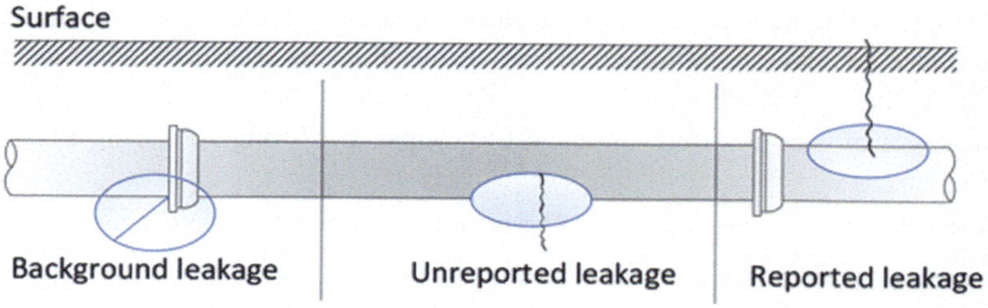

Fig. 6.1 Components of leakage [11]

volume of real losses is a function of awareness time (A), location time (L) and repair time (R) (ALR). As an example, estimated time and flow rates of reported bursts at 40 m average night pressure of any study area are depicted in Fig. 6.2 [12].

Awareness time (A), which is generally higher for background leakage, is the time that water utility becomes aware of the leak. Location time is duration of leakage localization, while repair time is the required time for repairment of leak [5, 12]. On contrary to common misunderstanding, cumulative volume of real losses resulting from background leakage is generally higher as awareness and localization of background leakage takes longer time periods.

6.2.2 Performance Indicators of Water Supply Services and Water Losses

Performance evaluation is a key factor for sustainable management of water supply services. Implementing suitable performance indicators (PIs) will be helpful for assessment of water supply services [4, 14–16]. IWA developed a total number of 170 PIs, which are based on six categories namely water resources (WR), personnel (Pe), physical (Ph), operational (Op), service quality (Qs), and economic and financial (Fi). Each category has groups and each group has subgroups, which are helpful to select specific indicators for different purposes [15].

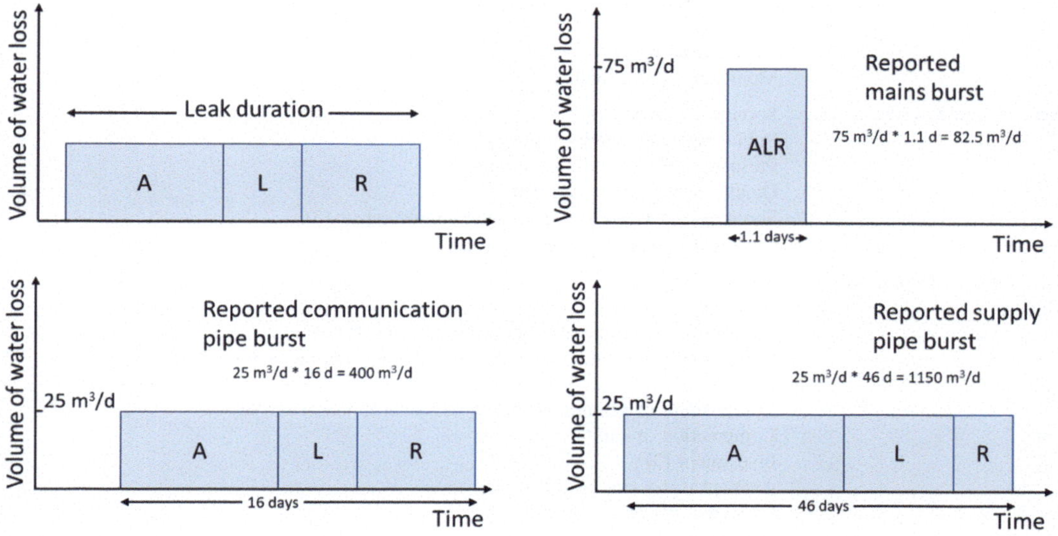

Fig. 6.2 Effects of time on cumulative volume of water losses [12, 13]

General categories, groups of indicators and total number of PIs are summarized in Table 6.2. According to the PIs developed by IWA, water losses can be assessed under the groups of operational PIs, and economic and financial PIs. Assessing performance of economic water losses based on PIs developed by IWA can be achieved using following PIs:

Table 6.2 Performance indicators of water supply services [15]

Category of indicators	Groups of indicators	Total number of PIs
Water resources (Wr)	Availability of water resources Inefficiency of use of water resources Own water resources availability Reuse supplied water	4
Personnel (Pe)	Total personnel-(2) Personnel per main function-(7) Technical services personnel per activity-(6) Personnel qualification-(3) Personnel training-(3) Personnel health and safety-(4) Overtime work-(1)	26
Physical (Ph)	Water treatment-(1) Water storage-(2) Pumping-(4) Valve, hydrant, and meter availability-(6) Automation and control-(2)	15
Operational (Op)	Inspection and maintenance-(6) Instrumentation calibration-(5) Electrical and signal transmission equipment inspection-(3) Vehicle availability-(1) Rehabilitation-(7) Water Losses-(7) Failure-(6) Water metering-(4) Monitoring water quality-(5)	44
Service quality (QS)	Service coverage-(5) Public taps and standpipes-(4) Pressure and continuity of supply-(8) Quality of supplied water-(5) Service connection, meter installation and repair-(3) Customer complaints-(9)	34
Economic and Financial (Fi)	Revenue-(3) Cost-(3) Composition of running costs per type of costs-(5) Composition of running costs per main function of the water utility-(5) Composition of running costs per technical function activity-(6) Composition of capital costs-(2) Investment-(3) Average water charges-(2) Efficiency-(9) Leverage-(2) Liquidity-(1) Profitability-(4) Economic water losses-(2)	47

- Fi 46—Non-revenue water by volume (%) is an indicator that represents the percentage of system input volume corresponding to non-revenue water [15].
- Fi 47—Non-revenue water by cost (%) is the percentage of system input volume corresponding to the appraisal of non-revenue water [15].

Performance indicators of operational water losses are summarized as follows [15]:

- Op 23—Water losses per connection (m^3/connection/year), which is suitable for urban water distribution systems, stands for total annual volume (apparent and real losses) of water losses per connection [15].
- Op 24—Water losses per mains (m^3/km/day) represents the total volume of annual water losses per mains' length per day. This indicator is suitable for bulk supply and water distribution systems with low density of service connection [15].
- Op 25—Apparent losses (%) is the percentage of water supplied to the system (difference between system input volume and exported water) corresponding apparent losses resulting from unauthorized consumption, and metering inaccuracies and data handling errors, system input volume. This indicator is suitable for urban water distribution systems [15].
- Op 26—Apparent losses per system input volume (%), which is a suitable indicator for bulk supply and water distribution systems with low density of service connection, stands for the percentage of water supplied to the system including exported water corresponding apparent losses [15].
- Op 27—Real losses per connection (L/connection/day) is an indicator that is suitable for urban water distribution systems and average daily volume of real losses per connection [15].
- Op 28—Real losses per mains length (L/km/day) is a suitable indicator for bulk supply and water distribution systems with low density of service connection and average daily volume of real losses per mains' length [15].
- Op 29—Infrastructure leakage index (ILI) is the ratio of current annual real losses (CARL) to unavoidable annual real losses (UARL), which can be calculated using an empirical equation based on system operating pressure, mains length, number and average length of service connection [15, 17]. ILI is aimed to eliminate some factors that are not related with physical condition of water distribution systems. However, using ILI in some water distribution systems with high pressure fluctuations and apartment blocks and individual apartment meters, is one of the disadvantages of this PI [15].

Each performance indicator of water losses should be selected and implemented considering advantages and disadvantages, and suitability to the water utilities.

6.3 Assessment of Water Losses

6.3.1 Top-Down Water Balance Assessment

Top-down water balance approach, which can be used for whole system, is one of the water loss assessment methods. This approach is based on assuming and/or estimating apparent losses to calculate real losses in water distribution systems [3, 18]. The first step in this approach is to determine the system input volume and billed authorized consumption, of which difference gives the volume of non-revenue water. Then, the volume of water losses can be calculated by subtracting the unbilled authorized consumption. For calculation of real losses, components of apparent losses should be assumed and/or estimated. Metering inaccuracies for different types of water meters at varying flow rates can be estimated by testing campaigns carried out by water utilities [19–22]. Data handling errors can also be presumed by analysis of historical billing data sets [23], whereas unauthorized consumption can be estimated through experiences of water utilities as illegal consumption depends on many factors such economic and sociocultural status of end-users. Real losses can then be calculated by subtracting apparent losses from total volume of water losses. Selecting and calculating suitable performance indicators for target setting and assessment of water losses, and strategies for water loss can be carried out by water utilities based on top-down water balance approach. Top-down water balance approach is a cost effective and pressure independent way to calculate the real losses. In this approach, real losses might be overestimated due to the assumptions of apparent losses [24]. However, more detailed methodologies should be carried out following calculation of total amount of real losses, as components of real losses should be determined. For this purpose, bottom-up approach and component analysis should be followed the top-down water losses assessment approach.

6.3.2 Bottom-Up Water Losses Assessment

Bottom-up water balance approach is a detailed methodology for assessing water losses occurring in water distribution systems. This method is based on estimation of real losses, then other components of water losses can be calculated [25]. This methodology can be utilized by analysis of minimum night flow (MNF). The period of MNF is generally for urban settlements between 2:00 am and 4:00 am, at what time there is no consumption or water consumption is at the level of minimum. The period of MNF can be different at some places due to tourism activities, industrial activities etc. The great amount of water during the period of MNF is due to leakage [5, 26, 27]. Dividing water distribution systems into district metered areas (DMAs), which are hydraulically isolated parts of networks with single or multiple inputs [5], is an efficient way to assess water losses during the period of the MNF, as determination of system input volume and consumption of end-users in DMAs is easy to determine. Dividing water distribution systems into DMAs should consider many factors such as geographic conditions, length of pipes, number of connections, elevation of the area, water quality. A generalized DMA setup is given in Fig. 6.3.

Volume of real losses during MNF period can be estimated by assessing water consumption of each end-user connected to the distribution system within DMAs. Around 6% of end-users are assumed to be active during MNF period in urban settlements [5, 28], and water consumption is generally associated with toilet use [5]. Night water consumption occurring during MNF period can be estimated based on these assumptions. Furthermore, exceptional night use within DMAs can also be determined through water meter readings. The volume of water calculated by subtracting night water consumption and exceptional water consumption from MNF is mostly due to leakage.

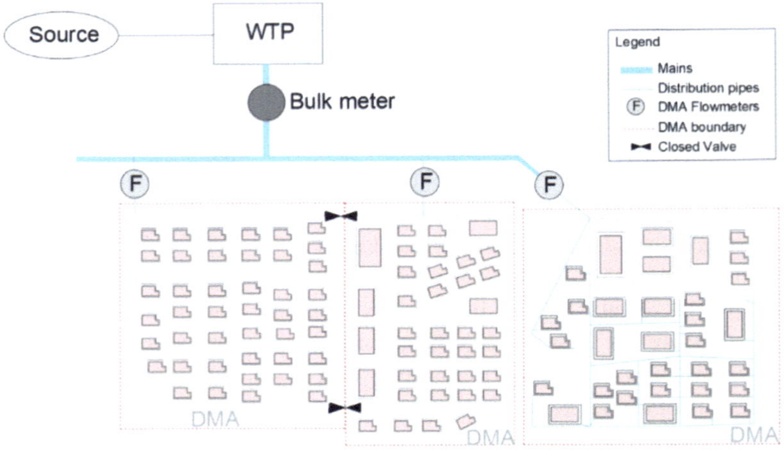

Fig. 6.3 A generalized DMA configuration of a water distribution system (Modified from [29])

6.3.3 Components Analysis of Real Losses

A model for analysis of components of real losses, known as burst and background estimation (BABE), has been developed considering that leakage occurs in three categories as background, reported and unreported leakage [13]. BABE approach assumes that real losses consist of unavoidable and avoidable real losses. Total volume of leakage resulting from pipe bursts is a function of flow rate and duration of repairment of individual bursts, and can be calculated by multiplying the average flow rate of water lost and time required for awareness and repairment. Annual volume of unavoidable real losses (UARL) at specific water pressures can be calculated using parameters given in Table 6.3 [17]. The volume UARL can also be calculated by the following Eq. (6.1) [17].

$$\mathrm{UARL(L/day)} = \left[(18 * L_m) + (0.80 * N_c) + (25 * L_p)\right] * P \qquad (6.1)$$

where L_m represents the length of mains (km), N_c stands for the number of service connections, L_p is the total length of service connections (km) between the edge of property and metering point, and P stands for the average operating pressure (m) [17].

Fixed and variable area discharges (FAVAD) concept, which aims to explain the relationship between pressure and leakage, was firstly introduced by May (1994) [30]. The concept assumes that leakage area on a pipe is a function of water pressure, and this assumption has also been confirmed by the studies existing in the literature [31].

Leaks in pipes can be considered as flows through orifices. The Torricelli's Eq. (6.2) expresses the flow rate through an orifice [33, 34]:

$$Q = C_d * A * \sqrt{2gh} \qquad (6.2)$$

where Q stands for the flow rate through an orifice, A is the area of an orifice, g stands for the acceleration of gravity, h is pressure head and Cd is the discharge coefficient due to loss of energy and jet contraction. However, the Torricelli's equation has been found to be unsuitable to describe pressure and leakage relationship in real case tests [33]. A general Eq. (6.3) for representing pressure-leakage relation has been implemented:

$$Q = C * h^{N1} \qquad (6.3)$$

Table 6.3 Required parameters for calculation of UARL [17, 32]

Infrastructure components	Background leakage	Reported leakage	Unreported leakage
Pipes	Length, pressure*, Losses rate/km	Number/year, pressure*, Ave. flow rate and duration	Number/year, pressure*, Ave. flow rate and duration
Service connections from pipe to the edge of property	Number, pressure*, Losses rate/connection	Number/year, pressure*, Ave. flow rate and duration	Number/year, pressure*, Ave. flow rate and duration
Service connections from edge of property to the water meter	Length, pressure*, Losses rate/km	Number/year, pressure*, Ave. flow rate and duration	Number/year, pressure*, Ave. flow rate and duration
Reservoirs	Leakage from water structures	Overflows Flow rate x duration	Overflows Flow rate x duration

* At a certain level of pressure

where C is the leakage coefficient. N1 is the leakage exponent, h stands for the pressure head and Q is the discharge flow rate through the orifice.

Leakage modelling has been widely utilized for component analysis of real losses and describing pressure—leakage relationship [7, 33–40]. Component analysis of real losses is an effective way for assessment of real losses. BABE and FAVAD approaches, providing an estimation for the volume of real losses occurring in different infrastructure components, should be integrated with top-down water balance assessment approaches.

6.4 Water Losses Control Methods and Intervention Tools

6.4.1 Intervention Tools and Control Real Losses

In any water distribution system, there is a level of water losses that cannot be totally eliminated, which is called as unavoidable annual real losses (UARL). Economic level of real losses is a certain level, at which further reduction is not feasible. This level of water losses is called as economic level of real losses. Water losses problem results in an economic loss for water utilities. On the other hand, reducing water losses is an economic issue. The main aim for utilities is to reduce current annual real losses to an economic level [41, 42]. IWA Water Loss Task Force recommends four basic intervention tools to control real losses in water distribution systems, as given in Fig. 6.4.

6.4.1.1 Pressure Measurement

Pressure management is an important issue to control real losses. The main objectives related to pressure management are reduction of the volume of real losses, frequency of new pipe bursts, water consumption resulting from excess pressure, and increase of economic life of infrastructure systems.

Water pressure flexes the volume of real losses. As pressure and leakage relationship discussed in the previous section, reducing water

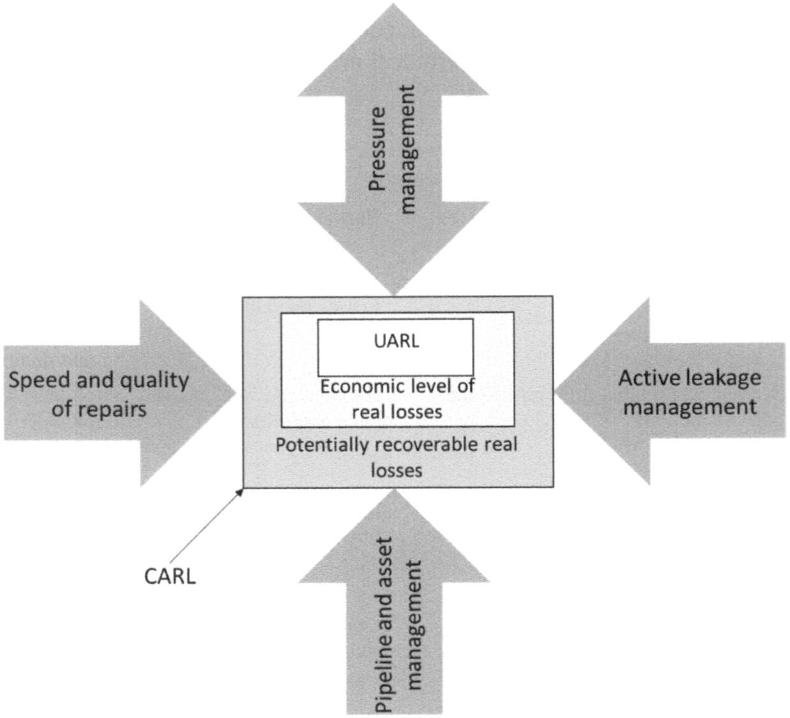

Fig. 6.4 Intervention tools to control real losses in WDSs [42]

pressure results in a decrease in the volume of real losses including background, reported and unreported leakages, whereas increasing water pressure causes higher rates of real losses [43]. Effects of excess pressure on frequency rates of new pipe bursts in water distributions systems are reported in the literature [42, 43]. Water pressure is not only the factor causing new bursts, but also a significant factor. Reducing water pressure helps in decreasing frequencies of new bursts and extending the economic life of infrastructure systems [42]. Pressure management also helps in reducing adverse hydraulic effects on components of water distribution systems such as valves. Water consumption of end-users might be affected by reduced water pressure, and reduction in residential consumption due to reduced pressure has been reported [44]. Reduction in water consumption of end-users might be considered as a revenue loss of water utilities. On the other hand, reduction in water consumption of end-users may be seen as a more cost-effective method than increasing supply to meet higher water consumption of end-users [5]. Pressure management in water distribution systems can be implemented by pressure reducing valves (PRVs), and implementing pressure management in DMAs with using hydraulic simulation models is very common [7]. Pressure management can be applied in many ways such as fixed outlet, time based, flow based, and remote node based [43]. Recently, pump as turbines (PATs) and micro turbines have been considered as an efficient way to reduce pressure while producing energy [45–49]. When implementing pressure management, concerns on fire flows should be considered. No universally accepted standard for water pressure in water distribution systems is available [50]. However, a minimum level of water pressure around 20–25 m [51] and an average level around 4–5.5 bar for operating conditions are common [52].

After detailed analysis of real losses components, appropriate methods to reduce real losses can be implemented. Details of intervention tools required for reduction in real losses are discussed in the following subsections.

6.4.1.2 Speed and Quality of Repairments

Time required for awareness, location and repairment of pipe bursts plays a key role for controlling real losses, as discussed in the previous section. Therefore, reducing time for ALR decreases the volume of real losses. Awareness time is a period required to become aware of an existing leakage in the system, and reduction of this period generally depends on availability of active leakage control. Awareness time for reported leaks is usually short, while awareness of unreported and background leakages takes more time [42]. Active leakage control, which can be implemented at least once a year, helps reducing awareness time for unreported leakages [5]. Increasing frequency of active leakage control results in a decrease in time required for awareness. DMAs help water utilities reduce awareness time, as flow rates can be continuously monitored, and anomalies can be detected in a short time period [53].

Leakages on pipes in the systems can be localized by hardware-based and software-based methods. Hardware based methods consist of acoustic methods such as leak correlators, noise loggers, listening rods, and non-acoustic methods such as ground penetrating radar and thermal infrared imaging [54]. Software-based methods for leakage localization is generally based on numerical and non-numeral modeling approaches including hydraulic modelling and data analysis [54]. Performance of leakage localization depends on the available tools and skills that water utilities have.

Time for repairment of a localized leakage depends on factors such as number of existing skilled personnel, equipment, etc. In addition to time required for repairment, quality of repairments plays an important role. Repairments with a poor quality may reoccur and result in new bursts on pipes [5].

6.4.1.3 Active Leakage Control

Leakage control in water distribution systems can be carried out by implementing proactive and reactive leakage detection methods. Reactive leakage detection, known as passive leakage control, is based on repairment of leaks that are reported by customers and/or become visible on the surface [55]. On the other hand, proactive leakage detection, also well-known as active leakage control, is a well-planned program to aware, to localize and to repair leaks [55]. Active leakage control is generally based on detection of leakages including unreported and background leakages [5, 8]. Active leakage control helps in reducing time required for ALR and is a helpful intervention tool for controlling real losses.

6.4.1.4 Asset Management

Assets of any system have an economic life, and ageing of components of the system is one of the main problems causing real losses. Corrosion in pipes, which vary depending on many factors such as water quality, cathodic protection in steel pipes, etc., causes real losses [56]. As any other assets, pipes age over time and need to be rehabilitated and/or replaced for some reasons such as higher leakage rates, hydraulic capacity [57]. Rehabilitation and replacement of ageing assets in any system can be carried out in many ways and help in controlling real losses.

6.4.2 Intervention Tools to Control Apparent Losses

Apparent losses in water distribution systems result from metering inaccuracies, unauthorized consumption and data handling errors including data transfer and data analysis errors. Apparent losses refer to the water used but not paid for it [24]. As described for real losses, there is an economic level of apparent losses that further reduction is not feasible. Each system has a certain level of apparent losses that cannot be reduced to zero, which is called unavoidable annual apparent losses (UAAL) [58]. The main objective for controlling apparent losses should be reducing the current annual apparent losses (CAAL) to an economic level. To control apparent losses, four basic intervention tools are recommended, as given in Fig. 6.5 [58].

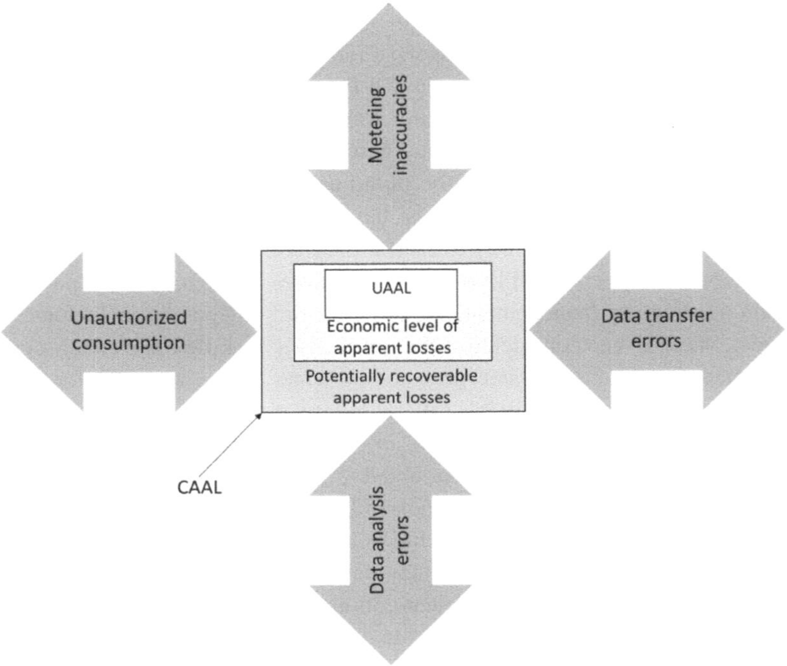

Fig. 6.5 Intervention tools to control apparent losses in WDSs [58]

Following a component analysis of apparent losses occurring in a system, required actions can be taken to control apparent losses.

6.4.2.1 Metering Inaccuracies

Metering errors are the main reason for apparent losses in well operated water distribution systems [59]. Inaccuracies of water meters depend on many factors such as type, size, accuracy class of meters, improper installation, water quality, ageing of water meters, and demand profile of end-users [19, 60]. Each water meter, even new meters, has an intrinsic measurement error. Selection of water meters is an important task for water utilities, as accurate metering is important for fair pricing [61], and determining inaccuracies of water meters is crucial for a proper water balance [3]. Apparent losses caused by metering accuracies can be minimized by implementing recommended techniques including selection criteria and replacement period of water meters [62]. For better management of apparent losses resulting from metering inaccuracies, water utilities should determine the inaccuracies of existing water meters and the selection criteria considering factors effecting water meter inaccuracies. Proper replacement periods of water meters should be determined and conducted.

6.4.2.2 Data Transfer and Analysis Errors

Manual readings of meters and failures of automatic meter reading systems are the main reasons of data transfer errors, while poor management of accounting is responsible for data analysis errors. Required actions can be determined and conducted by water utilities. The most important thing is to find out the reasons.

6.5 Conclusions

Water demand increases due to population increase, development and changing consumption patterns. On the other hand, water availability decreases. Some countries around the world have already been facing with water scarcity, while many others are projected to face with water stress and/or scarcity. Therefore, sustainable management of water resources has become a crucial task that should be tackled with. The problem of excessive water losses occurring in water distribution systems is a challenging task for sustainability. Some places around the world suffer from lack of water infrastructure systems, whereas other places, which have water supply systems, tackle with a serious problem—water losses threatening sustainable management of water resources. The first step to control water losses is to understand the concept of water losses. This chapter aims to provide a comprehensive review on understanding the concept of water losses, performance indicators, assessment of water losses components, and control and intervention tools of water losses management.

Acknowledgements Authors would like to thank Akdeniz University, Necmettin Erbakan University and EXCEED Swindon project funded by DAAD (German Academic Exchange Service).

References

1. Zhang XD. Conjunctive surface water and groundwater management under climate change. Front Environ Sci. 2015;3.
2. The World Bank Press Release: The World Bank and the international water association to establish a partnership to reduce water losses. September 2016.
3. Lambert A, Hirner W. Losses from water supply systems: standard terminology and recommended performance measures. IWA's Blue Pages. 2000. pp. 1–13.
4. Alegre H, Hirner W, Baptista JM, Parena R. Performance indicators for water supply services. 1st ed. London: IWA Publishing; 2000.
5. Thornton J, Sturm R, Kunkel GA. Water loss control. New York: McGraw-Hill; 2008.
6. McKenzie R, Seago C. Assessment of real losses in potable water distribution systems: some recent developments. In: Wilderer P, editor. 4th world water congress: innovation in water supply - reuse and efficiency. London: IWA Publishing; 2005. p. 33–40.
7. Karadirek IE, Kara S, Yilmaz G, Muhammetoglu A, Muhammetoglu H. Implementation of hydraulic modelling for water-loss reduction through pressure management. Water Resour Manag. 2012;26 (9):2555–68.
8. Aboelnga H, Saidan M, Al-Weshah R, Sturm M, Ribbe L, Frechen FB. Component analysis for optimal

leakage management in Madaba, Jordan. J Water Supply Res Technol-Aqua. 2018;67(4):384–96.
9. Karadirek IE. Non revenue water management: current trends and future prospects. Fresen Environ Bull. 2019;28(7):5226–33.
10. Ahopelto S, Vahala R. Cost-benefit analysis of leakage reduction methods in water supply networks. Water. 2020;12(1):15.
11. Tardelli FJ. Control of e ReduçãoPerdas. In Abastecimento de Água, 3rd. São Paulo: Departamento de Engenharia e HidráulicaSanitária, Polytechnic School of the University of São Paulo; 2006.
12. Fanner P, Thornton J. The importance of real loss component analysis for determining the correct intervention strategy. In: Conference the importance of real loss component analysis for determining the correct intervention strategy, Halifax, Nova Scotia, Canada.
13. Lambert A. Accounting for losses: the bursts and background concept. Water Environ J. 1994;8 (2):205–14.
14. Alegre H, Baptista JM, Cabrera JE, Cubillo F, Duarte P, Hirner W, et al. Performance indicators for water supply services, 2nd edn. IWA Publishing; 2006.
15. Alegre H, Baptista JM, Cabrera JE, Cubillo F, Duarte P, Hirner W, et al. Performance indicators for water supply services, 3rd edn. IWA Publishing;2017.
16. Haider H, Sadiq R, Tesfamariam S. Performance indicators for small- and medium-sized water supply systems: a review. Environ Rev. 2014;22(1):1–40.
17. Lambert AO, Brown TG, Takizawa M, Weimer D. A review of performance indicators for real losses from water supply systems. J Water Serv Res Technol-Aqua. 1999;48(6):227–37.
18. Farley M, Trow S. Losses in water distribution networks : a practitioner's guide to assessment, monitoring and control. 2003.
19. Karadirek IE. An experimental analysis on accuracy of customer water meters under various flow rates and water pressures. J Water Supply Res Technol-Aqua. 2020;69(1):18–27.
20. Arregui FJ, Cabrera E, Cobacho R, García-Serra J. Reducing apparent losses caused by meters inaccuracies. Water Pract Technol. 2006;1(4): wpt2006093-wpt.
21. Arregui FJ, Soriano J, Garcia-Serra J, Cobacho R. Proposal of a systematic methodology to estimate apparent losses due to water meter inaccuracies. Water Sci Technol-Water Supply. 2013;13(5):1324–30.
22. Arregui FJ, Balaguer M, Soriano J, Garcia-Serra J. Quantifying measuring errors of new residential water meters considering different customer consumption patterns. Urban Water J. 2016;13(5):463–75.
23. Mutikanga HE, Sharma SK, Vairavamoorthy K. Assessment of apparent losses in urban water systems. Water Environ J. 2011;25(3):327–35.
24. Al-Washali TS, Kennedy M. Methods of assessment of water losses in water supply systems: a review. Water Resour Manag. 2016;30(14):4985–5001.
25. Covas D, Jacob A, Ramos H. Bottom-up analysis for assessing water losses: a case study. In: Water distribution systems analysis symposium; 2006–2008. pp. 1–19.
26. Tabesh M, Yekta AHA, Burrows R. An integrated model to evaluate losses in water distribution systems. Water Resour Manag. 2009;23(3):477–92.
27. Alkasseh JMA, Adlan MN, Abustan I, Aziz HA, Hanif AM. Applying minimum night flow to estimate water loss using statistical modeling: a case study in Kinta Valley, Malaysia. Water Resour Manag 2013;27(5):1439–55.
28. Hamilton S, McKenzie R. Water management and water loss. London: IWA Publ; 2014.
29. Ferrari G, Savic D, Becciu G. Graph-theoretic approach and sound engineering principles for design of district metered areas. J Water Resour Plan Manag. 2014;140(12):13.
30. May J. Leakage, pressure and control. In: BICS international conference on leakage control investigation in underground assets, March 1994. London, UK.
31. Cassa AM, van Zyl JE, Laubscher RF. A numerical investigation into the effect of pressure on holes and cracks in water supply pipes. Urban Water J. 2010;7 (2):109–20.
32. Thornton J. Water loss control manual. Estados Unidos: McGraw-Hill; 2002.
33. van Zyl JE, Cassa AM. Modeling elastically deforming leaks in water distribution pipes. J Hydraul Eng-ASCE. 2014;140(2):182–9.
34. Schwaller J, van Zyl JE. Modeling the pressure-leakage response of water distribution systems based on individual leak behavior. J Hydraul Eng-ASCE. 2015;141(5):8.
35. van Zyl JE, Malde R. Evaluating the pressure-leakage behaviour of leaks in water pipes. J Water Supply Res Technol-Aqua. 2017;66(5):287–99.
36. Niebuhr D, Nsanzubuhoro R, van Zyl JE. Field experience with pressure-based leakage characterisation of bulk water pipelines. Urban Water J. 2019;16 (10):709–17.
37. Marzola I, Alvisi S, Franchini M. Analysis of MNF and FAVAD model for leakage characterization by exploiting smart-metered data: the case of the Gorino Ferrarese (FE-Italy) District. Water. 2021;13(5):15.
38. Fontana N, Giugni M, Marini G. Experimental assessment of pressure-leakage relationship in a water distribution network. Water Sci Technol-Water Supply. 2017;17(3):726–32.
39. Ferrante M, Massari C, Todini E, Brunone B, Meniconi S. Experimental investigation of leak hydraulics. J Hydroinform. 2013;15(3):666–75.
40. Ferraiuolo R, De Paola F, Fiorillo D, Caroppi G, Pugliese F. Experimental and numerical assessment of water leakages in a PVC-A pipe. Water. 2020;12 (6):16.

41. Abravani M, Saghi H. Introducing a novel flexible conjunction system to pressure control in water distribution networks. Water Resour Manag 2017;31 (13):4323–38.
42. Lambert AO, Tveit OA, Abdin NAZ, Lazzari L, Lorenze H, Lee H, Farley M, Masakat E, Suphani R, Esko H, Johnson K, Rapinat M, Dohnal P, McKenzie R, Manesc A, Weimer D, Lai SKS, Somos E, Monteir A, Davis S, Martinez F, Lo SL, Onep. International report: water losses management and techniques. In: Wilderer PA, Arvin E, Blackwell L, Hamoda MF, Mikkelsen PS, Mino T, Morgenroth E, Otterpohl R, Pons MN, Rauch W, Stephenson T, Ujang Z, Jianrong Z (Eds.) 2nd world water congress: water distribution and water services management. London: IWA Publishing; 2002. pp. 1–20.
43. Vicente DJ, Garrote L, Sanchez R, Santillan D. Pressure management in water distribution systems: current status, proposals, and future trends. J Water Resour Plan Manage-ASCE. 2016;142(2):13.
44. Bamezai A. Is system pressure reduction a valuable water conservation tool? Preliminary evidence from the Irvine Ranch water district. Conference is system pressure reduction a valuable water conservation tool? Preliminary evidence from the Irvine Ranch water district.
45. Carravetta A, Del Giudice G, Fecarotta O, Ramos HM. Energy production in water distribution networks: A PAT design strategy. Water Resour Manag. 2012;26(13):3947–59.
46. Carravetta AD, Del Giudice G, Fecarotta O, Ramos HM. Pump as turbine (PAT) design in water distribution network by system effectiveness. Water. 2013;5(3):1211–25.
47. Puleo V, Fontanazza CM, Notaro V, De Marchis M, Freni G, La Loggia G. Pumps as turbines (PATs) in water distribution networks affected by intermittent service. J Hydroinform. 2014;16(2):259–71.
48. Fontana N, Giugni M, Glielmo L, Marini G. Real time control of a prototype for pressure regulation and energy production in water distribution networks. J Water Resour Plan Manage-ASCE. 2016;142(7):9.
49. Muhammetoglu A, Karadirek IE, Ozen O, Muhammetoglu H. Full-scale PAT application for energy production and pressure reduction in a water distribution network. J Water Resour Plan Manage-ASCE. 2017;143(8):12.
50. Ghorbanian V, Guo Y, Karney BW. Field data-based methodology for estimating the expected pipe break rates of water distribution systems. J Water Resour Plan Manage-ASCE. 2016;142(10):11.
51. Monsef H, Naghashzadegan M, Farmani R, Jamali A. Pressure management in water distribution systems in order to reduce energy consumption and background leakage. J Water Supply Res Technol-Aqua. 2018;67(4):397–403.
52. Xu Q, Chen Q, Ma J, Blanckaert K, Wan Z. Water saving and energy reduction through pressure management in urban water distribution networks. Water Resour Manag. 2014;28(11):3715–26.
53. Ferrari G, Savic D. Economic performance of DMAs in water distribution systems. In: Ulanicki BK, Boxall J (Eds.) Computing and control for the water industry. Amsterdam: Elsevier Science Bv; 2015. pp. 189–95.
54. Li R, Huang HD, Xin KL, Tao T. A review of methods for burst/leakage detection and location in water distribution systems. Water Sci Technol-Water Supply. 2015;15(3):429–41.
55. Shammas NK, Al-Dhowalia KH. Effect of pressure on leakage rate in water distribution networks. J King Saud Univ Eng Sci. 1993;5(2):213–26.
56. Maddison LA, Gagnon GA, Eisnor JD. Corrosion control strategies for the Halifax regional distribution system. Can J Civ Eng. 2001;28(2):305–13.
57. Kanakoudis V, Tsitsifli S. Results of an urban water distribution network performance evaluation attempt in Greece. Urban Water J. 2010;7(5):267–85.
58. Rizzo A, Vermersch M, John SG, Micallef G, Riolo S, Pace R. Apparent water loss control: the way forward. Water. 2007;21(9):45–7.
59. Arregui FJ, Gavara FJ, Soriano J, Pastor-Jabaloyes L. Performance analysis of ageing single-jet water meters for measuring residential water consumption. Water. 2018;10(5):18.
60. Criminisi A, Fontanazza CM, Freni G, La Loggia G. Evaluation of the apparent losses caused by water meter under-registration in intermittent water supply. Water Sci Technol. 2009;60(9):2373–82.
61. Kanakoudis V, Tsitsifli S. Socially fair domestic water pricing: who is going to pay for the non-revenue water? Desalin Water Treat. 2016;57 (25):11599–609.
62. Arregui FJ, Soriano J, Cabrera E, Cobacho R. Nine steps towards a better water meter management. Water Sci Technol. 2012;65(7):1273–80.
63. Tabesh M, Delavar MR. Application of integrated GIS and hydraulic models for unaccounted for water studies in water distribution systems. Leiden: A a Balkema Publishers; 2003.
64. Van Zyl JE. Introduction to integrated water meter management. Gezina, South Africa: Water Research Commission (WRC). 2011.

Management Strategies for Minimising DBPs Formation in Drinking Water Systems

7

Nuray Ates, Gokhan Civelekoglu, and Sehnaz Sule Kaplan-Bekaroglu

Abstract

The disinfection process is one of the most critical processes in drinking water treatment plants to protect public health. Disinfection by-products are formed as a result of the reaction of chemical disinfectants with natural organic substances in water resources. Due to carcinogenic and mutagenic effects in case of long-term exposure to DBPs, some species of DBPs are restricted by legislation issued by the USEPA and WHO. Control strategies should be implemented both in the removal of organic matter and in the optimization of the disinfection process in order to prevent and to minimize DBP formations. In addition, optimization of treatment techniques, disinfection process, and improvement of operating conditions in the plant and distribution network need to be considered in controlling DBP formation. In order to understand the kinetics of the formation and to monitor of DBPs in the treatment plant and distribution network, various experimental and theoretical models have been developed. They are presented in this article.

Keywords

Disinfection · Disinfection by-products · Natural organic matter · Predictive models · Precursors

7.1 Introduction

Disinfection process is applied in water treatment system to eliminate microorganisms and to inhibit microbial growth in the pipe network. Chlorine, chloramine, chlorine dioxide, ozone and ultraviolet light are among the most commonly used disinfectants [1]. Disinfectants used in the disinfection process react with organic matter, dissolved organic nitrogen, anthropogenic pollutants and bromide/iodide salts naturally found in most of the source waters, resulting in the formation of disinfection by-products (DBPs). DBPs are undesirable in drinking water, as there is some evidence that long-term exposure may cause health risks [2–4]. The most common DBPs are trihalomethanes (THMs) and haloacetic acids (HAAs) in chlorinated and chloraminated waters, bromate in ozonation, and chlorite in chlorine dioxide waters [5–7]. Besides, chloral hydrate, haloacetamides,

N. Ates (✉)
Environmental Engineering Department, Erciyes University, Kayseri, Turkey
e-mail: nuraya@erciyes.edu.tr

G. Civelekoglu
Environmental Engineering Department, Akdeniz University, Antalya, Turkey

S. S. Kaplan-Bekaroglu
Environmental Engineering Department, Suleyman Demirel University, Isparta, Turkey

M. Bahadir and A. Haarstrick (eds.), *Water and Wastewater Management*,
Water and Wastewater Management, https://doi.org/10.1007/978-3-030-95288-4_7

haloacetonitriles, halopropanones, N-nitrosamines, aldehydes, ketoacids, carboxylic acids, and iodinated-DBPs are DBPs formed in chlorination and ozonation process [1, 3, 8]. The studies on DBPs formation speciation reported that more than 700 of the DBP species are formed in micro and nano amounts in disinfected drinking water [9]. However, more than half of these DBP species have not been quantified yet [8]. Moreover, the potential health effects of many DBPs caused by chemical disinfectants are not well known [2, 10].

The nature and quantity of DBPs formed depends on the type of disinfectant, dose, and the type of precursors or other constituents present in the water, treatment plant operation conditions, temperature of the water, and the residence time of the water in the distribution system [11]. In order to minimize the formation of DBPs and to ensure compliance with DBPs regulations in the treatment plant effluents and distribution systems, control strategies should be developed and implemented. Removing DBP precursors from water prior to disinfection, optimizing the disinfection process, and improving hydraulic properties in the distribution system are the main control strategies [12–14]. Since DBP formation in water resources is not only site specific, but DBP analysis requires advanced instrumentation, analytical techniques, and experienced personnel, is tedious and time consuming. Therefore, many researchers have made efforts to develop models for the prediction of DBP formation.

In this chapter, a brief literature review on formation, control, and modeling of DBPs in water treatment plants and distribution systems is provided. The chapter consists of 7 sub-chapter: (1) A general introduction, (2) Formation and health effects of DBPs and related legislation, (3) Control technologies to remove DBP precursors, (4) Alternative disinfectants and optimization of the disinfection process, (5) Improvement of operational conditions in treatment plants and in distribution systems management of DBPs formation, (6) Modeling of DBP formation, and (7) Summarize of the literature review and discussions.

7.2 Disinfection By-Products in Drinking Water

Disinfection process is mandatory for drinking water treatment systems to inactivate microorganisms (viruses, bacteria, protozoa, etc.) that cause infectious waterborne diseases, and to protect the distribution system. Chlorine, chloramines, chlorine dioxide, ozone and ultraviolet light are among the most applied disinfectants for disinfection purposes [1, 15]. Disinfectants react with organic matter, dissolved organic nitrogen, anthropogenic pollutants and bromide/iodide naturally found in the source waters and form disinfection by-products (DBPs) [15, 16]. The basic schematic illustration of DBPs formation by reaction of disinfectants with organic substances and inorganic ions is shown below [17]. The speciation and amount of DBPs are affected by the type of precursors (Fig. 7.1), type of disinfectant and dose, background ions present in water, operation conditions of the treatment plant, water temperature, and residence time of water in the distribution system [6, 18, 19]. The presence of high level of bromide ion (>100 μg/L) in natural sources causes DBP species to shift from chlorinated by-products to brominated ones [19, 20]

With the discovery of the gas chromatography device and the development of advanced extraction methods in the 1950s and 1960s, chlorinated compounds in water sources have become precisely measurable [21]. In the 1970s, Rook [22] detected higher amount of trihalomethanes (THMs) in chlorinated waters compared to raw surface waters, while Beller et al. [23] reported that higher amounts of chlorine leads to higher THM concentrations. Haloacetic acids (HAA) are nonvolatile chlorinated halides and the most abundant among DBPs following THMs in the chlorination/chloramination process [8, 24]. Nowadays, more than 700 DBP species are found in micro and nano amounts in disinfected drinking water [16]. More than half of the total organic halides formed during chlorination and more than half of the DBPs formed during the ozonation process

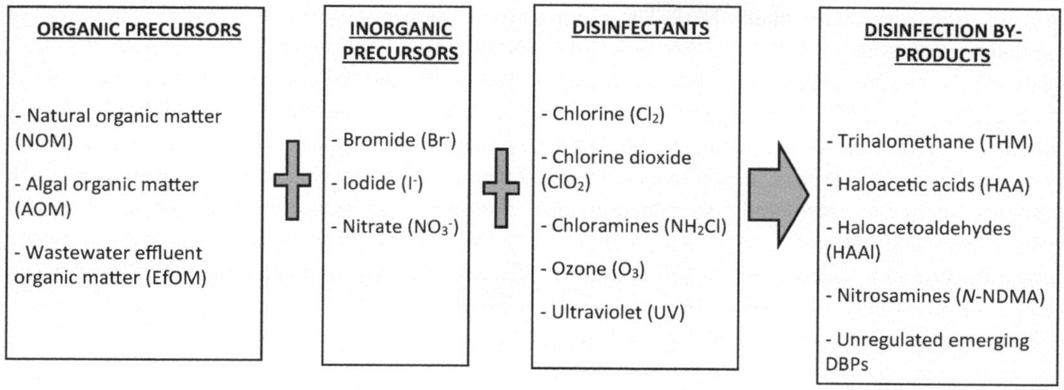

Fig. 7.1 Precursors and disinfectants

have not been identified yet [15]. Among the identified DBPs, THMs and HAAs are classified as carbonaceous DBPs (C-DBPs), and n-nitrosodimethylamine (*N*-NDMA), haloacetonitriles (HANs), and haloacetamides (HAcAms) are included in nitrogenous DBPs (N-DBPs) in disinfected waters [25].

Although the potential health effects of many DBPs caused by chemical disinfectants are not well known [2, 10], epidemiological studies indicate that long-term exposure to DBPs leads to increased cancer risk, and to potential genetic and mutagenic disorders [2, 26]. DBPs formed during the disinfection process, which cause serious risks for public health, are taken into the human body through digestion, respiration and dermal adsorption [3, 4, 27]. In the past 40 years, the studies on the health effects revealed that long-term exposure to DBPs is linked with the risk of brain cancer [28], bladder cancer [29, 30], and colon and rectal cancer [31]. Brominated and iodinated DBPs formed by chemical oxidation of bromine and iodine are more cytotoxic and genotoxic than their chlorinated analogues [3, 32, 33].

The first regulations for DBPs date back to the late 1970s. The first legislation for THMs was established under the interim total THM (TTHM) regulations by the US Environmental Protection Agency (USEPA) in 1979 [34]. According to the regulation, the total THM level was set as 100 μ/L maximum contaminant level (MCL) for water service systems serving more than 10,000 people based on annual average quarterly samples [34, 35]. Only 17 out of about 700 DBPs [16] including THMs (total and four species), HAAs (total and three species), HANs (dibromoacetonitrile and dichloroacetonitrile), bromate, chlorate, chlorite, *N*-NDMA, cyanogen chloride and 2,4,6 trichlorophenol are the most widely regulated ones by USEPA, WHO, European Countries, and some other countries [26, 36–38]. In Turkey, total THM was only DBP group regulated by Ministry of Health in 2005 with a limit of 100 μg/L [39].

7.3 Precursor Control Technologies

In order to control and to minimize DBP formation, responsible organic and inorganic precursors for DBP formation should be removed from source waters. The major precursor is NOM, which refers to a complex mixture of different organic compounds with both hydrophobic and hydrophilic moieties with different properties and molecular sizes such as humic substances, proteins, carboxylic acids, and carbohydrates [40, 41]. NOM concentration in natural waters range between few μg/L up to hundreds of mg/L [42]. Also, residual algae organic matter and micropollutants such as pesticides, pharmaceuticals and personal care products (PPCPs), and estrogens may produce DBPs upon reacting with disinfectants [43]. The main inorganic precursors of DBP are bromide and iodide ions [44]. Characterization and identifying of precursors of DBP in the source water is crucial for determining the strategy of DBP precursors' removal. Source water management strategies might be considered for reduction of DBP precursors. Blending various sources, alternating between sources, using the optimum intake and Riverbank filtration (RBF) are some of strategies to reduce DBP precursors. RBF was able to reduce 35–67% of dissolved DOC and THM- formation potential (FP), and HAA-FP reductions ranged from 57 to 73% and from 50 to 78%, respectively, in three full-scale RBF sites [45]. Removal of the DBP precursors is commonly used approach in order to reduce overall DBP formation. Enhanced coagulation, adsorption, ion exchange, biofiltration, advanced oxidation processes and membrane filtration are the main DBPs precursor control processes in drinking water treatment plants [12].

7.3.1 Coagulation/Enhanced Coagulation

The NOM removal mechanism through coagulation is a charge neutralization, adsorption, entrapment, and complexation process [46]. Enhanced coagulation is more effective for NOM removal compared to conventional coagulation methods [47]. Enhanced coagulation can include following operational processes such as lowering pH, increasing coagulant dose and adding coagulant aid [48]. Iron or aluminum-based coagulants are most widely used coagulants for NOM removal [46]. Typical dosage of iron-based or aluminum-based coagulants ranged between 5 and 150 mg/L depending on turbidity and NOM concentration. Optimum pH range for aluminum-based coagulants ranged 5.5–7.7, which is slightly higher than the optimum range for iron-based coagulants (4.5–7.0) [48]. 20–66% DOC removal is reported at several full-scale drinking water treatment plants [49]. HAA-FP and THM-FP reductions of water were higher than DOC removal, and these results proved that coagulation removed UV absorbing fractions of NOM [50]. Hydrophobic fractions are the preferentially removed components than hydrophilic fractions in enhanced coagulation process because of the higher charge density in the latter ones [51]. Type and dose of coagulant, pH, alkalinity, operation condition, characteristics of NOM, and the presence of anions and cations are the main factors for the efficiency of enhanced coagulation [52].

7.3.2 Adsorption

The adsorption process, mass transfer between solute and adsorbent, is one of the most preferred technologies because of ease of operation, economic feasibility and simplicity [53]. Activated carbon, carbon nanotubes, iron oxide particles, and other low-cost natural adsorbents have been tested so far for the removal of NOM with varying degree of success [49]. Activated carbon in form of powdered activated carbon (PAC) and granular activated carbon (GAC) is one of the most frequently used adsorbent for the removal of NOM with high removal efficiencies [54]. The adsorption process is also frequently used after coagulation to achieve additional removal of NOM and to increase the bio-stability of the water. Marais et al. [55] reported that approximately 20–30% additional NOM removal was achieved by GAC filtration after conventional water treatment. Characteristics of GAC, characteristics of NOM and process operation conditions are important parameters to determine effectiveness of activated carbon [56]. Reductions of DOC, HAA-FP and THM-FP and were at 80%, 89%, and 95%, respectively, for empty bed contact time (EBCT) of 21 min after 50 days operation, but the removal levels of DOC, HAA-FP and THM-FP were decreased in a full-scale trial to 42%, 71% and 40% after 250 days, respectively. In general, 10–15 min of EBCT is preferred for NOM reductions [49]. Fouling of GAC pores by NOM fractions and ineffectiveness for inorganic DBPs precursors are major disadvantages of GAC [57].

7.3.3 Ion Exchange Processes (IEX)

NOM removal by IEX is based on reversible exchange of ions between anionic functional groups on NOM and anions sorbed on the cationic resin surface. Magnetic ion exchange resin (MIEX) was designed specifically for removal of NOM as an alternative to enhanced coagulation. Results of full- and pilot-scale studies of MIEX showed that DOC removals ranged between 36 and 80% [58]. Bolto et al. [59] reported that 10–40% of NOM cannot be removed by ion exchange due to uncharged species in the NOM. Besides MIEX, suspended ion exchange and fluidized ion exchange have been tested for NOM removal as well. Additional 50–62% reductions in DOC, HAA-FP, and THM-FP were reported for suspended ion exchange process, as compared with conventional treatment which was based on coagulation and filtration [60]. Except of particulate NOM, a variety of NOM fractions were successfully removed by anion exchange resins [61]. Removal of hydrophobic, transphilic and hydrophilic fractions by MIEX ranged at 63–75%, 70–89% and 2–67%, respectively [49]. NOM characteristic, composition of raw water, properties of resins and operational conditions are the most important parameters for NOM removal by MIEX. Besides removal of NOM, the capability of inorganic precursors removal (e.g., bromide) and decrease in coagulant and oxidant/disinfectant demand are the main advantages of MIEX [61].

7.3.4 Membrane Processes

Membrane processes gained an increasing amount of attention for the removal of NOMs because of achieving higher rejection, lower disinfectant demand, easier operation control and less sludge generation. Microfiltration (MF) membranes are unable to remove NOM because of their large pore sizes. Removal of NOM with negatively charged ultrafiltration (UF) membranes have obtained rejections rates ranged from 30 to 50% [62]. NOM and DBPs precursors can be more effectively removed by nano filtration (NF) and reverse osmosis (RO) than UF membranes [63]. Many studies have been conducted on the removal of NOM using ceramic and polymeric NF membranes [64]. Sieving effects, electrostatic and hydrophobic interactions are the proposed mechanism for NOM removal. Over 90% of the formation potential of THM and HAA was eliminated using NF and RO membranes [65]. The major challenges in using membrane filtration are fouling, operational costs and lower removal performance for low molecular weight NOM. Therefore, NF and RO membranes require extensive pretreatment to control of fouling and scaling. DBPs precursors (organic and inorganic compounds) can be simultaneously removed by RO membranes, but high operational costs and concentrate management are current limitation factors for the widespread application in drinking water treatment plants [64].

7.3.5 Biological Treatment

The main biological treatment process for NOM removal is biologically active carbon (BAC) filters. BAC filters are one of the most promising, environmentally friendly and economic alternatives to overcome several limitations of other NOM control processes [66]. Biologically active filtration (BAF), multiple-function process is effective to remove a wide range of organic and inorganics compounds [67]. Pre-ozonation is used before BAC filter to transform NOM into biodegradable organic matter, however, BAF without pre-ozonation process is also used for removing DBP precursors. Brown et al. [68] reported that 38% of full scale BAF in USA do not perform ozone in the treatment. Extensive NOM removal studies at pilot/full scale plants have demonstrated that ozone/BAF successfully removed NOM to control DBP formation [67]. Removal of C-DBPs (THMs/HAAs) and N-DBPS (HNMs, HANs, and NDMA) precursors were 13–57% and 15–50%, respectively, in two full scale BAFs and two pilot scale BAFs [69]. Since hydrophilic fraction of NOM was removed preferentially by ozone/BAF, DOC could be reduced by 5–25% on pilot/full scale ozone/BAF [70]. Filter media, EBCT, concentration of attached biomass, ozone dose, temperatures, backwashing, and nutrient supplementation are the main factors affecting biofiltration performance. EBCT < 5 min, EBCT of 5–10 min, EBCT > 10 min in a survey of 40 full-scale BAF facilities were employed for 30%, 50%, 20% of biofilters in North America respectively [68].

7.3.6 Advanced Oxidation Processes (AOPs)

AOPs have gained interest in the last decades to remove NOM, algal organic matter, and effluent organic matter from drinking waters. AOPs generate hydroxyl radicals to achieve the transformation of organic matter into smaller molecular weight compounds [71]. Combination of ozone (O_3) and/or hydrogen peroxide (H_2O_2) with different catalysts and/or UV light are tested in AOPs for the NOM removal [72]. The removal efficiency of AOPs mainly depends on amounts of hydroxyl radicals and the water quality. The concentration of DOC, THM, and HAA by catalytic ozonation was reduced by 8.2–51.4%, 41.3–51.2% and 31.7–48.3%, respectively [73]. DOC, THM-FP and HAA-FP in O_3/UV treated waters decreased approximately 50%, 70%, 80%, respectively [74]. The main disadvantages of AOPs are the formation of oxidation by-products and operational costs due to high energy demand. In order to improve the overall removal efficiencies of NOM and other organic and inorganic DBP precursors, combinations of two or more treatment technologies are proven as promising options.

7.4 Alternative Disinfection to Minimise DBP Formations

Most of the identified and well-known DBPs are associated with chlorine, which is the most common disinfectant used worldwide. Therefore, operators should manage disinfection practices both to achieve microbial elimination and to comply with legal restrictions on disinfection by-products in water treatment facilities. While ozone, chlorine dioxide, chloramines or UV as alternative disinfectants have been used as primary disinfectants instead of chlorine to reduce the formation of disinfection by-products, chlorine or chloramines are applied as post-disinfectant to provide microbial safety in the water networks [3].

7.4.1 Alternative Disinfectants

Chloramines

The first alternative disinfectant applied for post-disinfection in drinking water treatment plants is chloramines, because it can be easily formed by only adding ammonia to the chlorine system, which complies with regulations by reducing DBPs by 90%, and provides residual disinfectant in distribution system [16]. The amount/formation of DBPs including chloroform, dichloropropanone, trichloropropanone and dichloroacetonitrile in waters containing 3 mg/L of NOM tested at different HOCl/NH_2Cl (5:0 to 0:5) ratios at pH 7 were remarkably decreased [75]. Huang et al. [76] observed I-DBPs for monochloramine and chlorine application in 5 mg/L disinfectant dose during 30 min contact time. The more polar I-DBP formations in chloraminated waters than chlorinated waters are attributed to oxidation of I^- to HOI/IO^-, which reacts with NOM to produce higher DBPs [76, 77].

Ozone

Ozone can eliminate microorganisms and degrade high molecular weight organic matter into smaller molecular weight organics due to its high oxidation potential [78]. Thus, it is used as pre-treatment before biological treatment processes for degradation of recalcitrant organics [79–81]. Combining O_3 and H_2O_2 before chlorination reduces the formations of HAA and brominated DBPs but enhances formation of THM with respect to sole ozone application [82, 83]. During ozone treatment followed by chlorination or chloramination, ozone can react with certain polymers (e.g., polyamine and poly-

DADMAC [poly(diallyldimethylammonium chloride]) and amine groups in NOM, and form *N*-NDMA and halonitromethanes (HNMs) [3, 84]. Besides, the reaction of ozone with organic substances and bromine produce bromate, which is regulated by USEPA as 10 μg/L [37].

Chlorine Dioxide

Chlorine dioxide (ClO_2) is preferred as an alternative to chlorine as it is a highly selective oxidant and does not react with NOM to form THM [85, 86]. Unlike chlorine, the occurrence of oxidation–reduction reactions between ClO_2 and NOMs instead of substitution reaction limits the formation of DBP [87]. When pre-chlorine dioxide followed by post-chlorination is applied, significant reductions were obtained in THM (–90%) and HAA formations (–87%) compared with pre-chlorination [88]. Moreover, about 50–70% of ClO_2 is reduced during the oxidation process, producing inorganic DBPs ClO_2^- and ClO_3^-, which are restricted by regulations [85, 87]. On the other hand, it is considered that ClO_2, when applied with chlorine or chloramines, leads to the formation of iodoacetic acid in water sources with high iodide levels and of tribromopyrrole, the latter being more cytotoxic than the regulated DBPs [15].

Ultraviolet Photolysis (UV)

Recently, UV technology gained interest as a promising alternative disinfection technique and widely used coupled with chlorination/chloramination to minimize DBP formations and to meet regulation limits. Besides, coupling UV with H_2O_2 and O_3 is suggested as a promising technique for the removal of organic micro-contaminants (e.g., pharmaceuticals and pesticides) [89]. However, recent studies showed that application of medium-pressure UV photolysis to post-chlorinated or post-chloraminated source waters has significant impact on increasing DBPs formations including formation of THMs, HAAs, HANs, HNMs, HAcAms, chloral hydrate, and cyanogen chloride [84, 90, 91].

7.4.2 Management of Disinfectants

In order to control DBPs formations in conventional treatment plants, shifting chlorine to its alternatives for pre- and/or post-disinfection can result in lower concentrations regulated or unregulated DBPs, but sometimes leading to more toxic and unknown species [92, 93]. In the case of disinfectant replaced by alternatives, besides regulated DBPs some other emerging DBPs including HNMs, nitrosamines, haloamides, halonitriles, halofuranones, iodo-acids, iodo-THMs, haloaldehydes, halopyrroles, haloquinones, haloketones, chlorate and iodate can be produced, depending on type and amount of organic matter, nitrogenous compounds, level of bromine or iodine, and water quality [3]. Although O_3 treatment prior to chlorination or chloramination significantly reduces the formation of THMs and HAAs [7], it causes bromate formation, especially in the presence of high levels of bromide salts in the source waters [3]. However, since water sources containing high iodide would contain high concentrations of bromide, great care must be taken when determining the ozone and chlorine dosage, contact time and operational conditions in order to control the formations of THM, HAA and bromate that exceed the regulation limits [7]. In the case of ClO_2 and chlorine applied together, adsorption or membrane processes are proposed to minimize regulated and emerging DBPs [3]. Since chloramine and chlorine dioxide form I-DBPs in water sources containing iodide, the application of these disinfectants should be avoided unless there is a pre-treatment process such as chlorination and ozonation that can oxidize iodide [7]. The order of the tendency towards formation of chlorinated and brominated by-products in conventional drinking water treatment plants based on the type of disinfectant is chlorine $\gg$ chloramines > chlorine dioxide. On the other hand, disinfectants can be put in order as chloramines > chlorine dioxide > chlorine > ozone in the tendency to form iodized by-products [7].

7.5 Operational Improvements and Adjustments for Water Treatment Plants and Distribution Systems

One of the most effective approaches to control and to minimize DBP formation in the treatment plant and the water distribution network is the optimization of the operating parameters that are effective in DBP formation. Therefore, some operational improvements and adjustments are required for an effective control of DBP formation. Although operating parameters can be optimized based on experiences gained so far, it should be considered to carry out DBP control on a plant-by-plant basis, since the characteristic of each source water is unique.

The formation of THMs and HAAs increase rapidly in treatment facilities by increasing contact time and chlorine dosage. While both DBP groups are formed before the contact time reached 5 h, 90% of the ultimate formation occurred in the 24-h period [94]. When the water is chlorinated in the primary disinfection process, it travels for a long time in the distribution system until it reaches the first user. This time is usually more than the minimum contact time of 20 min accepted as a requirement [95]. Although high contact times increase the formation of DBP, low contact times reduce disinfection efficiency. Therefore, it should be aimed to exactly determine the location of chlorine application, and if necessary, intermediate points may be located in the water distribution system to minimize DBP formation by performing secondary disinfection procedures.

The chlorine dose is adjusted according to the concentration of residual chlorine that must be provided in the distribution system, and the demand for chlorine increases in spring and autumn due to turbidity caused by storm water drainage. The increased chlorine dose increases the formation of DBPs [96]. It should be noted that the presence of bromide ion in chlorine solution causes the formation of brominated species [94]. Therefore, it is necessary to optimize the chlorine dosage during the period of highest THM or brominated species formation potential or highest chlorine demand. Although the removal of DBP precursors is the basic approach, as an alternative, reducing the dosage of chlorine without lowering the threshold residual chlorine level is a more appreciable cost-effective method [95].

It has been observed that the chlorine dose applied in disinfection together with ozone has a positive correlation with the formation of bromate. The increased concentration of bromide in water also increases the formation of bromate [97]. Although the value of CT (concentration x time) is an appropriate parameter in terms of disinfection with ozone, it may not be a good parameter for estimating the formation of bromate. This is because CT depends on residue concentration of ozone and the duration of detention in the ozone chamber, and the effects on bromate formation differ as these parameters change [97].

The use of UV and ozone in the disinfection process reduces the formation of halogenated DBPs [38]. Since residue disinfectant cannot be provided in UV oxidation, the use of this process as secondary disinfection by combining it with chlorine or chloramine can provide residual disinfectant in the water distribution system. Similarly, in an algae-rich water source with low specific UV absorbance (SUVA) values, UV irradiation has been reported to be effective in reducing the formation of THM and dichloroacetic acid during subsequent chlorination [98]. Another study reported that the high UV dosage applied during UV-chloramination resulted in a 24% reduction of haloketone, dichloroacetic acid and trichloroacetic acid [43].

Surface waters have a naturally low pH, which is suitable for disinfection with chlorine, and historical data showed that the pH adjustment had a little significant effect on THM formation [95, 99]. Besides, it has been reported that THM concentrations decrease as pH decreases and HAAs increase [100]. Furthermore, increasing pH tends to favor the formation of THMs (up to pH 9.5) and to decrease the formation of thrichloroacetic acid (TCAA) and total organic halogen (TOX) [94].

Since bromate formation generally increases at higher pH values [101], reducing the pH in ozonated water has proven to be an effective bromate control strategy [101, 102]. It has been stated when increasing pH value from 6.0 to 8.5 causes an increase of about 20 ppb bromate due to the transition from molecular ozone to hydroxyl radical oxidation [97]. In addition, it has been reported that the concentration of bromide required to create bromate decreases as the pH increases [103].

The contact time required for UV disinfection is quite low and is not affected by pH. It does not provide residues in the UV distribution system and therefore must be associated with chlorination or another type of disinfectant that provides residues. If the turbidity is high, the processes may be used for groundwater disinfection. It has been stated in water that UV radiation costs about twice the cost of chlorine disinfection [95].

The temperature both increases the reaction rate and the potential for THM formation [95]. For instance, increasing the temperature from 10 to 30 °C increases the formation of THM by 15–25% [94]. Due to the decay of organic matter, DOC concentration increases in the waters, which most likely leads to the formation of higher THM in the fall season [44, 104]. Although the process of HAA formation and decay is slightly different from THM's, the formation of HAA is similarly increased by increasing temperature [95].

Bromate formation is a concern in warmer waters due to its high reaction rate. Therefore, the rate of bromate formation is seasonally changed and increases in summer, if the water temperature rises above 15 °C [97]. Another study reported that the increase in water temperature from 20 to 30 °C, increased bromate formation by about 31% [103]. As the water temperature increases, ozone becomes more reactive, and its half-life decreases. Bu this also causes decreased ozone exposure at constant ozone doses [97].

Whole water entering a distribution system must contain a disinfectant concentration of monochloramine or free chlorine residue at a level of 0.5 mg/L or higher, after contact time in peak hourly flow for at least 20 min [95, 96, 105]. In the water distribution system, all the points should have a detectable free chlorine residue. A higher residual disinfectant may be required depending on the pH, temperature and other properties of water. Longer contact time in the distribution system and, therefore, reaction time often leads to higher residual chlorine consumption and leads to the formation of more THMs. Blended water has more stable chlorine residues than new or old water. Chlorine residues above 4 mg/L can cause known or expected health risks such as eye and nasal irritation, and stomach discomfort [95]. If the concentration of residual chlorine at the beginning of the distribution system exceeds 4 mg/L, the chlorine dose should be reduced, and other options should be explored to keep the amount of residual chlorine constant. Since there is a direct relationship between chlorine use and DBP levels, it is necessary to optimize the dose of chlorine.

Each engineering solution must be unique in the same way since each water source or water distribution system has its own characteristics. Therefore, a single-dimension solution to DBP problems does not work. Water age is the water retention time in the water distribution system and is affected by pipe length and water flow rates [13]. Water age is an important performance indicator and it increases when the pressure decreases in a water distribution system [106]. Therefore, higher duration of water in distribution system may lead to increased DBP concentrations [106]. According to the results of a survey covering more than 800 water supply networks in the U.S., the average age of water is about 1.3 days. The water distribution network is designed so that this value is not more than 3 days [13].

Since the changes in water quality depend on hydraulic and system operating conditions, hydraulic and water quality parameters should be measured at the same time. In order to reduce hydraulic retention times, different pressure zone can be defined with software and automation applications, or changing pressure adjustment points is possible. Thereby, it is possible to always maintain sufficient water pressure at all locations. Maintaining and repairing the water system for certain periods can help to keep the system clean and sediment-free. In addition, replacement of damaged pipelines in time, which are similarly known to contribute to the deterioration of water quality, is a method that will reduce the potential for DBP formation [107].

The location of water storage tanks in the distribution system can affect the amount of residual chlorine, water age and DBP levels. Tanks located at the end of the distribution system tend to increase the age of water in the tank and the distribution network, then increase the variability of the amount of residual chlorine throughout the system. If water storage tanks are located at the head of the distribution system, the overall age of water and the amount of residual chlorine in the distribution network tend to decrease. For water storage tanks with long residency periods, ventilation systems can be used to remove volatile DBP compounds from water. With the installation of a ventilation system for water storage tank, attention should be paid to the resulting loss of residual chlorine [95].

7.6 Models for Prediction and Management of DBP Formation in Drinking Waters

Formation and concentration of DBPs in drinking water depend on the characteristics of raw water, the conditions of operation of the treatment plant, the type of disinfectant, the water temperature, and the retention time of the water in the distribution system [43]. Models are designed to minimize DBP formation. Operation control can be used as a decision-making tool for distribution system design and maintenance [95]. At the same time, these models can be used to detect different water qualities and operating parameters on DBP formation [108].

The models to estimate DBP formation are generally developed using different approaches under variable water quality parameters and different operating conditions. Therefore, it is only possible to compare the performance of the models if water qualities and operating conditions are similar. Different performance indicators and statistical techniques such as the determination coefficient (R^2), the correlation coefficient (R), errors between the measured and predicted data (AE, MSE, RMSE, etc.) are used to estimate the capability of the models. Even if it has a good prediction performance, regression models need to be used in experimental boundary conditions to get relatively rational results [109].

Numerous models have been developed to predict the formation of DBP in drinking waters. The largest focus area of these models is the formation of THMs. The data based on the model were obtained from measurements performed in raw, pre-treated or synthetic prepared waters in the laboratory or field. Some models have been developed by analyzing databases containing historical data. In most DBP models, multiple linear regression analysis (MLR) was used to evaluate the empirical relationship between the measured variables and DBP concentrations. In some approaches, principal components analysis (PCA), a data preprocessing method that can be used to determine the number of variables in the model, has been applied [110]. DBP formation models developed on kinetic relationships are very limited in the literature [108].

The major responsible DBP precursor is NOM, and due to non-linear nature of NOM reactions, it is mostly difficult to express DBP formation with MLR models [111]. Artificial intelligence (AI) methods can define complex and nonlinear relationships and have been used frequently for DBP formation in recent years [112, 113]. AI models have a certain degree of interpretability. Therefore, the selection of appropriate analysis methods as a decision support mechanism for the formation of DBP has the potential to be used in the stages of determining forward-looking DBP control strategies based on water treatment processes and historical data [112].

Some DBP models selected from the literature developed with different techniques are summarized in Table 7.1. These models do not only predict the potential of DBP formation, but also determine the effect of different water quality and operating parameters on DBP formation in order to in-depth understanding of the DBP formation

Table 7.1 MLR and ANN models developed for predicting of different DBPs

Model ID	Model descriptions	R^2	References
MLR	$HAAs = -8{,}202 + 4{,}869(TOC) + 1{,}053(D) + 0{,}364(t)$	0.92	[114]
MLR	$DCAN = 3{,}567(D)^{1.03}(pH)^{-1.64}(R)^{0.18}t^{0.234}$	0.69	[115]
MLR	$THMs = 10^{-1{,}375}(t)^{0.258}(D/TOC)^{0.194}(pH)^{1{,}695}(T)^{0.507}(Br^{-})^{0.218}$	0.87	[116]
MLR + PCA	$BrO_3^{-} = (EC)^{0.46}(D)^{0.62}(t)^{0.50}$	0.77	[117]
ANN + PCA	3 input neurons in the input layer, 2 hidden neurons in each of the 2 hidden layers, and 1 neuron in the output layer was employed to predict BrO_3^{-}	0.97	[117]
ANN + PCA	7 input neurons in the input layer, 2 hidden layers, and 1 neuron in the output layer was employed to predict THMs, HAAs and TOX	0.98	[118]
ANN	5 input neurons in the input layer, 7 hidden neurons in 1 hidden layer, and 1 neuron in the output layer was employed to predict THMs	0.92	[110]
ANN + PCA	32 input neurons in the input layer, 5 hidden neurons in 1 hidden layer, and 3 neurons in the output layer was employed to predict HAAs	0.98	[119]

THMs: Total trihalomethanes; HAAs: Haloacetic acids; TOX: Total organic halide; BrO_3^{-}: Bromate; TOC: Total organic carbon (mg/L); D: Chlorine or ozone dose (mg/L); T: Temperature (°C); t: Reaction time (min/hr); Br^{-}: Bromide ion concentration (mg/L); R: Residual chlorine (mg/L); DCAN: Dichloroacetonitrile (μg/L); EC: Electrical conductivity (μS/cm)

mechanism [111]. It can be noted that most models are developed using MLR statistical techniques. As it is shown in Table 7.1, different DBP concentrations can be estimated with MLR and artificial neural network (ANN) models in the performance (R^2) range of 0.69–0.92 and 0.92–0.98, respectively. This suggests that ANN models accurately integrate complex relationships between precursors and DBPs. Although ANN cannot provide precise equations such as MLR and does not clearly show the relationship between variables, the general approach is that it is possible to understand the causal mechanism between the model inputs and outputs of the ANN structure. The main advantage of combining ANN and PCA is the simplification of the complex model structure and making a clear diagnosis of the relationship between variables. Combining and improving different AI methods can also improve the model prediction capabilities.

7.7 Conclusions

Controlling the formation of DBPs has been one of the major challenges for water treatment industry, since THMs were first discovered in drinking water in the early 1970s. Over the last 40 years, much research has been conducted to improve understanding of formation and control of DBPs. DBP formation depends on many factors such as properties of precursors, parameters of water quality, disinfection conditions, and operation conditions of treatment plant and specific characteristics of the distribution system. Predictive models are useful for selection of control strategies to minimize DBPs based on the statistical relationships. Continuous assessment and monitoring of source water are important to determine the concentration, characterization and reactivity of NOM, and to select the optimal strategy for DBPs control to mitigate the health risk posed by DBPs. Two approaches are mainly used for management strategies for minimizing DBPs formation in drinking water systems: (1) shifting chlorine to alternative disinfectants, and (2) precursor removal before disinfectant addition. The use of alternative disinfectants may not be always feasible due to formation of other disinfection by-products. Enhanced coagulation, adsorption, ion exchange, and membrane filtration are the main precursor removal processes available for DBP control. Also, biological methods, such as BAC filters have gained more attention in recent years. Combinations of two approaches have shown better results for the control of DBPs. Source-specific studies such as bench- and/or pilot-scale testing, are crucial to select the most effective control DBPs strategies.

References

1. Tang YL, Long X, Wu MY, Yang S, Gao NY, Xu B, Dutta S. Bibliometric review of research trends on disinfection by-products in drinking water during 1975–2018. Sep Purif Technol. 2020; 241. ARTN 116741/10.1016/j.seppur.2020.116741.
2. Richardson SD, Plewa MJ, Wagner ED, Schoeny R, DeMarini DM. Occurrence, genotoxicity, and carcinogenicity of regulated and emerging disinfection by-products in drinking water: a review and roadmap for research. Mutat Res-Rev Mutat. 2007;636 (1–3):178–242. https://doi.org/10.1016/j.mrrev.2007.09.001.
3. Richardson SD, Postigo C. Drinking water disinfection by-products. In: Emerging organic contaminants and human health. Springer, Berlin, Heidelberg; 2011. pp. 93–137.
4. Du YJ, Zhao L, Ban J, Zhu JY, Wang SW, Zhu X, Zhang YY, Huang ZH, Li TT. Cumulative health risk assessment of disinfection by-products in drinking water by different disinfection methods in typical regions of China. Sci Total Environ. 2021;770.
5. Krasner SW, Mcguire MJ, Jacangelo JG, Patania NL, Reagan KM, Aieta EM. The occurrence of disinfection by-products in United-States drinking-water. J Am Water Works Ass. 1989;81(8):41–53.
6. Singer PC. Control of disinfection by-products in drinking-water. J Environ Eng-Asce. 1994;120 (4):727–44.
7. Hua GH, Reckhow DA. Comparison of disinfection byproduct formation from chlorine and alternative disinfectants. Water Res. 2007;41(8):1667–78.
8. Krasner SW, Weinberg HS, Richardson SD, Pastor SJ, Chinn R, Sclimenti MJ, Onstad GD, Thruston AD. Occurrence of a new generation of disinfection byproducts. Environ Sci Technol. 2006;40(23):7175–85.
9. Richardson SD, Ternes TA. Water analysis: emerging contaminants and current issues. Anal Chem. 2018;90(1):398–428. https://doi.org/10.1021/acs.analchem.7b04577.
10. Brown D, Bridgeman J, West JR. Understanding data requirements for trihalomethane formation modelling in water supply systems. Urban Water J. 2011;8(1):41–56.
11. Hua GH, Reckhow DA. DBP formation during chlorination and chloramination: Effect of reaction time, pH, dosage, and temperature. J AWWA. 2007;100(8):82–95.
12. Wang F, Gao BY, Yue QY, Bu F, Shen X. Effects of ozonation, powdered activated carbon adsorption, and coagulation on the removal of disinfection by-product precursors in reservoir water. Environ Sci Pollut Res. 2017;24(21):17945–54.
13. Kourbasis N, Patelis M, Tsitsifli S, Kanakoudis V. Optimizing water age and pressure in drinking water distribution networks. Environ Sci Proc. 2020;51:1–9.
14. Tsitsifli S, Kanakoudis V. Disinfection impacts to drinking water safety—a review. Proceedings. 2018;2(11): 603.
15. Richardson SD, Postigo C. Drinking water disinfection by-products. In: Barceló D, editor. Emerging organic contaminants and human health. Berlin, Heidelberg: Springer; 2012. p. 93–137.
16. Richardson SD, Plewa MJ. To regulate or not to regulate? What to do with more toxic disinfection by-products? J Environ Chem Eng. 2020;8(4).
17. Krasner SW. The formation and control of emerging disinfection by-products of health concern. Philos T R Soc A. 2009;367(1904):4077–95.
18. Hong HC, Xiong YJ, Ruan MY, Liao FL, Lin HJ, Liang Y. Factors affecting THMs, HAAs and HNMs formation of Jin Lan Reservoir water exposed to chlorine and monochloramine. Sci Total Environ. 2013;444:196–204.
19. Hua GH, Reckhow DA. Evaluation of bromine substitution factors of DBPs during chlorination and chloramination. Water Res. 2012;46(13):4208–16.
20. Gould JP, Fitchorn LE, Urheim E. Formation of brominated trihalomethanes: extent and kinetics. In: Jolley RL, Brungs WA, Cotruvo JA, Cumming RB, Mattice JS, Jacobs VA (ed.) Water chlorination: environmental impact and health effects. Ann Arbor Sci. Publ., Ann Arbor, Michigan; 1983.
21. Kristiana I, Charrois JWA, Hrudey SE. Research overview, regulatory history and current worldwide status of DBP regulations and guidelines. In: Disinfection by-products and human health. London: International Water Association Publishing; 2012.
22. Rook JJ. Formation of haloforms during chlorination of natural waters. J Water Treat Examination. 1974;23:234–43.
23. Beller TA, Lichtenberg JJ, Kromer RC. The occurrence of organohalide in chlorinated drinking water. J Am Water Works Assoc. 1974;66:703–11.
24. Chen BY, Westerhoff P. Predicting disinfection by-product formation potential in water. Water Res. 2010;44(13):3755–62.
25. Qian YK, Chen YA, Hu Y, Hanigan D, Westerhoff P, An D. Formation and control of C- and N-DBPs during disinfection of filter backwash and sedimentation sludge water in drinking water treatment. Water Res. 2021;194.
26. WHO: World Health Organization (WHO). Guidelines for drinking-water quality, world health organization, 9241544805, vol. 3, 4th ed. Geneva, Switzerland; 2011.

27. Lee SC, Guo H, Lam SMJ, Lau SLA. Multipathway risk assessment on disinfection by-products of drinking water in Hong Kong. Environ Res. 2004;94(1):47–56.
28. Cantor KP, Lynch CF, Hildesheim ME, Dosemeci M, Lubin J, Alavanja M, Craun G. Drinking water source and chlorination byproducts in Iowa. III. Risk of brain cancer. Am J Epidemiol. 1999;150 (6):552–560.
29. Diana M, Felipe-Sotelo M, Bond T. Disinfection byproducts potentially responsible for the association between chlorinated drinking water and bladder cancer: a review. Water Res. 2019;162:492–504.
30. Freeman LEB, Cantor KP, Baris D, Nuckols JR, Johnson A, Colt JS, Schwenn M, Ward MH, Lubin JH, Waddell R, Hosain GM, Paulu C, Mccoy R, Moore LE, Huang AT, Rothman N, Karagas MR, Silverman DT. Bladder cancer and water disinfection by-product exposures through multiple routes: a population-based case control study (New England, USA). Environ Health Persp. 2017;125(6).
31. Rahman MB, Cowie C, Driscoll T, Summerhayes RJ, Armstrong BK, Clements MS. Colon and rectal cancer incidence and water trihalomethane concentrations in New South Wales, Australia. BMC Cancer. 2014;14.
32. Ates N, Kaplan-Bekaroglu SS, Dadaser-Celik F. Spatial/temporal distribution and multi-pathway cancer risk assessment of trihalomethanes in low TOC and high bromide groundwater. Environ Sci-Proc Imp. 2020;22(11):2276–90.
33. Regli S, Chen J, Messner M, Elovitz MS, Letkiewicz FJ, Pegram RA, Pepping TJ, Richardson SD, Wright JM. Estimating Potential increased bladder cancer risk due to increased bromide concentrations in sources of disinfected drinking waters. Environ Sci Technol. 2015;49(22):13094–102.
34. USEPA: U.S. Environmental Protection Agency (USEPA), National interim primary drinking water regulations; control of trihalomethanes in drinking water. Federal Register, 68624; 1979.
35. Vidić RD. Control of disinfection by-products in drinking water: regulations and costs. In: Maksimović Č, Calomino F, Snoxell J, editors. Water supply systems, vol. 15. Berlin, Heidelberg: Springer; 1996. p. 259–73.
36. EC: European Comission (EC). The quality of water intended for human consumption, council directive 2020/218, December 16, The European Parliament and of The Council, Brussels; 2020.
37. USEPA: United States Environmental Protection Agency (USEPA). National primary drinking water regulations: stage 2 disinfectants and disinfection byproducts rule developed by United States environmental protection agency, Washington, DC; 2006.
38. Yang LY, Chen XM, She QH, Cao GM, Liu YD, Chang VWC, Tang CY. Regulation, formation, exposure, and treatment of disinfection by-products (DBPs) in swimming pool waters: a critical review. Environ Int. 2018;121:1039–57.
39. THOM: Turkish Ministry of Health (TMOH). Regulation concerning water intended for human consumption, Official News Paper, No. 25730, Ankara, Turkey; 2005.
40. Matilainen A, Gjessing ET, Lahtinen T, Hed L, Bhatnagar A, Sillanpaa M. An overview of the methods used in the characterisation of natural organic matter (NOM) in relation to drinking water treatment. Chemosphere. 2011;83(11):1431–42.
41. Tak S, Vellanki BP. Natural organic matter as precursor to disinfection byproducts and its removal using conventional and advanced processes: state of the art review. J Water Health. 2018;16(5):681–703. https://doi.org/10.2166/wh.2018.032.
42. Wall NA, Choppin GR. Humic acids coagulation: influence of divalent cations. Appl Geochem. 2003;18(10):1573–82.
43. Chaukura N, Marais SS, Moyo W, Mbali N, Thakalekoala LC, Ingwani T, Mamba BB, Jarvis P, Nkambule TTI. Contemporary issues on the occurrence and removal of disinfection byproducts in drinking water—a review. J Environ Chem Eng. 2020;8(2).
44. Gilca AF, Teodosiu C, Fiore S, Musteret CP. Emerging disinfection byproducts: a review on their occurrence and control in drinking water treatment processes. Chemosphere. 2020;259.
45. Weiss WJ, Bouwer EJ, Ball WP, O'Melia CR, Lechevallier MW, Arora H, Speth TF. Riverbank filtration—fate of DBP precursors and selected microorganisms. J Am Water Works Ass. 2003;95 (10):68–81.
46. Matilainen A, Vepsalainen M, Sillanpaa M. Natural organic matter removal by coagulation during drinking water treatment: a review. Adv Colloid Interfac. 2010;159(2):189–97.
47. Bolto B, Dixon D, Eldridge R, King S. Removal of THM precursors by coagulation or ion exchange. Water Res. 2002;36(20):5066–73.
48. Sillanpaa M, Ncibi MC, Matilainen A, Vepsalainen M. Removal of natural organic matter in drinking water treatment by coagulation: a comprehensive review. Chemosphere. 2018;190:54–71.
49. Bond T, Goslan EH, Parsons SA, Jefferson B. Treatment of disinfection by-product precursors. Environ Technol. 2011;32(1):1–25.
50. Liang L, Singer PC. Factors influencing the formation and relative distribution of haloacetic acids and trihalomethanes in drinking water. Environ Sci Technol. 2003;37(13):2920–8.
51. Mikola M, Ramo J, Sarpola A, Tanskanen J. Removal of different NOM fractions from surface water with aluminium formate. Sep Purif Technol. 2013;118:842–6.
52. Saxena K, Brighu U, Choudhary A. Parameters affecting enhanced coagulation: a review. Environ Technol Rev. 2018;7(1):156–76.

53. Bhatnagar A, Sillanpaa M. Removal of natural organic matter (NOM) and its constituents from water by adsorption—a review. Chemosphere. 2017;166:497–510.
54. Bond T, Goslan EH, Parsons SA, Jefferson B. A critical review of trihalomethane and haloacetic acid formation from natural organic matter surrogates. Environ Technol Rev. 2012;1(1):93–113.
55. Marais SS, Ncube EJ, Msagati TAM, Mamba BB, Nkambule TTI. Comparison of natural organic matter removal by ultrafiltration, granular activated carbon filtration and full scale conventional water treatment. J Environ Chem Eng. 2018;6(5):6282–9.
56. Golea DM, Jarvis P, Jefferson B, Moore G, Sutherland S, Parsons SA, Judd SJ. Influence of granular activated carbon media properties on natural organic matter and disinfection by-product precursor removal from drinking water. Water Res. 2020;174.
57. Erdem CU, Ateia M, Liu C, Karan T. Activated carbon and organic matter characteristics impact the adsorption of DBP precursors when chlorine is added prior to GAC contactors. Water Res. 2020;184.
58. Singer PC, Boyer T, Holmquist A, Morran J, Bourke M. Integrated analysis of NOM removal by magnetic ion exchange. J Am Water Works Ass. 2009;101(1):65–73.
59. Bolto B, Dixon D, Eldridge R. Ion exchange for the removal of natural organic matter. React Funct Polym. 2004;60:171–82. https://doi.org/10.1016/j.reactfunctpolym.2004.02.021.
60. Metcalfe D, Rockey C, Jefferson B, Judd S, Jarvis P. Removal of disinfection by-product precursors by coagulation and an innovative suspended ion exchange process. Water Res. 2015;87:20–8.
61. Levchuk I, Marquez JJR, Sillanpaa M. Removal of natural organic matter (NOM) from water by ion exchange—a review. Chemosphere. 2018;192:90–104.
62. Xu DL, Bai LM, Tang XB, Niu DY, Luo XS, Zhu XW, Li GB, Liang H. A comparison study of sand filtration and ultrafiltration in drinking water treatment: removal of organic foulants and disinfection by-product formation. Sci Total Environ. 2019;691:322–31.
63. Ates N, Yilmaz L, Kitis M, Yetis U. Removal of disinfection by-product precursors by UF and NF membranes in low-SUVA waters. J Membr Sci. 2009;328(1–2):104–12. https://doi.org/10.1016/j.memsci.2008.11.044.
64. Zazouli MA, Kalankesh LR. Removal of precursors and disinfection by-products (DBPs) by membrane filtration from water—a review. J Environ Health Sci. 2017;15.
65. Sentana I, Rodriguez M, Sentana E, M'Birek C, Prats D. Reduction of disinfection by-products in natural waters using nanofiltration membranes. Desalination. 2010;250(2):702–6.
66. Korotta-Gamage SM, Sathasivan A. A review: potential and challenges of biologically activated carbon to remove natural organic matter in drinking water purification process. Chemosphere. 2017;167:120–38.
67. Liu C, Olivares CI, Pinto AJ, Lauderdale CV, Brown J, Selbes M, Karanfil T. The control of disinfection byproducts and their precursors in biologically active filtration processes. Water Res. 2017;124:630–53.
68. Brown J, Upadhyaya G, Carter J, Brown T, Lauderdale C. North American biofiltration knowledge base. In: Water Research Foundation Report. Denver, Colorado; 2016.
69. Selbes M, Brown J, Lauderdale C, Karanfil T. Removal of selected C- and N-DBP precursors in biologically active filters. J Am Water Works Ass. 2017;109(3):E73–84.
70. Selbes M, Amburgey J, Peeler C, Alansari A, Karanfil T. Evaluation of seasonal performance of conventional and phosphate-amended biofilters. J Am Water Works Ass. 2016;108(10):E523–32.
71. Matilainen A, Sillanpaa M. Removal of natural organic matter from drinking water by advanced oxidation processes. Chemosphere. 2010;80(4):351–65.
72. Sillanpaa M, Ncibi MC, Matilainen A. Advanced oxidation processes for the removal of natural organic matter from drinking water sources: a comprehensive review. J Environ Manage. 2018;208:56–76.
73. Chen KC, Wang YH, Chang YH. Using catalytic ozonation and biofiltration to decrease the formation of disinfection by-products. Desalination. 2009;249(3):929–35.
74. Chin A, Berube PR. Removal of disinfection by-product precursors with ozone-UV advanced oxidation process. Water Res. 2005;39(10):2136–44.
75. Liu Z, Xu B, Zhang TY, Hu CY, Tang YL, Dong ZY, Cao TC, El-Din MG. Formation of disinfection by-products in a UV-activated mixed chlorine/chloramine system. J Hazard Mater. 2021;407.
76. Huang Y, Zhang YY, Zhou Q, Li AM, Shi P, Qiu JF, Pan Y. Detection, identification and control of polar iodinated disinfection byproducts in chlor(am)inated secondary wastewater effluents. Environ Sci-Wat Res. 2019;5(2):397–405.
77. Zhu XH, Zhang XR. Modeling the formation of TOCl, TOBr and TOI during chlor(am)ination of drinking water. Water Res. 2016;96:166–76.
78. Rice RG, Wilkes JF, Miller GW, Hill AGM. Use of ozone-bromide reactions. J Am Water Works Ass. 1981;85:63–72.
79. Carlson G, Silverstein J. Effect of ozonation on sorption of natural organic matter by biofilm. Water Res. 1997;31(10):2467–78.
80. Chang CN, Ma YS, Zing FF. Reducing the formation of disinfection by-products by pre-ozonation. Chemosphere. 2002;46(1):21–30.

81. Siddiqui MS, Amy GL, Murphy BD. Ozone enhanced removal of natural organic matter from drinking water sources. Water Res. 1997;31(12):3098–106.
82. Irabelli A, Jasim S, Biswas N. Pilot-scale evaluation of ozone vs. peroxone for trihalomethane formation. Ozone-Sci Eng. 2008;30(5):356–366.
83. Mosteo R, Miguel N, Martin-Muniesa S, Ormad MP, Ovelleiro JL. Evaluation of trihalomethane formation potential in function of oxidation processes used during the drinking water production process. J Hazard Mater. 2009;172(2–3):661–6.
84. Reckhow DA, Linden KG, Kim J, Shemer H, Makdissy G. Effect of UV treatment on DBP formation. J Am Water Works Ass. 2010;102(6):100.
85. Korn C, Andrews RC, Escobar MD. Development of chlorine dioxide-related by-product models for drinking water treatment. Water Res. 2002;36(1):330–42.
86. Gordon G, Rosenblatt AA. Chlorine dioxide: the current state of the art. Ozone-Sci Eng. 2005;27(3):203–7.
87. Yang B, Fang H, Chen B, Yang S, Ye Z, Yu J. Effects of reductive inorganics and NOM on the formation of chlorite in the chlorine dioxide disinfection of drinking water. J Environ Sci-China. 2021;104:225–32.
88. Badawy MI, Gad-Allah TA, Ali MEM, Yoon Y. Minimization of the formation of disinfection by-products. Chemosphere. 2012;89(3):235–40.
89. Kolkman A, Martijn BJ, Vughs D, Baken KA, van Wezel AP. Tracing nitrogenous disinfection byproducts after medium pressure UV water treatment by stable isotope labeling and high resolution mass spectrometry. Environ Sci Technol. 2015;49(7):4458–65.
90. Hua ZC, Li D, Wu ZH, Wang D, Cui YL, Huang XF, Fang JY, An TC. DBP formation and toxicity alteration during UV/chlorine treatment of wastewater and the effects of ammonia and bromide. Water Res. 2021;188.
91. Shah AD, Dotson AD, Linden KG, Mitch WA. Impact of UV Disinfection combined with chlorination/chloramination on the formation of halonitromethanes and haloacetonitriles in drinking water. Environ Sci Technol. 2011;45(8):3657–64.
92. Richardson SD, Ternes TA. Water analysis: emerging contaminants and current issues. Anal Chem. 2011;83(12):4614–48.
93. Goslan EH, Krasner SW, Bower M, Rocks SA, Holmes P, Levy LS, Parsons SA. A comparison of disinfection by-products found in chlorinated and chloraminated drinking waters in Scotland. Water Res. 2009;43(18):4698–706.
94. Hrudey SE. Chlorination disinfection by-products (DBPs) in drinking water and public health in Canada. In: A primer for public health practitioners reviewing evidence from over 30 years of research. National Collaborating Centre on Environmental Health, Canada; 2008.
95. Dawe P. Best management practices for the control of disinfection by-products in drinking water systems in Newfoundland and Labrador. Department of Environment and Conservation Water Resources Management Division. Canada; 2009.
96. Al-Zahrani MA. Optimizing dosage and location of chlorine injection in water supply networks. Arab J Sci Eng. 2016;41(10):4207–15.
97. Young KB. Development of operatic strategies to our district bromate formation in the Moorhead Water Treatment Plant. M.Sc. Thesis, North Dakota State University; 2014.
98. Chen S, Deng J, Li L, Gao NY. Evaluation of disinfection by-product formation during chlor(am)ination from algal organic matter after UV irradiation. Environ Sci Pollut R. 2018;25(6):5994–6002.
99. AWWARF: AWWA Research Foundation (AWWARF). Use of chlorine dioxide and ozone for control of disinfection by-products, American Water Works Association. Denver, USA; 2004.
100. Hung YC, Waters BW, Yemmireddy VK, Huang CH. pH effect on the formation of THM and HAA disinfection byproducts and potential control strategies for food processing. J Integr Agr. 2017;16(12):2914–23.
101. Williams MD, Coffey BM, Krasner SW. Evaluation of pH and ammonia for controlling bromate during cryptosporidium disinfection. J Am Water Works Ass. 2003;95(10):82–93.
102. Neemann J, Hulsey R, Rexing D, Wert E. Controlling bromate formation during ozonation with chlorine and ammonia. J Am Water Works Ass. 2004;96(2):26–9.
103. Siddiqui MS, Amy GL. Factors affecting DBP formation during ozone bromide reactions. J Am Water Works Ass. 1993;85(1):63–72.
104. Guilherme S, Rodriguez MJ. Occurrence of regulated and non-regulated disinfection by-products in small drinking water systems. Chemosphere. 2014;117:425–32.
105. Zhang CQ, Lu JR. Optimizing disinfectant residual dosage in engineered water systems to minimize the overall health risks of opportunistic pathogens and disinfection by-products. Sci Total Environ. 2021;770.
106. Mazhar MA, Khan NA, Ahmed S, Khan AH, Hussain A, Rahisuddin CF, Youse M, Ahmadi S, Vambol V. Chlorination disinfection by-products in municipal drinking water—a review. J Clean Prod. 2020;273.
107. USEPA: United States Environmental Protection Agency (USEPA). Effects of water age on distribution system water quality. Washington DC, USA; 2002.
108. Chowdhury S, Champagne P, McLellan PJ. Models for predicting disinfection byproduct (DBP) formation in drinking waters: a chronological review. Sci Total Environ. 2009;407(14):4189–206.

109. Montgomery DC, Runger GC. Applied statistics and probability for engineers. 4th ed. New York, USA: Wiley; 2007.
110. Singh KP, Gupta S. Artificial intelligence based modeling for predicting the disinfection by-products in water. Chemometr Intell Lab. 2012;114:122–31.
111. Li L, Rong SM, Wang R, Yu SL. Recent advances in artificial intelligence and machine learning for nonlinear relationship analysis and process control in drinking water treatment: a review. Chem Eng J. 2021;405.
112. Guo K, Pan YP, Yu HY. Composite learning robot control with friction compensation: a neural network-based approach. Ieee T Ind Electron. 2019;66(10):7841–51.
113. Ike IA, Karanfil T, Ray SK, Hur J. A comprehensive review of mathematical models developed for the estimation of organic disinfection byproducts. Chemosphere. 2020;246.
114. Serodes JB, Rodriguez MJ, Li HM, Bouchard C. Occurrence of THMs and HAAs in experimental chlorinated waters of the Quebec City area (Canada). Chemosphere. 2003;51(4):253–63.
115. Kolla RB. Formation and modeling of disinfected ion by-products in Newfoundland Communities. M. Sc. Thesis, Memorial University of Newfoundland; 2004.
116. Hong HC, Liang Y, Han BP, Mazumder A, Wong MH. Modeling of trihalomethane (THM) formation via chlorination of the water from Dongjiang River (source water for Hong Kong's drinking water). Sci Total Environ. 2007;385(1–3):48–54.
117. Civelekoglu G, Yigit NO, Diamadopoulos E, Kitis M. Prediction of bromate formation using multi-linear regression and artificial neural networks. Ozone-Sci Eng. 2007;29(5):353–62.
118. Kulkarni P, Chellam S. Disinfection by-product formation following chlorination of drinking water: artificial neural network models and changes in speciation with treatment. Sci Total Environ. 2010;408(19):4202–10.
119. Ceto X, Saint C, Chow CWK, Voelcker NH, Prieto-Simon B. Electrochemical fingerprints of brominated trihaloacetic acids (HAA3) mixtures in water. Sens Actuat B-Chem. 2017;247:70–7.

8 Water Sensitive Planning and Design

Hoda Soussa

Abstract

Water sensitive planning became emerging as a designing approach recently, which involves sustainable management of water resources and uses from building lot to catchment area. This approach does not only rely on acclimating peak flows, minimizing water pollution and wastewater reuse, but it extends to use and to maintain natural water resources safely and efficiently. Climate change challenges, the efficiency of the existing urban water networks, and posing the need to cope with the new patterns in both floods and droughts situations. New management practices and solutions will be presented in this chapter to increase the efficiency of existing urban water network and/or to planning sustainable solution. Measures to minimize the environmental impacts of urbanization in terms of water supply and pollution threats to natural water resources should be adopted (Allison et al. in Environ Manage 42:344–359, 2008 [1]). Four main steps for developers are needed to apply Water Sensitive Design. First, develop an efficient plan to treat the ambient water, to decrease its pollution, and protect it from any future pollution or over-exploitation. Second, increase public awareness to efficiently use of available water and to minimize expected pollution, and to secure public support of wastewater reuse based on purpose (agriculture and landscaping). Third, collect urban storm water and to safeguard its quality for discharge to water bodies. Finally, to present sustainable solutions for holistic management of water resources by mimicking the natural system to minimize negative impacts on water cycle and receiving waterways, recharge groundwater, and bays. This chapter will present water sensitive design of best practices design and management at developed and developing countries.

Keywords

Urban water supply · Water distribution · Innovative solutions · Water challenges · Impacts of climate change · Water quality · Storm water quality · Integrated water management · water sensitive planning and design

H. Soussa (✉)
Faculty of Engineering, Ain Shams University, Cairo, Egypt
e-mail: hoda_soussa@eng.asu.edu.eg

M. Bahadir and A. Haarstrick (eds.), *Water and Wastewater Management*,
Water and Wastewater Management, https://doi.org/10.1007/978-3-030-95288-4_8

8.1 Introduction

The design concepts for water sensitive cities are challenging problems especially in old cities due to aging of the infrastructure, unavailable as-built maps for water and wastewater networks, and degraded water ways. To transform cities to sustainable urban water cities, or to Water Sensitive Cities, major renovation of the hydro-social system should be effectuated to conventional approaches [1]. In new cities, challenges are less but still adapting practices for environmental management and integration with the whole hydrologic cycle presenting a challenge to the designer especially if there are no guidelines forcing the implementation of such an approach. Governance plays a major role in applying this approach by considering the urban water system as a holistic environment from the planning phase of the city. The concepts of socio-technical approaches, context design based on responses, and nature-based solutions were developed and framed in much research. Figure 8.1 combines different concepts to show the urban water transition framework, as stated by Ashly et al. [2].

Managing water resources efficiently as a part of hydrologic cycle will safeguard not only the environment and human health, but it will also increase the economic value of the city. Water sensitive city means a good infrastructure, clear and clean waterways, and safe life against extreme climate change events by adopting different strategies of accommodate peak flows. Selecting key factors of ecosystem functioning for policy and management purposes using the pentatope model as a holistic framework, and a combined natural and socio-economic valuation scheme were applied to reduce the water system complexity [3].

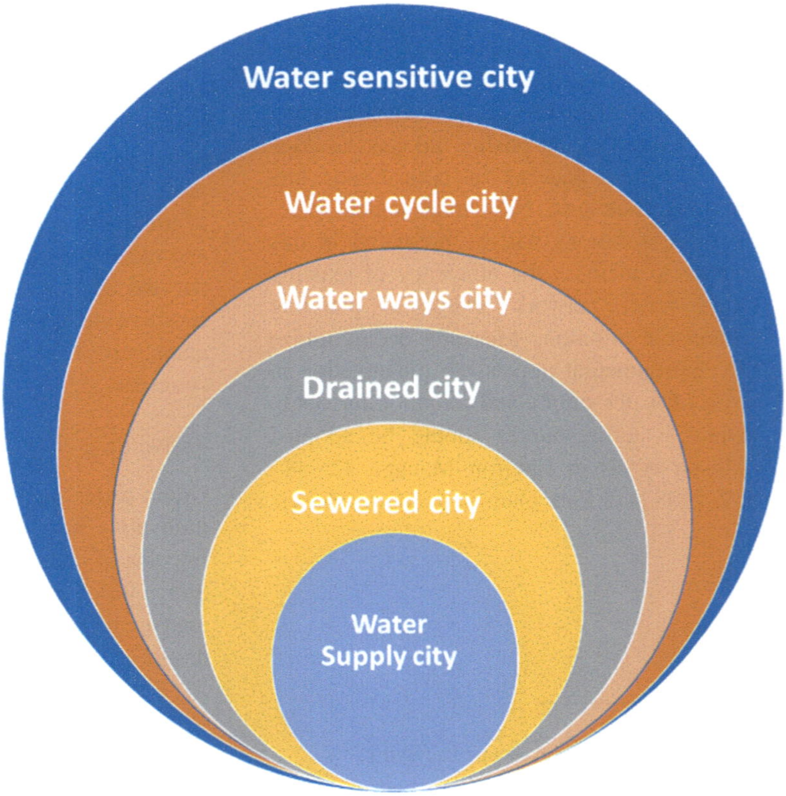

Fig. 8.1 Urban water transitions framework (after [2])

8.2 Egyptian National Approach

8.2.1 Efficient Plan to Treat the Ambient Water and to Minimise Pollution

Protecting water resources from pollution leads not only to preserving its quality, but it decreases the costs needed to treat wastewater by minimizing its quantity from the origin. For example, greywater reuse for irrigating gardens and recreational areas could save 80% of wastewater treatment costs, if all these domestic wastewaters would go directly to wastewater treatment plants after separation of greywater. In addition, adequately treating industrial wastewater at the factory before delivering to drains is presenting a must. However, it is still considered as a challenging problem especially for riparian zones in cities. Urbanization has invaded many areas, leading to reduce ecosystems, public health, and vitality. Citizens in both urban and rural areas in Egypt used to dump solid and liquid wastes in waterways and lakes (Fig. 8.2).

They also eliminate the vegetation in the area, which contribute to absorb polluted gases from the ecosystem. Many informal settlements are constructed in the riparian lands, away from the governmental supervision boosting the risk of massive urban floods. Nearly none of all the construction laws and regulations consider the existence of the poor communities, even if they are living in slums. In rural areas, bad agricultural practices are adding pollution to the riparian areas due to their extensive uses of herbicides, pesticides, and chemical fertilizers. These practices adversely affect the quality of water bodies. As a step to protect surface water bodies, many local authorities started to remove encroachments and to imply vegetative buffers. These buffers help in filtering sediments and contaminants from the runoff and in providing soil stability from erosion. Figure 8.3 shows a main flood wadi at Borg El Arab city, north Egypt, with local plants filtering the sediments within the flash flood events. During storms and floods, these waterways and swales provide a path for distribution of water to prevent urban flooding,

Fig. 8.2 Solid wastes thrown in the main water wadi, Borg El Arab city, Egypt, *(taken by the author)*

Fig. 8.3 Food wadi at Borg El Arab city, Egypt *(photo taken by the author)*

and provide opportunities for recreational space, garden waterfront areas.

8.2.2 Increase Public Awareness for Wastewater Reuse

The main challenge facing wastewater reuse is the lack of public awareness. In developing countries, many farmers use drains of untreated wastewater illegally to satisfy their irrigation needs. Farmers use to clean their stuff and wash their fresh foods at the nearest surface water available due to lack of potable water at their houses. Drinking water networks are not covering all rural areas, or in some cases, the infrastructure exists but the service is only available for few hours daily and with a bad quality. Solving these problems is in two folds: First, to secure water infrastructure (complete coverage of drinking water and sewage networks) in parallel with sewage networks. Many countries are advancing fast in drinking water coverages without securing sewage networks; Second, to increase public awareness to build a common understanding of water issues and to create shared values on how water resources should be used and managed. This can be realized by continuous meetings and workshops with people explaining water quality and quantity management and its impacts on public health and ecosystem. Conducting and announcing water quality analysis and crops analysis results may help them to realize the risks of bad water practices. Nevertheless, giving incentives to farmers using safe fertilizers and modern irrigation may encourage the rest to follow the same. Government and private companies may also encourage organic farming by purchasing and marketing these products. The aim of raising public awareness for water issues is to engage people in water conservation, clean water uses and protecting the ecosystems. Thoroughly, public awareness should be realized through interaction of many active stakeholders capable of influencing each other and providing social control on public and private companies to safely use water and to reinforce strict actions on water pollution.

8.2.3 Urban Cities are Catchments for Supplying Water

Collecting and treating urban stormwater to meet water quality standards for discharge to surface water present a sustainable solution for flash floods protection. In addition, storing stormwater in urban landscape could be used to form artificial lakes and bio-swales that could present recreational attractiveness. Old cities are suffering from coping with the rapid urban migration and growth. New cities are encroaching on flood-prone areas. Due to climate change in the last few years, the flash floods events are increasing in quantity and frequency. This is not a trend only in developing countries, floods in US, Europe and China highlighted the risks of development in environmentally sensitive and low areas [4]. A severe flood in Beijing in 2012 affected more than 1.6 million persons and caused about 10 billion Yuan (US$1.6 billion) damages and at least 8200 homes had been destroyed [5]. In Egypt, most of new cities suffer in flash floods periods, the main streets are stuck with water and a complete chaos on the city's transportation systems occurs.

Environmental sustainability is associated with development activities without any negative impacts on ecosystem. Such goals may seem impossible to reach, as they set challenges to realize wide ranging benefits to environment, society, and economics towards the goal of sustainability.

8.2.4 Institutional Conflicts and Governance

De Jonge et al. [6] perceived that it is still unclear for many stakeholders, how technology can help to avoid breaching of sustainable levels of ecosystem utilization. Lack of information on how the ecosystem functions, make it hard to formulate regulations to maintain natural stocks and flows that provide ecological goods and services. The allocation of ecosystem resources is believed to be an important step of sustainable use, following the steps of realizing the dependency of humans on Ecosystem Goods and Services (ES G&S), the quantification of resource availability and uses, and finally defining sustainable use, which means setting limits to the use of natural resources on the one hand, and on the environmental impacts on the other hand, [7]. The concept of ES G&S allows exploring the economic and non-economic values of natural resources, depending on the internalization of goods and impacts. Decision makers usually consider environmental services as an economic externality. Although the United Nation/World Bank High-Level Panel on Water developed an Action Plan based on a more holistic social-technical approach to addressing complex water issues, the Action Plan claims that "*sustainable solutions require integrated approaches, addressing technical, institutional, financial, social, and environmental issues simultaneously.*".

This is a multifold approach, which consists of institutional regulations and governance, design concepts and management strategy (the governance), and finally public involvement. Institutional directive is the main driver to realize a water sensitive city. It also presents a main constraint in such approach due to the institutional conflicts between different players, as water management is common factor, between public and private sectors, different ministries, and local authorities. So, presenting an integrated framework with compulsory standards may help solving these conflicts. For example, in Egypt the Ministry of Housing, Utilities and Urban Communities is the sole party to issue guidelines (codes of practice) for civil designs, i.e., code for smart cities and code for storm water design systems.

8.2.5 International Examples on WSUD

In the following section, different success stories are discussed to show the transformation of some areas towards green and water sensitive cities. These cities within their urban water management planning are considering the influences of social and technical issues of urban livability.

8.2.5.1 Australia's Water-Sensitive Cities

In Australia, the Millennium Drought (1997–2009) stimulated the implementation of the principle of water sensitive cities. Many cities and towns started to change towards this objective. Many coastal cities included desalination plants to their water supply systems, although it represents an expensive solution to construct and even to maintain. However, with this water supply security was achieved, decision makers started to search for more sustainable and cost-effective alternatives with longer time to implement. They started with wastewater reuse, runoff harvesting and treating wastewater based on use purposes. They also built new infrastructure that serves multiple purposes. Different runoff harvesting projects using sustainable solutions by redefining urban landscapes and streetscapes in many cities, based on water-sensitive urban design approach (WSUD) in Australian cities.

Melbourne Water formulated comprehensive guidelines for WSUD cities [8] for councils on the southern and eastern fringe of Melbourne. These guidelines describe the process of planning and design starting from the early planning and site assessment, moving towards concept design, and ending by the detailed design, construction, and maintenance. WSUD combines urban water management with urban planning and design. Its main goal is to offer a sustainable alternative to the conventional conveyance approach of rainfall and flood management by acting at the source, so reducing the required size of the storm water drainage infrastructure, and filter rainfall water to remove pollutants. The Australian portfolio of sources executes the minimum cost, containing environmental impacts and different outwardness, including managed groundwater aquifers, recharge schemes, urban storm water, rainwater, recycled wastewater, and desalinated water.

8.2.5.2 Seoul Star Rainwater City

The Korean succeeded to apply multipurpose and proactive rainwater management to ensure sustainability and to meet the Millennium Development Goals (MDG) for developing countries, based on the design of decentralized solutions located near flooded areas and involving local activities, which exactly fit the uneven nature of rainfall in Seoul.

Seoul is an old city with six hundred years history and a population over ten million. The climate change is a very effective factor on the city rainfall pattern that raises its trend in the last five years compared to the late thirty years. Decision makers were thinking first to increase the sewer networks capacity to contain the increase in water flows, however, this solution was very expensive. So, they sought alternatives to cope with the new floods flows by managing the water within its watershed and storing in local areas. As shown in Fig. 8.4, three thousand cubic meters tank capacity was designed in three compartments, the first 1000 m^3 tank to collect the rainwater from unpaved areas, the second tank to collect rain from roofs and to use for flushing and gardening, and the remaining 1000 m^3 tank is used as an emergency stock for drinking water and other domestic uses. This designed tank system was awarded the International Water Association, 2010 Project Innovation Award. This system was applied in all new public buildings and new public facilities (parks, parking lots and school), and recommended for private buildings. Operation and maintenance of this decentralized systems required education of the public, school children and the army. A rainwater piggy bank microcredit project was designed and installed at a house level to promote the idea (Fig. 8.5).

8.2.5.3 China's Sponge Cities

In 2014, the Chinese government began to implement a plan to construct sponge cities to holistically tackle climate change and its associated water quantity and quality problems. The sponge city initiative even though theoretically was designed with appropriate principles, different implementation challenges appeared, extending from technological complexity to lack of governance, financing, public awareness, and participatory. The objective of sponge city is to fundamentally shift traditional Chinese water management approach to an integrated

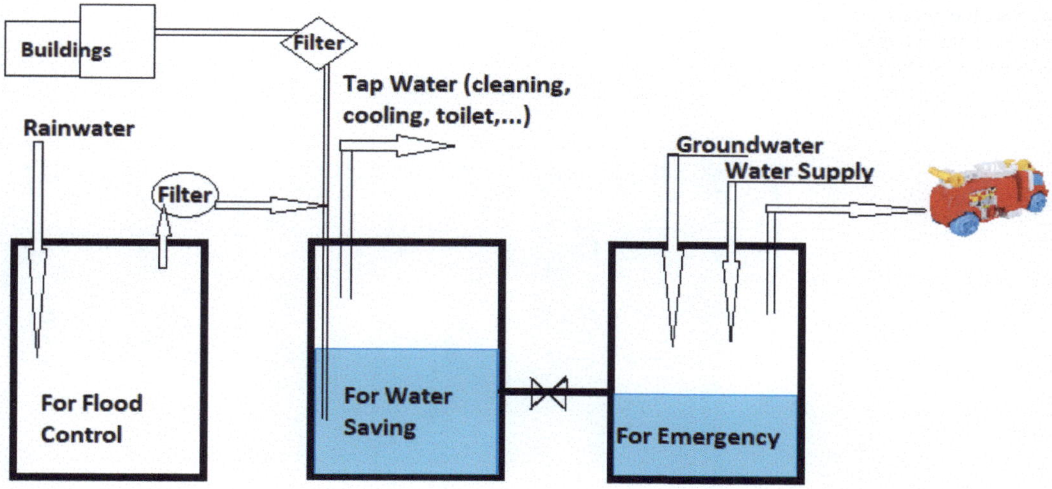

Fig. 8.4 Example of multipurpose rainwater tank design in the Star City, Seoul, Korea [7]

Fig. 8.5 Rainwater piggy bank microcredit system [7]

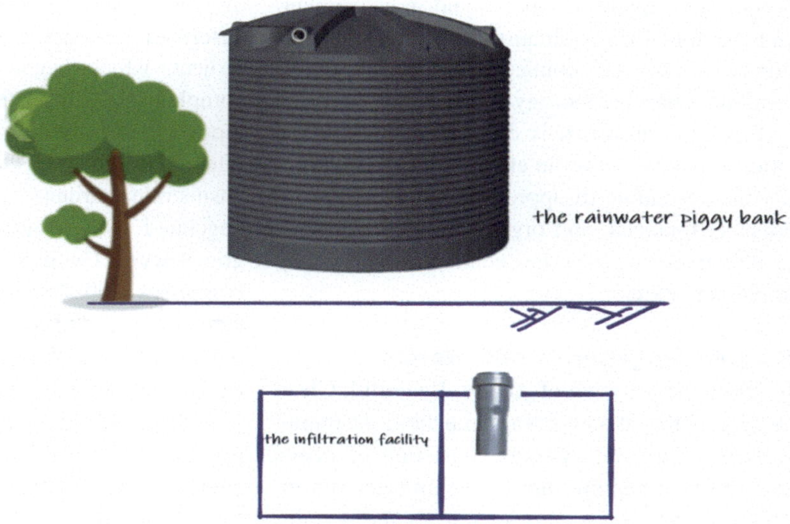

sustainable urban planning and design approach that benefits of ec000ological function to decrease the diverse, interrelated urban water problems [9]. To realize a good design of sponge city, huge database is to be established, which includes information on urban planning data, underground pipe networks, digital elevation data, and hydrological station networks. Integrated technologies should be adapted, including low-impact development, urban drainage, waterlogging defense, ecosystem protection, prevention of urban surface pollution and rainwater usage.

Shao et al. [9] applied the sponge city urban data integration plan with an empirical application on Fenghuang County, in Hunan Province of China, and evaluated the outcome of this application. Figure 8.6 integrates the main technologies used to formulate a framework for an interlinked unified and coordinated sponge city construction system. Integrating data and technologies stimulated the sponge city's planning,

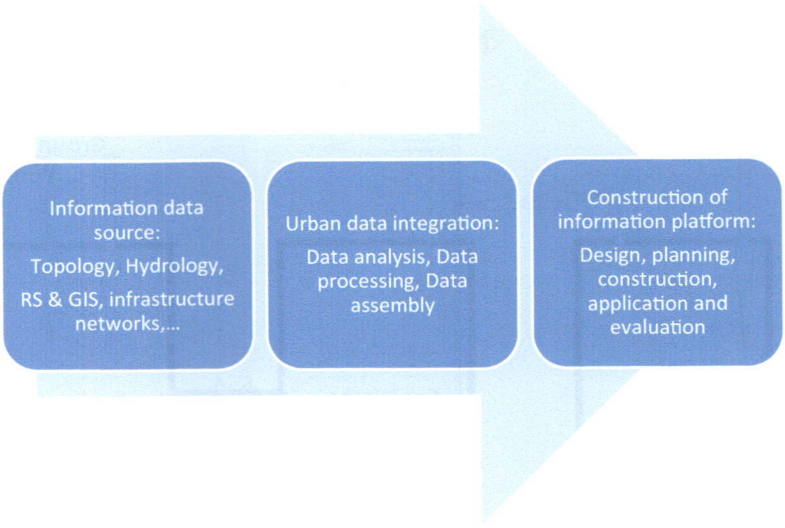

Fig. 8.6 Integrated framework for sponge city construction system [11]

design, construction, operational management, and the functions evaluation [10]. This integration can also be connected to smart cities, bringing smart technology to the sponge city.

From the above, it is concluded that sponge cities initiative can be an effective approach only, if China commits to appropriate technical, governance, financial, and organizational 0measures to effectively address the challenges for policy implementation.

8.2.5.4 Singapore's ABC Waters

In 2006, the program of Active, Beautiful, Clean Waters (ABC Waters) was launched, designing features of natural systems consisting of plants and soil that capture and treat rainwater runoff. This program forms a major part of the sustainable urban drainage system (SuDS) in Singapore; however, its effectiveness has not been studied or documented [12]. Its main function is to improve water quality and other ecosystem services such as providing social and recreational spaces for the public or residents in development. Types of ABC Waters design features include the bioretention basins or rain gardens, bioretention swales, vegetated swales, constructed wetlands and sedimentation basins, while ABC Waters design features are like Sustainable urban Drainage Systems (SuDS), but have a stronger focus on the cleansing function. Kuei-Hsien Liao described the construction of the waterway ridges project, which began in April 2012 and was completed in April 2017, [12]. It contains four types of ABC Waters design features that are integrated in the design: 1. bioretention basins/rain gardens, 2. bioretention lawns, 3. vegetated swales, and 4. vegetated swales with gravel layer. The innovative design implemented comprises 400–750 mm thick detention gravel storage layers below these four features, receiving runoff from about 60% of the total site area during rainfall event (Fig. 8.7a–c).

Yau et al. [13] applied a 1D SWMM model for the Waterway Ridges pilot project. The analysis revealed that ABC Waters design features reduced peak flow and runoff coefficient during storm events [13]. The reduction in peak flow (and effective runoff coefficient) is 33% and 47% for the 10-year and 3-month design storms, respectively, when compared with the scenario no treatment or detention is performed.

8.2.5.5 The United States' Low Impact Development

The low impact development (LID) approach has been recommended as an alternative to traditional storm water design in the US. Allison et al. [1] identified seven obstacles facing sustainable

Fig. 8.7 (**a**) Vegetated swales in Waterway Ridges [12], (**b**) Bio-retention lawns in Waterway Ridges [12], (**c**) Bio-retention lawns in Waterway Ridges [12]

storm water management: (1) performance uncertainties, (2) insufficient engineering guidelines, (3) institutional conflicts, (4) lack of institutional capacity, (5) lack of legislative mandate, (6) lack of funding and effective market incentives, and finally (7) resistance to change. Solutions to each of the seven obstacles were addressed to encourage implementation of WSUD with watershed-based goals to save human health and stream ecosystems.

Whelans et al. [14] indicated that the way of protecting urban cities had been improved over time from the 1960, concentrated on water quantity, passing through recreation and aesthetics convenience towards new 21th goal of sensitive city design.

8.2.5.6 Egypt's Water Sensitive Urban Practices

Egypt is among the most vulnerable countries to the potential impacts of climate change (increased average temperatures, more erratic precipitation, and sea level rise). The big cities in Egypt faced during the last decades fast urbanization activities related to huge increase of population. Old cities are difficult to accommodate the new WSUD system; however, Nassar et al. [15] selected an urban residential area in Port Said coastal city. In this work, researchers implemented a proposed framework from WSUD with urban rehabilitation standards to improve the quality.

The implementation of WUSD is much easier, if it is effectuated from the preliminary phase of execution, which is done in the new cities. The new administrative capital (NAC) city is currently under construction in the eastern desert region of Greater Cairo. Between the NAC and Cairo is another recent city development, named New Cairo city, which faced flooding in 2018 and 2020. Flash floods in these areas have enormous impact on buildings, chaos to the street network and traffic flow, and splits in the urban structure to create various isolated entities, in addition to flooding basements and causing massive economic losses to the owners. Figure 8.8 shows photos in these areas from the March 2020 rainfall event. The cause of this problem was construction of buildings within the flood prone areas, resulting from the Minister of Housing selling off building plots between 2003 and 2009. Walid et al. [16] highlighted the importance of considering flash floods risk, while the masterplan of the city is in the development phase.

The researchers combined different analytical methods to identify urban zones subjected to flood risk and propose solutions. Figure 8.9 shows the different hazards degrees associated to flood zones. A remedy plan was suggested to place WSU features in the NAC composed of two folds: (1) identify the essential locations for retention reservoirs and gardens, and (2) widen

Fig. 8.8 Flooded streets in the New Cairo *(photo taken by the author)*

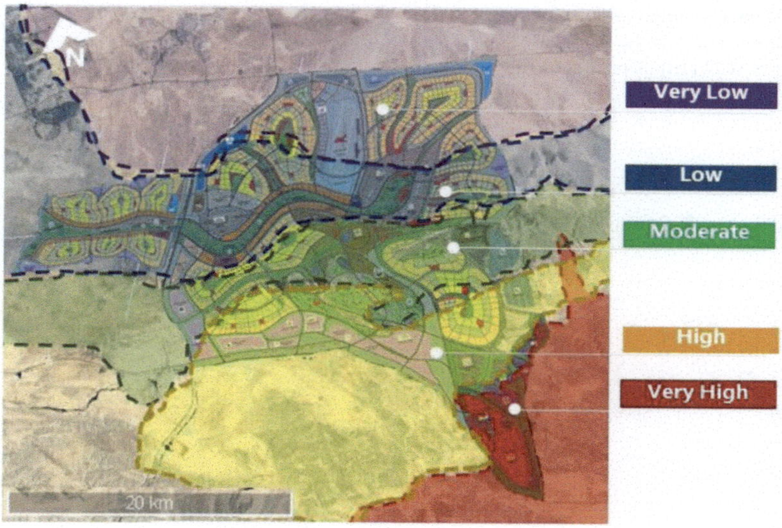

Fig. 8.9 Hazard degrees associated to flood zones [16]

the green valley (green belt) to serve as a bio-retention swale.

It is recommended to consider the remedy plan during the construction phases as well as to develop regulations to ban building in flood prone risk zones to minimize ecosystem and economic losses.

Reservoirs and bio retention gardens should be added to the implemented phase as shown in Fig. 8.10 at the cross-point between the existing flood channels and the beginning of the master-plan from the northern side and the southern side. Moreover, several culverts and tanks should be added to the city's infrastructure and the drainage system through the natural flood channels. Feasibility studies of flood risks and flooding in masterplan should be conducted. The generated urban morphology for a new urbanism should be in parallel to the flood orientation. Space syntax represents a successful tool to test the ability of the city's urban configuration to hold connection against existing flood torrents.

8.3 Conclusions

Over the last decade, climate change was increasing the frequency and the intensity of occurrence of the rain events. Analysis of existing development strategies and solutions to storm water management and flood controls worldwide reveal resolving the actual causes and effects. Consequently, problems remain constant while making cities vulnerable to impacts. Therefore, alternative approaches are needed to enhance flood adaptation and to optimize a flood and storm water resilient urban environment.

This review addresses the need for a combined approach that offers water sensitive urban design components in landscape planning at the catchment scale. Investigating the effectiveness of different WSUD approaches in several cities, coupled with adaptive management, is crucial to advance research to develop appropriate guidelines. The short timeframes of research projects

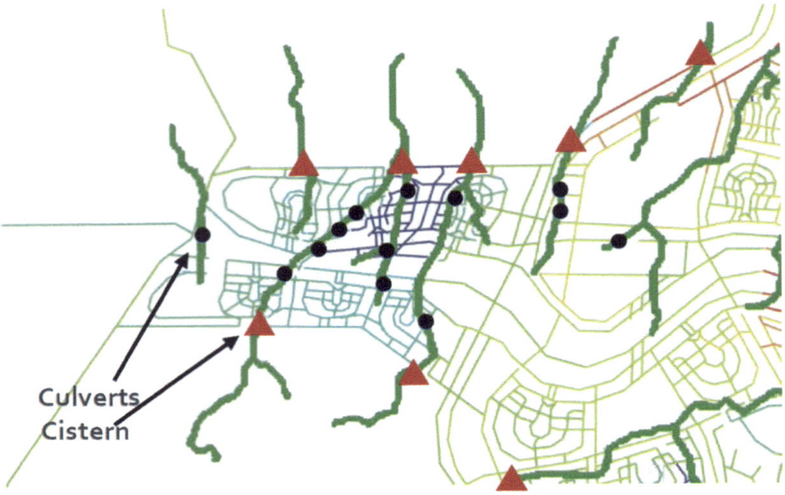

Fig. 8.10 Proposed locations for reservoirs in the first phase of NACC [16] as read in [17]

disinclined general approaches. So, financial supports tangible ecological practices in a watershed will demonstrate the benefits of WSUD and guarantee community and political support for pervasive implementation. Consequently, having varying technologies to address the problem of storm water runoff, it is time to use this information to move towards environmental sustainability.

Acknowledgements The author would like to thank Dr. Tamir El-Khouly, Ph.D. (The Bartlett, UCL), for his kind understanding and support by giving the author permission to use two figures of his publication, that helped to clarify the author idea. In addition, I am thankful to PUB, Singapore's National Water Agency for their understanding and sharing their pictures with sensitive cities experiences. At the end, the author is grateful for EXCEED network, Prof. Bahadir, and Prof. Haarstrick for their support in publishing this chapter in this final form.

References

1. Allison HR, Wenger SJ, Fletcher TD, Walsh CJ, Ladson AR, Shuster WD, Thurston HW, Brown RR. Impediments and solutions to sustainable, watershed-scale urban stormwater management: lessons from Australia and the United States. Environ Manage. 2008;42:344–59.
2. Ashley R, Lundy L, Ward S, Shaffer P, Walker L, Morgan C, Saul AJ, Wong T, Moore S. Water-sensitive urban design: opportunities for the UK. Proc Inst Civil Eng Municipal Eng. 2013;166(ME2):65–76.
3. Vannevel R, Goethals PLM. Identifying ecosystem key factors to support sustainable water management. Sustainability. 2020;12(3):1148.
4. Wong THF, Brown RR. The water sensitive city: principles for practice. Water Sci Technol. 2009;60 (3):673–82.
5. Morning Post, Beijing. "Beijing officials have yet to release latest flood casualtdata". EasyWeb News (163.com) (in Chinese). Retrieved 25 July 2012.
6. De Jonge VN, Pinto R, Turner RK. Integrating ecological, economic and social aspects to generate useful management information under the EU Directives 'ecosystem approach.' Ocean Coast Manage. 2012;68:169–88.
7. Han M. Progress of multipurpose and proactive rainwater management in Korea. Environ Eng Res. 2013;18(2):65–9.
8. Wong THF, Rogers BC, Brown RR. Transforming cities through water-sensitive principles and practices. One Earth. 2020;3(4):436–47.
9. Shao W, Zhang H, Liu J, Yang G, Chen X, Yang Z, Huang H. Data integration and its application in the sponge city construction of China. Proc Eng. 2016;154:779–86.
10. Cates EL, Westphal MJ, Cox JH, Calabria J. Field evaluation of a proprietary storm-water treatment system: removal efficiency and relationships to peak flow, season, and dry time. J Environ Eng ASCE. 2009;135:511–7.
11. Jianga Y, Zevenbergena C, Ma Y. Urban pluvial flooding and stormwater management: a contemporary review of China's challenges and "sponge cities" strategy. Environ Sci Policy. 2018;80:132–43.
12. PUB: Singapore's National Water Agency, https://www.pub.gov.sg/.
13. Yau WK, Radhakrishnan M, Liong SY, Zevenbergen C, Pathirana A. Effectiveness of ABC waters

design features for runoff quantity control in urban Singapore. Water. 2017;9:577.
14. Whelans C, Maunsell HG, Thompson P. Planning and management guidelines for water sensitive urban (Residential) design. Report prepared for the Department of Planning and Urban Development of Western Australia, ISBN 0 64615 468 0; 1994.
15. Nassar UA, El-Samaty H, Waseef A. Water sensitive urban design: a sustainable design approach to reform open spaces in low-income residential rehabilitation projects in Egypt. UPLanD J Urban Plan Landscape Environ Des. 2017;2(3):123–48.
16. Abdeldayem WS, El-Khouly T. A macroscopic view of water management of the new administrative capital city of Egypt. In: Proceedings of the 12th space syntax symposium, Beijing, China; 2019.
17. Eliwa HH. A summary of the advisory study on flood studies: the work of the regional studies required for the administrative capital site with an area of 170,000 feddan, for the detailed stage of an area of 30,000 feddan, and for a priority area of 10,500 feddan. National Authority for Remote Sensing and Space Science, Ministry of State for Scientific Research, Egypt; 2016.

Part IV

Wastewater Management (Technologies)

Sewerage Systems and Wastewater Treatment

9

Eyup Debik, Kubra Ulucan-Altuntas, and Neslihan Manav-Demir

Abstract

In residential areas and industry, wastewater should be collected through some engineered structures (sewerage systems) and treated with an appropriate treatment method for the intended reuse, recovery and/or final disposal. By providing that it does not harm human health and natural life. Wastewater reaching the treatment facility should be treated in treatment facilities designed according to the pollution of the wastewater. While domestic wastewater generally has higher biological content and is suitable for treatment by biological treatment methods, industrial wastewater may also contain recalcitrant pollutants that cannot be treated by biodegradation. Various treatment methods, i.e., mechanical and/or chemical processes, conventional activated sludge plants, trickling filter systems, and biodiscs are used as primary and secondary treatment for domestic and agricultural wastewater, and wastewater from various industries. Wastewater treatment for water reclamation by advanced treatment technologies is among the subjects that research has focused on in recent years. For this purpose, oxidants such as UV irradiation and ozone as well as biological treatment technologies such as anaerobic treatment, natural treatment, and advanced treatment technologies such as membrane filtration and electrochemical processes are used. When applying these treatment technologies, the presence of persistent organic pollutants and their residual by-products in treated water can pose a significant problem in water reuse. In this section, the items summarized above are explained in detail, considering the current technologies and developments.

Keywords

Reuse of treated wastewater · Sewage sludge management · Sewerage systems · Wastewater collection · Wastewater treatment

9.1 Introduction

As a result of the increasing intensity of industrial and agricultural activities and the lack of proper water management policies, the demand for freshwater is gradually increasing, leading to the global water crisis and freshwater scarcity. The increase of water demand for domestic and industrial purposes, which is the most important resource for maintaining human health and

E. Debik (✉) · K. Ulucan-Altuntas · N. Manav-Demir
Department of Environmental Engineering, Yildiz Technical University, Istanbul, Turkey
e-mail: debik@yildiz.edu.tr

K. Ulucan-Altuntas
Department of Chemical Sciences, University of Padova, Padova, Italy

M. Bahadir and A. Haarstrick (eds.), *Water and Wastewater Management*,
Water and Wastewater Management, https://doi.org/10.1007/978-3-030-95288-4_9

welfare as well as biological diversity and all kinds of vital activities, has further increased the importance of ensuring water security [1]. Considering these, establishment of wastewater collection systems with proper engineering, delivery of wastewater to treatment facilities, and appropriate treatment and discharge it to receiving water bodies appear significant issues regarding the sustainability of environmental and human health. But, it is not enough to meet the discharge standards, and the reuse of treated water becomes more important for the efficient use of resources.

9.2 Sewerage Systems—Wastewater Collection

All wastewater collected from residential, commercial and industrial sources in a city is transported to wastewater treatment plants through the sewer system [2]. It is of great importance to establish a well-designed wastewater collection system for appropriate planning and execution of the treatment process. Numerous research studies have hitherto been carried out on the factors affecting the operation of sewerage systems, mitigating odor emissions, and reducing the pressure on the sewerage system due to precipitation. Investigations are also ongoing to minimize environmental pollution due to short-term planning of sewerage systems by local governing bodies. In particular, unplanned urbanization and industrialization cause environmental problems before the infrastructure systems are completed [3]. For this reason, long-term strategic plan studies should be carried out and published by ministries and local government bodies. Sewerage systems consist of a great network of wastewater channels and carry numerous pollutants originating from the city to the wastewater treatment plant. For this reason, sewerage systems should be designed and structured so that they carry the produced wastewater to the disposal point without harming public health.

Sewerage systems can be classified in three different ways according to their construction: (1) separate systems, (2) combined systems, and (3) semi-separate systems. Each system has its advantages and disadvantages, and the design is usually carried out depending on the parameters that are effective in system selection. Effective parameters in selecting the system can be listed as costs, topographic conditions, groundwater level, discharge location and existing infrastructure facilities. Separate systems are preferred in residential areas having high population density and/or steep slopes, and/or located along a stream and/or a possibility of flooding the basements if a combined system is used. In old but still developing residential areas, new settlement areas should be designed according to the separate system [4]. In recent years, the trend has been to plan for separate systems [5]. The most important reasons for this are that the rainwater is less polluted than the wastewater, and when heavy rains occur, it is difficult to treat the resulting high volume of combined wastewater and rainwater causing overflows [6].

Generally, wastewater is desired to be transmitted by gravity, but pressurized systems are also used if necessary. Wastewater channels are designed to provide non-uniform flow conditions and to prevent the settling of solids. Reducing investment costs by designing sewerage systems depending on hydraulic and operational criteria is still an important issue [7]. Sewerage systems are designed depending on wastewater characteristics, minimum velocity, minimum slope, diameter and peak flow characteristics. The materials, diameters, minimum and maximum velocities, and peak flow properties of the pipes are determined according to the characteristics of the wastewater. Manning equation (Eq. 9.1), Kutter tables, and software approaches can be employed in design [8].

$$V = \frac{1}{n} * R^{\frac{2}{3}} * J^{\frac{1}{2}} \tag{9.1}$$

where; V is the mean flow velocity (m/s), n is a non-dimensional roughness coefficient, R is the hydraulic radius (m), and J is the slope of the hydraulic gradient (m/m) [9].

The placement of manholes is another important issue. Manholes are built-in intersections, places with changing flow direction and when the length between two manholes exceeds certain values. Manholes also prevent anaerobic conditions in the pipelines and provide easier access to open the blocked areas. Design criteria to be considered in the construction of sewerage systems and manholes are shown in Table 9.1.

Sewerage systems can be well designed with remote sensing systems such as GIS and GPS. While various software approaches like SewerGEMS are gaining importance for the efficient design of sewer networks, various innovative methods (i.e., remote sensing systems) have started to improve the planning and control of sewerage systems nowadays. Using available software, modelling, and hydraulic simulation of sewer systems could also provide future projections based on several scenarios [11]. The most common problem encountered in sewer systems is leakage, which can cause underground water pollution, and system renewal brings great costs [12]. Considering these, three different maintenance approaches, namely preventive, routine and emergency, gain importance for sewerage systems. Preventive and routine maintenance is carried out regularly in order to prevent a fault in the system. In this way, the reliability required in the sewer system is made more economically [13].

9.3 Wastewater Treatment

Biological treatment is the most commonly used method for wastewater treatment because of its advantages, including low operating costs, high treatment efficiencies, and lower management requirements. Suspended growth processes and attached growth processes constitute the two most important categories of biological processes used for the treatment of wastewater. Although numerous processes have been proposed, including suspended growth processes (i.e., oxidation ditches, sequencing batch reactors and activated sludge processes), attached growth processes (i.e., trickling filters and biofilm processes), and hybrid processes (i.e., rotating biocontactors, activated biofilters, solid contact chambers, and filters followed by an activated sludge process), the conventional activated sludge process is the most commonly used one. One of the most important differences between suspended growth and attached growth processes is that suspended growth systems are suitable for

Table 9.1 Design parameters of gravity-flow sewerage systems [10]

Pipe size	Slope			Sewerage pipe size	Required velocity	Pipe size	Manhole distance	Manhole size
mm	n = 0.013	n = 0.015		mm	m/s	mm	m	m
200	0.0033	0.0044	Min	200	0.61	≤ 600	<100	1.2
250	0.0025	0.0033	building connections	100–150	–	700–1200	<200	[c]
300	0.0019	0.0026						
375	0.0014	0.0019	Max	–	2.5–3.0	>1200	[b]	[c]
450	0.0011	0.0015						
525	0.0009	0.0012						
600	0.0008	0.0010						
675	0.0007[a]	0.0009						
750	0.0006[a]	0.0008[a]						
900	0.0004[a]	0.0006[a]						

[a] Minimum slope for sewerage pipes is about 0.0008 m/m
[b] Depending on locations such as street intersections, distance between manholes may be higher
[c] Larger manhole bases should be constructed for sewers larger than 600 mm, provided that the size of the manhole barrel is same

high flowrates of wastewater, while attached growth processes are used usually for smaller flow rates [14].

9.3.1 Suspended Growth Process

Suspended growth processes employ responsible microorganisms, which are suspended in wastewater. This suspension formed suspended solids and microorganisms in wastewater are referred to as mixed liquor suspended solids (MLSS). In these processes, aeration is used to perform aerobic reactions and takes place with the help of bottom diffusers or mechanical mixers. Ratios of return activated sludge (RAS) and internal recycle (IR) are two of the most influential operating parameters. RAS ratio affects the process in such a way that it is used to maintain MLSS concentration in the reactors at desired levels, while IR ensures that nitrate, which is formed in the aerated reactor, is transported to a non-aerated reactor, where the recycled nitrate is used as electron acceptor for the oxidation of substances causing chemical oxygen demand (COD) [15]. In a conventional activated sludge process, the removal of COD, biochemical oxygen demand (BOD), and ammonium nitrogen are accomplished under aerobic conditions, on the one hand. On the other hand, anoxic conditions are used to remove nitrite and nitrate nitrogen, and an anaerobic reactor is also incorporated to remove phosphorus in various combinations. In a biological wastewater treatment system, while the organic matter is removed by heterotrophic bacteria, and ammonia-oxidizing (*Nitrosomonas*) and nitrite-oxidizing bacteria (*Nitrobacter*) oxidize the ammonia to nitrite and nitrate. This process is called nitrification. Then nitrite and nitrate are reduced to nitrogen gas by the following denitrification. Unlike nitrogen removal, biological phosphorus removal is accomplished in a three-step process by a facultative aerobic group of bacteria, called phosphorus accumulating organisms (PAOs), In the first step, under strict anaerobic conditions, PAOs store volatile fatty acids and releases phosphate. The second step involves the uptake of vast amounts of phosphate by PAOs under aerobic or anoxic conditions. Then, phosphorus removal is completed by removing the bacteria from the system as waste activated sludge [16].

Many activated sludge processes have been developed so far, and the parameters considered in the design of these processes usually are (1) process loading, (2) amount of waste sludge to be generated, (3) oxygen requirement in and air supply to the aerobic reactor, (4) reactor dimensions and configuration, (5) mixing of reactors and energy requirement, and (6) the design of secondary sedimentation tank. Many process types have been developed so far, and in general, important parameters in the design and operation of all suspended biological processes are sludge age (SRT) (3–40 days), food to microorganism ratio (F/M) (0.04–2.0 kg_{BOD} per kg_{MLVSS} per day), organic loading rate (L_{org}) (0.1–2.4 kg_{BOD} per m^3 per day), mixed liquor suspended solids (MLSS) (1,000–8,000 mg/L), hydraulic retention time (HRT) (1–40 h) and return activated sludge flowrate (Q_R) (25–150%). Several systems for carbon, nitrogen, and phosphorus removal (biological nutrient removal) are listed as A^2O, Bardenpho, UCT (standard and modified), VIP, Johannesburg, and sequencing batch reactors (SBRs). Process flow charts for some of these systems are shown in Fig. 9.1 and design criteria are shown in Table 9.2. The main difference of UCT and VIP processes from others is that RAS enters the anoxic step instead of the anaerobic step in these processes, through which recycling of nitrate to anaerobic step is prevented and ensured strict anaerobic conditions effectively. These processes have, further, their advantages and disadvantages.

9.3.2 Attached Growth Process

Attached growth processes as an alternative to suspended growth processes are classified under three groups as (1) non-submerged attached growth processes, (2) suspended growth processes with fixed-film packing, and (3) submerged attached growth processes. In attached growth processes (trickling filters, rotating

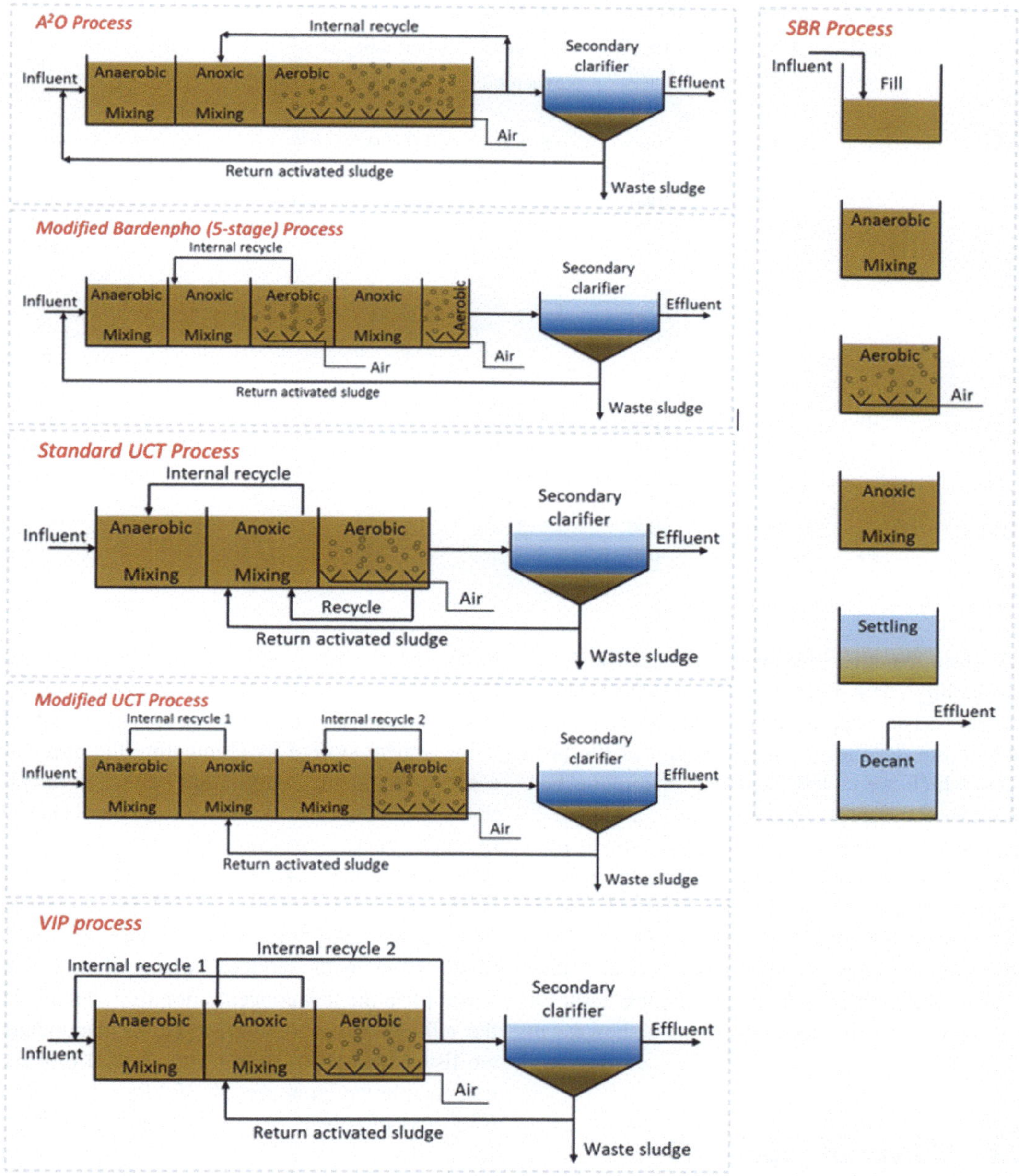

Fig. 9.1 Suspended growth processes for biological nutrient removal

biological contactors, etc.), microorganisms responsible for the wastewater treatment grow attached on a support material, and these processes are usually referred to as biofilm processes. Disks, rock, gravel, sand, wood, plastics, synthetic materials and membranes can be used as support materials [17]. Sufficient surface area for microbial growth is considered as the most critical parameter for a support material [18]. Sufficient surface area provides reduced required volumes for the process, which is the most important advantage of attached growth

Table 9.2 Typical design criteria for suspended growth processes for biological nutrient removal [2]

Process	SRT (day)	F/M ratio (1/day)	MLSS (mg/L)	Hydraulic retention time (h)			RAS (%)	Internal recycle (%)
				Anaerobic step	Anoxic step	Aerobic step		
A^2O	5–25	0.1–0.25	3000–4000	0.5–1.5	1.0–3.0	4–8	25–100	100–400
UCT	10–25	0.1–0.2	3000–4000	1–2	2–4	4–12	80–100	200–400 Anoxic 100–300 Aerobic
VIP	5–10	0.1–0.2	2000–4000	1–2	1–3	4–6	80–100	100–200 Anoxic 100–300 Aerobic
Bardenpho (5-stage)	10–20	0.1–0.2	3000–4000	0.5–1.5	1–3 Anoxic1 2–4 Anoxic2	4–12 Aerobic1 0.5–1 Aerobic2	50–100	200–400
SBR	20–40	0.1–0.2	3000–4000	1.5–3	0.5–2	2–4	–	–

processes. Another important consideration is the availability and cost of the support/packing material. The support media can be stationary or moving depending on the design of the process, which are called fixed bed or moving bed processes, respectively [14].

The main advantage of attached growth processes over activated sludge processes is their cost-effectiveness [19]. Besides, attached growth processes offer higher tolerance against high hydraulic loadings and shock loadings, and have smaller footprint [20]. Typical design criteria for attached growth processes are shown in Table 9.3.

9.4 Bioreactor Types

Recently, membrane bioreactor (MBR) processes, which combine activated sludge and membrane filtration by facilitating solid–liquid separation [21], and aerobic granular sludge (AGS) [22] processes are emerging technologies as biological treatment methods. Membrane bioreactors (MBRs) are a new process consisting of a combination of membrane technology and biological processes, and have been accepted as a high-tech application in twenty-first century [23]. These processes reduce the investment cost of the treatment system by eliminating the need for a secondary sedimentation tank. But, the operation and maintenance costs can be higher because of high energy consumption and membrane clogging/fouling [24]. The MBR processes can be operated at extremely high MLSS concentrations (5,000–20,000 mg/L) as well as long SRT (5–20 days) with reduced waste sludge and accomplish the low concentration of pollutants in the effluent. Several other operating parameters are listed for MBRs as flux (600–1,000 L/m^2h), transmembrane pressure (4–35 kPa), HRT (4–6 h), and dissolved oxygen concentration (0.5–1.0 mg/L).

Aerobic granular sludge is a promising technology that reduces the area for the clarifier by up to 33%. Dense biofilm layer and good settling properties of aerobic granules provide comparable nitrification capacity with respect to the conventional activated sludge processes even at sludge retention times as low as 2.5 days [25]. Thanks to the layered structure of microbial granules in these processes, aerobic, anoxic, and

Table 9.3 Typical design criteria for the popular attached growth processes [2]

Trickling filter	Low rate with nitrification	Standard low rate	Intermediate to high rate	Super rate	Roughing
Type of packing	plastic or rock	rock or slag	rock or slag	plastic	plastic
Hydraulic loading (m^3/m^2 day)	5–16	1–4	4–40	15–80	40–100
Organic loading ($kgBOD_5/m^3$ day)	0.08–0.48	0.08–0.32	0.24–2.4	0.8–4.8	1.6–6
Recirculation ratio	1–2	0–1	0.1–2	1–12	0–2
Depth (m)	2.5–12 (plastic) 1.0–2.5 (rock)	1.0–2.5	2.0–2.5	2.5–12	1.0–10
BOD_5 removal efficiency (%)	85–95	80–90	80–90 single-stage 90–95 two-stage	70–80 single-stage 80–90 two-stage	40–70
Effluent BOD_5 (mg/L) Effluent NH_4 (mgN/L)	<20 <3	<25 <5	<30 <5	<30 <5	<40 No nitrification
Ventilation	Forced air	Natural	Forced air	Forced air	Forced air
RBC		Unit	Treatment level		
			BOD removal	BOD and nitrification	Separate nitrification
Hydraulic loading		m^3/m^2 day	0.08–0.16	0.03–0.08	0.04–0.10
Organic loading		$gBOD_5/m^2$ day	8–20	5–16	1–2
Maximum 1st-stage organic loading		$gBOD_5/m^2$ day	24–30	24–30	–
NH_3 loading		gN/m^2 day	–	0.75–1.5	–
Hydraulic retention time		hour	0.7–1.5	1.5–4	1.2–3
Effluent BOD		mg/L	15–30	7–15	7–15
Effluent NH_3–N		mg/L	–	< 2	1–2

anaerobic microorganisms can coexist, providing simultaneous nitrification, denitrification, and phosphorus removal for municipal wastewater treatment [26].

Besides, processes such as sequencing batch reactors (SBR), up-flow anaerobic sludge bed (UASB) followed by a polishing pond (PP) or final polishing unit (FPU), downflow hanging sponge system (DHS), and moving bed biofilm reactor (MBBR) can also be listed among nutrient removal processes that employ granulated biomass for wastewater treatment. Several new processes such as ANAMMOX (anaerobic ammonium oxidation), CANON (completely autotrophic nitrogen removal over nitrite), denitrified phosphorus removal (DPR), and reverse/advanced osmosis membrane filtration has also been developed in order to reduce the energy consumption for nitrogen and phosphorus removal from wastewaters [27].

9.5 Reuse of Reclaimed Wastewater

Due to climate change, a significant reduction in freshwater resources is expected. Additionally, problems depending on the use of a high percentage of the existing freshwater resources for irrigation and the increasing population will bring the world to face water scarcity. Reusing reclaimed wastewater has emerged as a priority goal besides preserving limited water resources and the quality of receiving bodies. Reuse strategies for reclaimed wastewater are often seeking a solution to water scarcity without increasing environmental problems. For this reason, it is crucial to investigate inexpensive treatment approaches, which do not require a high level of treatment. In recent years, both laboratory and full-scale research studies have been dedicated to wastewater reuse (REF). The studies are aimed both to develop advanced treatment processes to be applied depending on the water quality required in the reuse process and to investigate the possible effects on soil, plant, groundwater and public health after reuse.

9.6 Environmental and Public Health and Economic Impacts

Another issue investigated for the reuse of reclaimed wastewater is its impact on the environment after use. The effects on soil, plants, and crops are mainly investigated for irrigation wastewater, as well as the effects on public health. Studies on soil generally examine soil pH, salinity, nutrient and heavy metal accumulations, sodium adsorption, and changes in the microbial structure of the soil. The studies on public health are performed on the absence of microbial contamination, potential accumulation of heavy metal and emerging pollutants on edible parts of plants. Some of these substances can also pose a health risk for the farmers, who can be affected by inhalation.

Furthermore, a significant number of studies reported that emerging compounds such as personal care products, endocrine disrupting compounds and microplastics were detected in wastewater effluents after secondary treatments. Using secondary effluents directly for irrigation can accumulate these compounds on soil and enter the food chain through uptake by plants [28, 29]. Most of the biological wastewater treatment plants are primarily ineffective in removing antibiotics, and the usage of secondary effluent can result in antibiotic-resistant bacteria and affect microbiota on the applied soil [30, 31]. In this regard, the tertiary treatment should be appropriately managed by the authorities.

9.7 Treatment Technologies for Wastewater Reuse

In general terms, the treatment technologies for wastewater reuse aim at removing the remaining nutrients, suspended solids, and microorganisms after primary treatment. They are mostly called tertiary, advanced or reclamation processes and based on chemical (i.e., coagulation-flocculation), physicochemical (i.e., desalination systems such as membrane processes), biological processes (i.e., fixed biofilms technologies such as wetlands) and combinations (such as membrane bioreactors (MBRs)). The selection of these treatment technologies mostly depends on (1) the water quality specified by regulations or institutions, (2) the current state of the technology, (3) the characteristics and amount of the wastewater, (4) concentration of the pollutants to be treated, and (5) the potential by-products after disinfection. For example, if the treated wastewater is planned for irrigation purposes (agricultural or recreational), efficient disinfection of the wastewater treatment plant's effluent may be sufficient after secondary treatment. However, if the treated water is to be used as process water, advanced treatment methods such as membrane technologies (i.e., reverse osmosis, ultrafiltration, nanofiltration), electrodialysis systems, activated carbon, advanced oxidation techniques processes, and MBRs can procure further treatment of excess pollutants and partial disinfection before applying final disinfection processes. Ensuring microbiological quality for the reuse of treated water is controlled by legal restrictions, and several types of disinfection processes are applied for this purpose. Before disinfection, the initial concentration of many pollutants can affect the amount and variety of by-products to be formed. For this reason, the effectiveness of the treatment technology to be applied before disinfection is essential.

Tertiary treatments used for reuse to meet the quality requirements of reclaimed wastewater are generally applied to provide the necessary microbial safety by chemical, physical or radiation disinfection [32]. Chlorine, ozone and hydrogen peroxide are mainly used in chemical oxidation processes that provide reclaimed wastewater suitable for agricultural reuse by generating less waste [33]. In selecting the disinfectant to be used, economical, operator safety, environmental effects, and ease are the decisive factors. Emerging compounds (i.e., pesticides, antibiotics) present in the recycled water may also affect the choice of disinfectant.

Chemical disinfectants are applied conventionally in the disinfection of both wastewater and drinking water. Generally, halogen-based chemicals are used for reclaimed wastewater disinfection to form strong oxidative species. Chlorine, sodium hypochlorite, calcium hypochlorite and chlorine dioxide are the principal compounds used in chlorine-based disinfection processes. Chlorine can be found as a liquid or gaseous form and is one of the most common disinfectants worldwide. The main disadvantage of chlorine usage in reclaimed wastewater is the potential production of disinfection by-products by reacting with organic substances. The long-term effects are still not known when it is discharged to the environment. In addition, the possible chlorine residue present in wastewater disinfected with it is toxic to aquatic organisms even at low concentrations. Therefore, it will be necessary to treat with reducing agents (i.e., sulphur dioxide, sodium bisulphite and sodium thiosulfate) or filter with activated carbon to remove free or bound chlorine residues from wastewater [34]. Besides chlorinated compounds, hydrogen peroxide and potassium permanganate can also be used for odour and colour control during reuse. Moreover, while peracetic acid, a mixture of acetic acid and hydrogen peroxide, is not widely applied due to the addition of organic matter to wastewater, disinfection by-products are not formed [35].

The use of ozone in the disinfection of reclaimed wastewater gains interest due to its capability to reduce trace components and disinfection. Ozone is an oxidising agent that is soluble in water and has high oxidation potential. Organic compounds can react with ozone directly and indirectly through free radicals such as hydroxy radicals by controlling the pH and temperature. The oxidation by ozone occurre through reaction of oxygen atoms with organic compounds and mineralisation/destruction into shorter chain by-products. Ozone does not form chlorinated disinfection by-products; however, it forms other by-products such as aldehydes and brominated compounds where bromide exists in reclaimed wastewater [36]. In the presence of bromide, UV irradiation could form bromate, a probably toxic substance [37]. In addition, the use of ozone is more suitable for the inactivation of viruses, while UV irradiation is more effective for controlling inactive bacteria [38]. UV light can be classified as physical disinfection rather than chemical disinfection. The effectiveness of UV irradiation can vary depending on the characteristics of reclaimed water and suspended particles. The dissolved natural organic matter can absorb UV light, accumulate on UV lamps, and cause fouling. In addition, the hardness of reclaimed wastewater can cause the deposition of minerals on quartz tubes.

Secondary treatment systems may fail to remove many compounds from wastewater for meeting the water quality standards for water reuse. These compounds include organic and inorganic substances such as nitrogen, phosphorus, soluble organics, hydrocarbons, phenolic compounds, and suspended organic matter. The suspended organic matter can result in low effective disinfection by adsorption of disinfectant and shelter for bacteria. In addition, many of these compounds cause toxicity to fish and taste and odour problems in water resources. For this reason, tertiary treatments such as coagulation, filtration and advanced oxidation processes are applied before disinfection to remove excess pollutants and suspended matter. While membrane processes can remove these substances with high efficiency, more economical technologies such as adsorption, ion exchange, chemical coagulation, advanced oxidation and advanced biological treatment can also be applied alone or in combination with each other. The basic goal in adsorption applied for reclaimed wastewater is to remove recalcitrant compounds, inorganic substances, and odour by gathering these compounds in a solid form.

Air stripping, coagulation, and flotation processes are widespreadly used to remove compounds such as phosphorus, ammonia, and volatile organics that the secondary treatment cannot achieve. Chemicals such as aluminum sulphate, iron sulphate, ferric chloride are used for coagulation, which can help to flocculate phosphorus and to remove it from wastewater through precipitation. For removing volatile organic compounds that cause odour and taste problems, adsorption on granular substances are also widely applied [39]. While many substances (i.e., persistent organic pollutants, pharmaceuticals) are retained by the adsorption process before using UV, it provides removal of particulate matter to prevent the reduction on the effect of UV. The most used material in the adsorption process is activated carbon, because it is cheap and easy to find [40], but ion exchange resins and membrane processes can also be referred to as adsorption processes. Generally, granular adsorption is a process that works by preparing a bed of granular materials to adsorb contaminants and/or to filtrate colloidal particles and passing water to be purified through it. For filtration purposes, materials such as fine sand and diatomaceous earth can also be used. By performing backwashing, the solids accumulated in the filter bed are periodically removed, and by this filters can be used many times. Activated carbon adsorption is effective in removing soluble refractory organics from water. Most of them are organometallic compounds, pesticides, and chlorinated compounds that conventional secondary treatment and tertiary filtration cannot remove. This process takes place in trapping organic compounds in the pores of the large surface area of the activated carbon by van der Waal forces and weak bonds. Afterwards, the powder activated carbon (PAC) or granular activated carbon (GAC) is saturated with the compound to be removed, regenerated, or the filter bed is replaced with new materials. Researchers are currently investigating innovative materials with higher adsorption capacity than activated carbon.

Membranes are the most advanced materials in filtration systems, acting as selective barriers that allow the passage of elements of a specific size. Depending on the selected pore size, they can selectively separate TSS (total suspended solids), turbidity and microorganisms as well as ionized salts. Depending on the purpose of reclaimed wastewater reuse, membrane processes such as nanofiltration (NF) or reverse osmosis (RO) may be preferred following sand filtration, microfiltration (MF), or ultrafiltration (UF) processes. When classifying membrane processes, (1) pore size, (2) driving force, (3) working range, (4) removal efficiency, and (5) membrane material are essential [41]. Membranes are classified according to their pore sizes as microfiltration, ultrafiltration, nanofiltration, reverse osmosis and electrodialysis. Accordingly, those with a pore size greater than 50 nm are named microfiltration, those with a mesopore of 2–50 nm pore diameter ultrafiltration or nanofiltration, and those with a pore size of less 2 nm reverse osmosis [42]. MF and UF are alternatives for reclaimed wastewaters that cannot be removed from their solids by settling and filtration [43].

MF treatment can remove solids causing TSS, turbidity, colloids, protozoa, limited bacteria and viruses. UF systems can hold more bacteria, viruses and large organic molecules than MF. NF and RO systems can also be used depending on the reuse purpose of the reclaimed wastewater. NF systems can hold smaller molecular organics and many viruses. Generally, NF systems are used before RO systems, removing smaller molecules than NF, ions, and hardness. The main disadvantages of membrane systems are the problem of brine disposal and membrane clogging. The presence of solids in the water or precipitates of ions such as Fe, Mn, natural organic substances, accumulation of microorganisms, and strong oxidants such as chlorine can lead to clogging. Unlike conventional membrane systems, electrodialysis is a demineralization process, in which electrical potential is used as driving force, and ions are transferred to ion-selective membranes [44]. Another process, in which membrane systems are used successfully, is the combination with biological treatment. Membrane bioreactors (MBRs) are processes, in which MF or UF membranes are placed submerged in the reactor, where biological growth occurs. MBRs have gained significant attention for reclaimed wastewater treatment because of the membranes' function, which ensures that the sludge required for biological treatment remains in the reactor and limits the solid amount of the effluent.

Apart from the aforementioned technologies, treatment processes that can be used actively are classified as advanced oxidation processes (AOPs). Generally, AOPs are the processes, in which oxidising chemicals are added to form oxidative species such as hydroxyl radicals. These species have unpaired electrons on their valence orbital, which can react with the bonds of recalcitrant compounds and degrade them into less toxic compounds. Chlorination, UV irradiation, ozone is also classified as AOPs, mainly used in combinations to degrade recalcitrant organic pollutants. In addition, Fenton like processes, photocatalytic processes, ultrasonic treatment, electrooxidation, and cold plasma technologies can be used according to the type of pollutant (i.e., phenol-based, long-chain). The reaction to produce active species (i.e., HO•, HOO•, O_3^{-}•) via ozone, H_2O_2 + UV, Fenton, TiO_2 photocatalytic reactions, ultrasonic, persulfate are given in Eqs. (9.2)–(9.11), (9.6)–(9.7), (9.8)–(9.10), (9.11), (9.12), respectively [45, 46].

$$O_3 + OH^- \rightarrow HO_2^- + O_2 \tag{9.2}$$

$$O_3 + HO_2^- \rightarrow HO_2\cdot + O_3^-\cdot \tag{9.3}$$

$$O_3^-\cdot + H^+ \rightarrow HO_3\cdot \tag{9.4}$$

$$HO_3\cdot \rightarrow \cdot OH + O_2 \tag{9.5}$$

$$H_2O_2 + UV \rightarrow 2\cdot OH \tag{9.6}$$

$$H_2O_2 + \cdot OH \rightarrow HO_2\cdot + H_2O \tag{9.7}$$

$$Fe^{2+} + H_2O_2 \rightarrow Fe^{3+} + \cdot OH + OH^- \tag{9.8}$$

$$Fe^{3+} + H_2O_2 \rightarrow Fe^{+2} + HOO\cdot + H^+ \tag{9.9}$$

$$H_2O_2 \rightarrow \cdot OH + HOO\cdot + H_2O \tag{9.10}$$

$$TiO_2 + UV \rightarrow e^- + h^+ \tag{9.11}$$

$$H_2O + US \rightarrow \cdot OH + H\cdot \tag{9.12}$$

9.8 Sewage Sludge Management

Sewage sludge is formed as a by-product in biological wastewater treatment plants in form of slurry or semisolid and requires appropriate handling and disposal. Sewage sludge is generally classified as primary and secondary sludge, generated from primary and secondary settling tanks, respectively. The primary sludge commonly comprises inorganic sedimentations and chemical precipitations, while secondary sludge is generally composed of organic compounds, nutrients, pathogens, and many other contaminants, and requires sophisticated treatment and management strategies [47]. Recently, rapid urbanisation has led to an increase in the number of treatment facilities required, resulting in a large volume of sewage sludge. The general layout of a sludge treatment system from biological wastewater treatment plants includes thickening, digestion, dewatering and drying processes.

Thickening is the first process applied in sludge management, where the solid settles mostly by gravity to form a thicker sludge because of its easiness and ability to lower the cost of other possible processes. Raw sludge collected in wastewater treatment plants contains a large amount of water. Therefore, the sludge thickening method is generally used before sending the sludge to further units such as the stabilization and conditioning units. Sludge with low solids content (about 4–6%) from primary and secondary treatment are kept in a cone type tank, and the solids are compressed to obtain a sludge with increased solids content of approx. 10% at the bottom [41]. Generally, gravity sludge thickening units are designed as circular, and sludge is fed from the middle and has three distinct zones: (1) supernatant zone, (2) settling zone, and (3) compaction zone [41]. The clear water is gathered via tanks' weirs and sent back to the entrance of the treatment plant in the supernatant region. In the settling zone, the sludge is conveyed towards the compaction zone. The sludge in this zone is mixed gently, so the gases formed and water trapped are carried to the surface. Finally, the compaction zone is compressed by gravity via the mass of the sludge itself, and compressed sludge is taken from the lowest centre point of the gravity thickening tank. In addition to commonly used gravity thickeners, Dissolved Air Flotation Thickening (DAF) is a process used to precipitate small-sized particles that generally have a very long settling time (i.e., sludges coming from coagulation-flotation and suspended growth nitrification processes). As a general expression, thin air bubbles are introduced into the liquid in DAF, causing the solids to float. Except for gravity thickeners and DAF units, centrifugal thickeners, gravity belt thickeners, and rotary drum thickeners are the generally applied processes for sludge thickening before sending to sludge stabilization and conditioning processes.

The sludge coming out of the thickening unit is stabilized by reducing the offensive smell, pathogenic bacteria and biodegradable organic materials. The methods used for sludge stabilization can be summarized as (1) biological methods (aerobic and anaerobic digestion), (2) chemical methods (chlorination, stabilization with lime), and (3) physical methods (thermal stabilization). Primary and secondary sludge, which contains a high amount of organic solids, are mostly treated to decompose by digestion. Digestion can disinfect the sludge from pathogens and also lower the solid mass. Aerobic digestion is mainly used in small plants and is similar to the activated sludge process. The thickened sludge is kept for about 40–60 days [Sludge Retention Time (SRT)] until biologically stable solids are obtained [41]. Cell tissues are highly oxidized in aerobic digesters. The most significant disadvantage of these facilities can be the cost due to the necessity of continuously air supply. In anaerobic digestion, microorganisms degrade the sludge under anaerobic conditions, in which methane gas is produced and used for energy recovery. The sludge from anaerobic digestion is stable, has no odour, and contains only low amounts of pathogenic bacteria. For this reason, anaerobically digested sludge is suitable for use as soil conditioner in the agriculture. Anaerobic degradation takes place in 3 phases involving a series of biochemical reactions: (1) the hydrolysis phase, (2) the acidogenic phase, and (3) the methanogenic phase [41]. In the hydrolysis phase, complex carbohydrates, proteins and lipids are degraded into, e.g., monosaccharides, amino acids and long-chain fatty acids. In the acidogenic phase, the compounds produced in the hydrolysis phase are degraded into short-chain organic acids, i.e., acetic acid, butyric acid, etc.. Finally in the methanogenic phase, these short-chain organic acids are converted to methane and carbon dioxide.

Another stabilization method used in small wastewater treatment plants is the chemical stabilization. In this process, chlorine, hydrogen peroxide and ozone are used to reduce odours and microbial activity. However, it is not applied in large scale facilities, because the chemical cost increases when the amount of sludge increases.

The fine particles in the sludge have electrostatic charges and cannot coalesce into bigger particles due to their zeta potential. Therefore, sludge conditioning methods are applied to bring the particles closer together to form flocs with lower content water in the sludge and are divided into chemical and physical processes. Lime, various coagulants and anionic/cationic polymers are used in chemical processes. In physical processes, applications such as heat treatments, ultrasonic vibrations and microbial fuel cells are applied [48–50]. However, the most widely used method is chemical conditioning. In chemical conditioning processes, where metal salts are used, metal hydroxides are formed by neutralizing negatively charged sludge solids, thus disinfecting the sludge and reducing odours [51]. Among the sludge conditioning chemicals, fly ash provides dewatering of the sludge but significantly increases the amount of sludge cake [52]. In addition, cationic polymers are widely used among organic polymers, since sludges mostly have negatively charged particles. In case of polymers, an increase of the sludge amount is not observed, as in the case of using metal salts and ashes.

The disposal of the sludge is done after its free water is removed. To transport these sludges with trucks, composting and burning them, sludge dewatering methods are applied. In natural dewatering methods, sludge after conditioning is spread on the land and dried under the sun in small-scale treatment plants with low amounts of sludge. These methods are called sludge drying beds or lagoons. Conventional sand sludge drying beds consist of layers of gravel and sand with a perforated pipe on the underside. The free water of the sludge is filtered and collected from these pipes. Sludge drying beds, which have been developed in recent years, are dried with solar energy [53]. Sludge drying lagoons are cost-effective and consist of shallow soil basins, in which supernatant is discharged from the surface. Disadvantages of sludge drying lagoons are the odour emission and potential groundwater contamination. In large-scale plants, mechanical dewatering systems are used. The most often applied methods are belt filter press and centrifugal filter press.

Therefore, selecting these processes mentioned above for sludge management depends on sludge amount, treatment costs, and disposal method. There are several options for the treatment and disposal of the sludge, and these options should be adapted to local conditions, where the treatment plant is established. Further, thermal and hybrid treatment processes such as incineration, pyrolysis, gasification, microwave treatment, ultrasonic destruction, enzyme hydrolysis process, and high-pressure homogenization are also used for sewage sludge treatment and help to reduce the volume of sludge to be handled. Available options for disposal include landfilling, composting, incineration, land application, and production of construction materials [54, 55].

9.9 Institutional Structures

In residential areas and industry, wastewater should be collected through sewerage systems and treated with an appropriate treatment method for the intended reuse, recovery and final disposal by providing that it does not harm human health and natural life. Although local governments usually manage these operations, treatment is carried out by private organisations and supervised by the local government and central government in some cities. Thus, private organisations are critical to assist the local government in managing treatment facilities with several advantages, such as providing local employment [56]. Nevertheless, the transfer of responsibility to the private sector can increase environmental pollution in the absence of adequate regulatory controls. Therefore, private organisation participation in the design, construction, and operation of wastewater treatment plants is reliable with government regulations and sanctions.

The agricultural reuse of treated wastewater is controlled by regulations and guidelines determined by International Organization for Standardization, World Health Organization, European Commission, and the countries themselves. These regulations and guidelines control the quality parameters of treated waters by categorising them with substances affecting human health (such as chemicals and disease-causing microorganisms), agronomic parameters (i.e., salinity, toxicity, pH) and physicochemical properties. For treated wastewaters that meet specified water quality criteria for irrigation, approval must be obtained from local, regional and national authorities. However, these regulations are often insufficient and differ significantly from country to country regarding emerging pollutants [57].

9.10 Final Remarks

The use of sewerage systems is of great importance in terms of population growth in urban and suburban areas, the diversity of industrial areas, and the transportation of wastewater from these regions and their transmission to treatment facilities in a way that does not threaten public health. It is essential to design the sewerage systems according to the region, where they will be installed or updated, according to the corrosive sewage waters, and adapting the developing satellite technologies to the sewerage systems and minimizing the installation and operating costs. In addition, sewerage systems should be operated in a way that does not damage the sewerage system and impair the efficiency of the treatment plants at the end of the sewer network. For this purpose, legal arrangements should be regulated. Thus, efficient operation of treatment plants should be ensured and, if possible, advanced treatment technologies should be used for the reuse of wastewater. New techniques should be developed to minimize advanced treatment costs, and the reuse of wastewater treated with these techniques should be encouraged not only for irrigation purposes but also for various industries.

Acknowledgements Dr. K. Ulucan-Altuntas is on unpaid leave from Yildiz Technical University and supported by Horizon 2020 Research and Innovation Programme under the Marie Skłodowska-Curie Action with grant agreement No. 898422 in University of Padova.

References

1. Liao Z, Chen Z, Xu A, Gao Q, Song K, Liu J, Hu HY. Wastewater treatment and reuse situations and influential factors in major Asian countries. J Environ Manage. 2021;282. https://doi.org/10.1016/J.JENVMAN.2021.111976.
2. Qasim SR, Zhu G. Wastewater treatment and reuse: theory and design examples. Princip Basic Treat. 2017;(1). https://doi.org/10.1201/b22368.
3. Dikici M. Büyükşehirlerin Kanalizasyon Hatlarının Etkili İşletmesi ve İstanbul Örneği, ALKÜ Fen Bilim. Derg. 2021;3:29–43. https://doi.org/10.46740/ALKU.841422.
4. Todeschini S, Papiri S, Ciaponi C. Placement strategies and cumulative effects of wet-weather control practices for intermunicipal sewerage systems. Water Resour Manag. 2018;328(32):2885–900. https://doi.org/10.1007/S11269-018-1964-Y.
5. Mannina G, Viviani G. Separate and combined sewer systems: a long-term modelling approach. Water Sci Technol. 2009;60:555–65. https://doi.org/10.2166/WST.2009.376.
6. Kim HB, Guerra Y. A Study on the installation of a sewage separator pipe inside an existing combined sewer system for CSO control. J Wetl Res. 2021;23:85–93. https://doi.org/10.17663/JWR.2021.23.1.85.
7. Tan E, Sadak D, Ayvaz MT. Optimum design of sewer systems by using differential evolution algorithm. Tek Dergi. 2020;31:10229–10250. https://doi.org/10.18400/TEKDERG.541507.
8. Sudha J, Prasad N, Ariramachandran, Nagarajan, Prashanth K. An evaluation and analysis of a sewerage system for Tondiarpet, Chennai—a case study. Mater Today Proc. 2020. https://doi.org/10.1016/J.MATPR.2020.09.358.
9. Tuozzolo S, Langhorst T, de Moraes Frasson RP, Pavelsky T, Durand M, Schobelock JJ. The impact of reach averaging Manning's equation for an in-situ dataset of water surface elevation, width, and slope. J Hydrol. 2019;578:123866. https://doi.org/10.1016/J.JHYDROL.2019.06.038.
10. Metcalf & Eddy, Tchobanoglous G. Wastewater engineering: collection and pumping of wastewater. McGraw-Hill; 1981.
11. Kruszyński W, Dawidowicz J. Computer modeling of water supply and sewerage networks as a tool in an integrated water and wastewater management

system in municipal enterprises. J Ecol Eng. 2020;21:261–266. https://doi.org/10.12911/22998993/117533.

12. De La Fuente A, Pons O, Josa A, Aguado A. Multi-criteria decision making in the sustainability assessment of sewerage pipe systems. J Clean Prod. 2016;112:4762–70. https://doi.org/10.1016/J.JCLEPRO.2015.07.002.
13. Sewer systems. In: Manual sewerage and sewage treatment part B operation maintenance, Central Public Health and Environmental Engineering Organization; 2012.
14. Musa MA, Idrus S. Physical and biological treatment technologies of slaughterhouse wastewater: a review. Sustain. 2021;13:4656. https://doi.org/10.3390/SU13094656.
15. Manav Demir N, Demir S. Model-based analysis of the effects of recycle ratios on the performance of an A2O process. Sigma J Eng Nat Sci. 2020;38:1741–1751.
16. Damalerio R, Pausta CM, Eusebio RC, Promentilla MA, Orbecido A, Patacsil L, Beltran A. Key sectors perspective in selecting optimal biological nutrient removal technologies for sewage treatment in the Philippines. ASEAN Eng J. 2021;11:1–13. https://doi.org/10.11113/AEJ.V11.16673.
17. Khalil M, Liu Y. Greywater biodegradability and biological treatment technologies: a critical review. Int Biodeterior Biodegradation. 2021;161. https://doi.org/10.1016/J.IBIOD.2021.105211.
18. Valipour A, Taghvaei SM, Raman VK, Gholikandi GB, Jamshidi S, Hamnabard N. An approach on attached frowth process for domestic wastewater treatment. Environ Eng Manage J. 2014;13:145–52.
19. Dias J, Bellingham M, Hassan J, Barrett M, Stephenson T, Soares A. Impact of carrier media on oxygen transfer and wastewater hydrodynamics on a moving attached growth system. Chem Eng J. 2018;351:399–408. https://doi.org/10.1016/J.CEJ.2018.06.028.
20. Budgen J, Le-Clech P. Assessment of brewery wastewater treatment by an attached growth bioreactor. H2Open J. 2020;3:32–45. https://doi.org/10.2166/H2OJ.2020.023.
21. Verhuelsdonk M, Glas K, Parlar H. Economic evaluation of the reuse of brewery wastewater. J Environ Manage. 2021;281. https://doi.org/10.1016/J.JENVMAN.2020.111804.
22. Zhao T, Qiao K, Wang L, Zhang W, Meng W, Liu F, Gao X, Zhu J. Isolation and characterization of a strain with high microbial attachment in aerobic granular sludge. J Environ Sci. 2021;106:194–203. https://doi.org/10.1016/J.JES.2021.01.019.
23. Shi Y, Wang Z, Du X, Gong B, Jegatheesan V, Haq IU. Recent advances in the prediction of fouling in membrane bioreactors. Membranes. 2021;11:381. https://doi.org/10.3390/MEMBRANES11060381.
24. Gao T, Xiao K, Zhang J, Zhang X, Wang X, Liang S, Sun J, Meng F, Huang X. Cost-benefit analysis and technical efficiency evaluation of full-scale membrane bioreactors for wastewater treatment using economic approaches. J Clean Prod. 2021;301. https://doi.org/10.1016/J.JCLEPRO.2021.126984.
25. Nguyen Quoc B, Wei S, Armenta M, Bucher R, Sukapanpotharam P, Stahl DA, Stensel HD, Winkler MKH. Aerobic granular sludge: Impact of size distribution on nitrification capacity. Water Res. 2021;188:116445. https://doi.org/10.1016/J.WATRES.2020.116445.
26. He Q, Xie Z, Fu Z, Wang H, Chen L, Gao S, Zhang W, Song J, Xu P, Yu J, Ma J. Effects of phenol on extracellular polymeric substances and microbial communities from aerobic granular sludge treating low strength and salinity wastewater. Sci Total Environ. 2021;752. https://doi.org/10.1016/J.SCITOTENV.2020.141785.
27. Rashid SS, Liu YQ, Zhang C. Upgrading a large and centralised municipal wastewater treatment plant with sequencing batch reactor technology for integrated nutrient removal and phosphorus recovery: Environmental and economic life cycle performance. Sci Total Environ. 2020;749. https://doi.org/10.1016/J.SCITOTENV.2020.141465.
28. Xiao C, Wang L, Zhou Q, Huang X. Hazards of bisphenol A (BPA) exposure: a systematic review of plant toxicology studies. J Hazard Mater. 2020;384. https://doi.org/10.1016/J.JHAZMAT.2019.121488.
29. Lesmeister L, Lange FT, Breuer J, Biegel-Engler A, Giese E, Scheurer M. Extending the knowledge about PFAS bioaccumulation factors for agricultural plants—a review. Sci Total Environ. 2021;766. https://doi.org/10.1016/J.SCITOTENV.2020.142640.
30. El Hadki A, Ulucan-Altuntas K, El Hadki H, Ustundag CB, Kabbaj OK, Dahchour A, Komiha N, Zrineh A, Debik E. Removal of oxytetracycline by graphene oxide and Boron-doped reduced graphene oxide: a combined density function theory, molecular dynamics simulation and experimental study. FlatChem. 2021;27. https://doi.org/10.1016/J.FLATC.2021.100238.
31. Syafiuddin A, Boopathy R. Role of anaerobic sludge digestion in handling antibiotic resistant bacteria and antibiotic resistance genes—a review. Bioresour Technol. 2021;330. https://doi.org/10.1016/J.BIORTECH.2021.124970.
32. Foglia A, Andreola C, Cipolletta G, Radini S, Akyol Ç, Eusebi AL, Stanchev P, Katsou E, Fatone F. Comparative life cycle environmental and economic assessment of anaerobic membrane bioreactor and disinfection for reclaimed water reuse in agricultural irrigation: a case study in Italy. J Clean Prod. 2021;293. https://doi.org/10.1016/j.jclepro.2021.126201.
33. Egbuikwem PN, Mierzwa JC, Saroj DP. Evaluation of aerobic biological process with post-ozonation for treatment of mixed industrial and domestic wastewater for potential reuse in agriculture. Bioresour Technol. 2020;318. https://doi.org/10.1016/j.biortech.2020.124200.
34. Yeom Y, Han J, Zhang X, Shang C, Zhang T, Li X, Duan X, Dionysiou DD. A review on the degradation

efficiency, DBP formation, and toxicity variation in the UV/chlorine treatment of micropollutants. Chem Eng J. 2021;424. https://doi.org/10.1016/J.CEJ.2021.130053.
35. Domínguez Henao L, Turolla A, Antonelli M. Disinfection by-products formation and ecotoxicological effects of effluents treated with peracetic acid: a review. Chemosphere. 2018;213:25–40. https://doi.org/10.1016/J.CHEMOSPHERE.2018.09.005.
36. Jahan BN, Li L, Pagilla KR. Fate and reduction of bromate formed in advanced water treatment ozonation systems: a critical review. Chemosphere. 2021;266. https://doi.org/10.1016/J.CHEMOSPHERE.2020.128964.
37. Fang J, Zhao Q, Fan C, Shang C, Fu Y, Zhang X. Bromate formation from the oxidation of bromide in the UV/chlorine process with low pressure and medium pressure UV lamps. Chemosphere. 2017;183:582–8. https://doi.org/10.1016/J.CHEMOSPHERE.2017.05.136.
38. Kong J, Lu Y, Ren Y, Chen Z, Chen M. The virus removal in UV irradiation, ozonation and chlorination. Water Cycle. 2021;2:23–31. https://doi.org/10.1016/j.watcyc.2021.05.001.
39. Pivokonsky M, Kopecka I, Cermakova L, Fialova K, Novotna K, Cajthaml T, Henderson RK, Pivokonska L. Current knowledge in the field of algal organic matter adsorption onto activated carbon in drinking water treatment. Sci Total Environ. 2021;799. https://doi.org/10.1016/J.SCITOTENV.2021.149455.
40. Soni R, Bhardwaj S, Shukla DP. Various water-treatment technologies for inorganic contaminants: current status and future aspects. Inorg Pollut Water. 2020;273–295. https://doi.org/10.1016/B978-0-12-818965-8.00014-7.
41. Qasim SR, Zhu G, CP&Y I. Wastewater treatment and reuse: theory and design examples. Volume 2, Post-treatment, reuse, and disposal, 1st ed. Florida; 2018.
42. Ulbricht M. Advanced functional polymer membranes. Polymer (Guildf). 2006;47:2217–62. https://doi.org/10.1016/J.POLYMER.2006.01.084.
43. Muralikrishna IV, Manickam V. Wastewater treatment technologies. In: Environ Manage, Butterworth-Heinemann; 2017. pp. 249–293. https://doi.org/10.1016/B978-0-12-811989-1.00012-9.
44. Ilhan F, Kabuk HA, Kurt U, Avsar Y, Sari H, Gonullu MT. Evaluation of treatment and recovery of leachate by bipolar membrane electrodialysis process. Chem Eng Process Process Intensif. 2014;75:67–74. https://doi.org/10.1016/J.CEP.2013.11.005.
45. Ibrahim KEA, Şolpan D. Removal of carbofuran in aqueous solution by using UV-irradiation/hydrogen peroxide. J Environ Chem Eng. 2019;7. https://doi.org/10.1016/j.jece.2018.102820.
46. Ulucan-Altuntas K, Ilhan F. Enhancing biodegradability of textile wastewater by ozonation processes: optimization with response surface methodology. Ozone Sci Eng. 2018. https://doi.org/10.1080/01919512.2018.1474339.
47. Abdel Wahaab R, Mahmoud M, van Lier JB. Toward achieving sustainable management of municipal wastewater sludge in Egypt: the current status and future prospective, Renew. Sustain Energy Rev. 2020; 127:109880. https://doi.org/10.1016/J.RSER.2020.109880.
48. Avsar Y, Kurt U. Thermotechnical comparison of conventional heating and microwave radiation method for dewatering of sewage sludge. Desalin Water Treat. 2017. https://doi.org/10.5004/dwt.2017.20425.
49. Ayol A, Biryol İ, Taşkan E, Hasar H. Enhanced sludge stabilization coupled with microbial fuel cells (MFCs). Int J Hydrogen Energy. 2021;46:29529–40. https://doi.org/10.1016/J.IJHYDENE.2020.10.143.
50. Bian C, Ge D, Wang G, Dong Y, Li W, Zhu N, Yuan H. Enhancement of waste activated sludge dewaterability by ultrasound-activated persulfate oxidation: operation condition, sludge properties, and mechanisms. Chemosphere. 2021;262. https://doi.org/10.1016/J.CHEMOSPHERE.2020.128385.
51. Abu-Orf M, Muller CD, Park C, Novak JT. Innovative technologies to reduce water content of dewatered municipal residuals. J Residuals Sci Technol. 2004;1.
52. Aboulfotoh A. Effect of flocculation time on the performance of fly ash as sludge conditioners. J Mater Environ Sci. 2019;10:1296–1303. http://www.jmaterenvironsci.com. Accessed 8 Aug 2021.
53. Poblete R, Painemal O. Improvement of the solar drying process of sludge using thermal storage. J Environ Manage. 2020;255. https://doi.org/10.1016/J.JENVMAN.2019.109883.
54. Chen G, Zhang R, Guo X, Wu W, Guo Q, Zhang Y, Yan B. Comparative evaluation on municipal sewage sludge utilization processes for sustainable management in Tibet. Sci Total Environ. 2021;765. https://doi.org/10.1016/J.SCITOTENV.2020.142676.
55. Chu L, He W. Toxic metals in soil due to the land application of sewage sludge in China: spatiotemporal variations and influencing factors. Sci Total Environ. 2021;757. https://doi.org/10.1016/J.SCITOTENV.2020.143813.
56. Davis R, Hirji R. Water resources and environment technical note D. 1 water quality; 2003.
57. Shoushtarian F, Negahban-Azar M. Worldwide regulations and guidelines for agricultural water reuse: a critical review. Water. 2020;12:971. https://doi.org/10.3390/W12040971.

Near-Nature Wastewater Treatment Methods

10

Elina Domscheit

Abstract

Constructed wetlands (CWs) are non-conventional treatment methods that have continuously been developed into a secure wastewater treatment technology for domestic, agricultural and industrial wastewater as well as for runoff and leachate waters. These nature-based systems have some advantages in comparison to conventional wastewater treatment plants. Furthermore, CWs are of special interest when it comes to the reduction of climate change impacts. The main disadvantage of CWs is the comparatively high surface requirement. According to their properties, contaminants get removed via several physical, chemical and biological removal processes. Furthermore, CWs perform sufficient removal of emerging contaminants. CW types can be classified according to the vegetation type and the flow pattern within the wetland. In this paper, basic methods and information about CW design are given. Further, a substantial literature review on CWs used for river water treatment is conducted. Generally, satisfying removal efficiencies of standard parameters and emerging contaminants are reported, and many authors stated that water quality could be improved with feasible use of CW systems. Even though there is comparatively less maintenance required, the most important management activities are assumed. Finally, the case study of a pre-design FWS-CW for river water treatment is briefly described.

Keywords

Ecological treatment · Emerging organic contaminants · Natural and constructed wetlands · Nutrient removal · Wetland types

10.1 Introduction

Sufficient wastewater treatment is important when it comes to hygienic and environmental aspects. Different systems were developed over time to clean up polluted water. Other than the conventional wastewater treatment methods, non-conventional systems have also been set up. A prominent example of non-conventional wastewater treatment methods are constructed wetlands (CWs). In comparison with conventional wastewater treatment systems, CWs generally convince due to their little environmental impact, strong adaptability [1], low sludge generation [2], and good self-purification capacity [3]. In addition, they provide ecosystem services and improve ecosystem health. As CWs are

E. Domscheit (✉)
Lower Saxony Water Management, Coastal and Nature Protection Agency (Niedersächsischer Landesbetrieb für Wasserwirtschaft, Küsten- und Naturschutz), Norden, Germany
e-mail: elina.domscheit@gmx.de

M. Bahadir and A. Haarstrick (eds.), *Water and Wastewater Management*,
Water and Wastewater Management, https://doi.org/10.1007/978-3-030-95288-4_10

water treatment systems that require low-cost, low-maintenance and low-energy, they are of special interest for implementation in developing countries and rural areas [4]. Moreover, CWs are of special interest when it comes to the reduction of climate change impacts [5]. Due to water regulation, they are able to improve adaption to extreme weather conditions such as local floods and droughts [6], which are more likely to occur with ongoing climate change [7]. When it comes to floods, CWs function as water reservoirs and buffers [8, 9]. In case of droughts, which may lead to water shortage and thus affect water supply and agricultural yields [10], CWs can improve water security and access to water [6], as they can provide water for non-potable purposes such as irrigation. Reusing treated water saves valuable drinking water [10]. Furthermore, CWs capture carbon like natural wetlands, which store the main part of global soil carbon due to low decomposition rates in anaerobic soils [11, 12]. In the context of global warming, CWs have a positive effect on local climate parameters such as precipitation, temperature and humidity [13]. Vegetation in CWs perform evaporative landscape cooling [14], resulting into a decrease of the overall temperature in CWs for about 2 °C [15]. Many CWs have the ability to function as temperature buffer [14]. However, in contrast to conventional watewater treatment plants, CWs require a comparatively high surface. A minor disadvantage is that the choice of plant species is geographically limited. Even though it can be prevented by proper management, CWs may support the breeding of disease producing organisms and insects and may generate odours [16]. Furthermore, adjustment of certain conditions as oxygen concentrations is less precise than in conventional treatment systems.

This paper provides an overview about natural and constructed wetlands. Different ways of pollutant removal in CWs are described with particular focus on emerging contaminants. The categorization of the different CW types is further explained and the basic design parameters are demonstrated. The various range of use for CW systems is pointed out and highlights the utilization of CWs for river water treatment.

10.2 Natural Wetlands

Globally, 12.1 million km^2 are covered by inland and coastal wetlands. Wetlands are defined as "areas of land where water covers the soil". Almost 54% of the world's wetlands are permanently flooded, while 46% are flooded seasonally. The water covering wetlands can have different properties such as static or flowing as well as fresh, brackish or saline. The term natural wetland includes rivers and streams, natural lakes, peatlands, marshes and swamps for inland wetlands, and estuaries, mangroves seagrass beds, coral reefs, coastal lagoons, kelp forests, coastal karst, and caves for coastal wetlands [11]. Wetlands are highly productive ecosystems that stand out due to their combination of terrestrial and aquatic habitats [17]. Therefore, they support a wide range of biodiversity and are habitat for water birds [17, 18]. Wetlands play a key role for migratory species, as they are used as feeding, breeding and stop-over grounds [17]. With a view to growth of adapted plants and the development of characteristic wetland soils are supported as a result of present water [19]. Typically, emergent aquatic vegetation such as cattails, rushes and reeds can be found in wetlands [16]. Furthermore, wetlands play an important role in the water cycle. Flows are regulated, water is received, stored and released [11], and groundwater gets recharged [17]. These regulatory properties can help protecting against extreme weather conditions such as floods and droughts [17, 20]. Additionally, wetlands serve many ecosystem services, which can be divided into cultural, provisioning and regulating services. Cultural services refer mainly to aesthetical and recreational aspects [20], while provisioning services enclose mainly resourcing of food and energy. Regulating services are mainly provided in the sense of carbon sequestration, water purification and water regulation [11].

10.3 Constructed Wetlands

Constructed wetlands are defined by EPA as "engineered or constructed wetlands that utilize natural processes involving wetland vegetation, soils, and their associated microbial assemblages to assist, at least partially, in treating an effluent or other water source" [21]. The vegetation of CWs is similar to the one of natural wetlands and can be divided according to their growth behaviour referring to the water surface into the three main categories of submerged, emergent and floating plants (Fig. 10.1) [14]. Due to the ability of controlling and adjusting hydraulic parameters and conditions in a constructed wetland, the positive qualities of natural wetlands can be adopted and used in an effective way [16]. CWs offer a secure and sustainable opportunity to treat domestic, agricultural, and industrial wastewater as well as runoff and river waters [2, 22].

10.3.1 Pollutant Removal by CWs

Contaminants get removed in CWs via several physical, chemical and biological removal processes according to their properties. Mechanisms that contribute to the removal of pollutants are microbial mediated processes, chemical networks, volatilization, sedimentation, sorption, photo degradation, plant uptake, vertical diffusion in soils and sediments, transpiration flux, seasonal cycles and accretion [14]. On average, wetlands can remove 60–95% of known pollutants from industry and households [23].

Physical removal mechanisms include sedimentation, filtration and adsorption, and take place mainly when water passes through substrate or root masses and in settings, where gravitational settling can take place. Adsorption occurs on substrate and plant surfaces [24], as cited in [16]. Additionally, volatile organic compounds like pesticides can be removed from the water by volatilization [23, 25].

Precipitation and chemical decomposition are the main processes of chemical removal. These are the main processes responsible for removal of nutrients as nitrogen and phosphorus as well as removal of heavy metals and pathogens. Decomposition happens as UV radiation, oxidation and reduction [24], as cited in [16]. Aerobic processes are supported by provision of oxygen through the root-system of the plants [26].

Removal due to biological processes can be categorized into bacterial and plant metabolism, plant adsorption, predation and natural die-off [25]. Plant metabolism and uptake removes refractory organics. Bacteria and viruses may be eliminated by toxic root excretions. Nutrients as

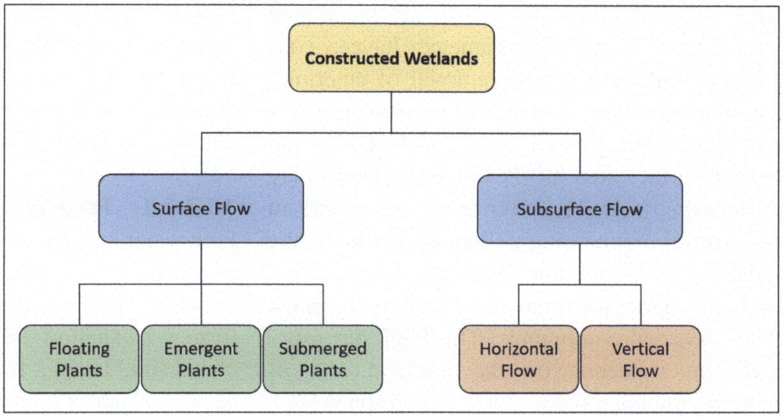

Fig. 10.1 Classification of CW types according to flow pattern within the wetland and vegetation type [14]

nitrogen and phosphorus as well as heavy metals and refractory organics can be taken up by plants [24], as cited in [16]. Predation mainly refers to removal of total suspended solids and pathogens [24, 25]. Removal due to natural die-off is only relevant for pathogens [24], as cited in [16]. Nevertheless, bacterial metabolism is the main factor, when it comes to removal of biodegradable compounds such as colloidal solids, BOD_5, nitrogen and refractory organics in CWs [16, 27].

In case of emerging contaminants, such as pesticides or pharmaceuticals, the removal efficiency depends on different factors. Often there is a various set of physical, chemical, and biological processes involved, which can be influenced by the CW design and the selected operational parameters [28]. Generally, oxygen concentration is shown to be one of key factors, since the best performance took place at of aerobic pathways [29]. According to Ding et al. [28] and Kadlec and Wallace [14], possible removal factors of antibiotics in livestock wastewater and pesticides in CWs are adsorption to soil particles and organic matter, sedimentation of particles, photo degradation, plant uptake, and biodegradation as well as physicochemical degradation. Especially "old" pesticides, such as DDT, are very persistent in the environment and partition to a relevant amount to particulate matter. In this case, CWs might rather act as a trap for the particulate matter than provide any very effective mechanisms for degradation. In contrast, modern pesticides degrade faster, and studies have shown that CWs generally reduce concentrations of many of these compounds [14]. As mentioned before, the CW design parameters have a considerable influence on the removal of emerging contaminants. For example, vegetation, primary treatment, loading mode (e.g., batch mode) and specific surface area (m^2/Person Equivalent) play an important role. However, a key parameter in the removal of emerging pollutants in CWs is the hydraulic retention time (HRT); the greater the HRT, the higher the removal efficiencies for most of the selected compounds. The HRT has especially an influence on the removal rate of hydrophobic compounds such as hormones [28].

The most relevant removal factors can be differentiated between the different types of emerging contaminants. Non-steroidal anti-inflammatory drugs (NSAID) such as Ibuprofen and Diclofenac, which are the most common drugs used in humans, are negatively charged at environmental pH. Due to that, the process of sorption was found to be negligible. Instead, their removal in CWs can mainly be explained by biodegradation next to photo degradation. Consequentially, oxygen is a key parameter affecting the removal rates next to water depth. For lipid regulator drugs and anti-epileptic agents such as Carbamazepine and Clofibric acid, HRT has the greatest influence on removal efficiencies [29]. Masi et al. [30] found up to 100% removal rates of oestrogens in hybrid systems. Removal mechanisms for these compounds in CWs are mainly associated with sorption by organic matter due to its high hydrophobicity, next to biofilm interaction [29, 30]. According to Matamoros and Bayona [29], removal of oestrogen could be improved by an increase in HRT, since it might increase the interaction time.

10.3.2 CW Types

The different CW types are divided into surface and subsurface flow CWs [22]. Surface flow CWs can further be grouped due to the vegetation type, while subsurface flow CWs can additionally be classified according to the flow direction into horizontal and vertical subsurface flow CWs (Fig. 10.1). These CW types can be joined with each other in hybrid systems in order to combine the particular advantages of each CW type [31]. In few cases, floating treatment wetlands are used [32].

10.3.2.1 Free Water Surface CWs

Free water surface CWs (FWS-CWs) are defined as "wetland systems, where the water surface is exposed" [33]. Typically, FWS-CWs are performed as shallow basins or channels with vegetated soil [16, 25]. Implementation for treatment of river water or channels can be operated as off-

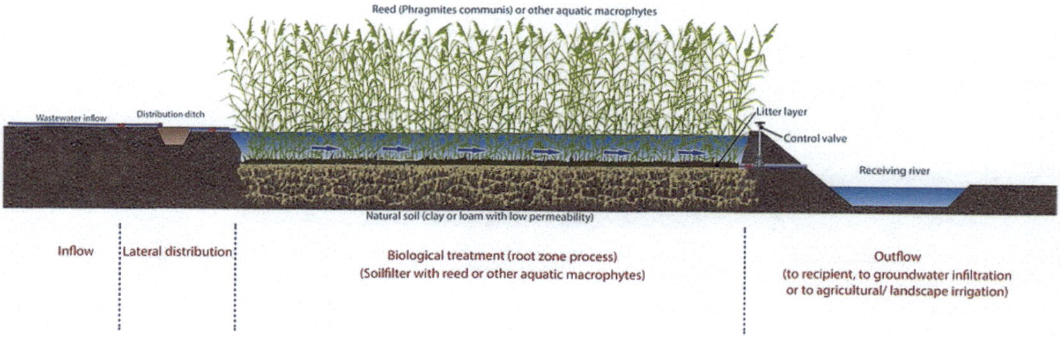

Fig. 10.2 Cross section of a free water surface constructed wetland planted with reed [34]

stream or on-stream system. Subsurface barriers like silty soils prevent seepage and keep water above the soil [6, 16]. The water to be treated flows in a horizontal pattern through the vegetation and top soil from an inlet to an outlet point (Fig. 10.2) [6, 25]. In very few cases, there is no effluent due to evapotranspiration and infiltration within the wetland [25]. Short circuiting is minimized by the shallow water depth, low flow velocity and the presence of plants [16]. FWS-CWs display the CW type that mimics the hydrologic regime of natural wetlands the most [25]. As this treatment system has its highest efficiency during warmer periods [23], FWS-CWs fit best for warm climates [31].

FWS-CWs are distinguished by their process stability and ability to tolerate fluctuating water levels and nutrient loads. They perform effective treatment with e.g., high reduction of BOD_5 and solids and moderate removal of pathogens [31]. Due to the passive treatment, mechanical and technical equipment as well as energy and skilled operator needs can be minimized [33]. Besides, the use of chemicals like coagulants is not needed. Additionally, FWS-CWs can be built with local materials [31] and are less expensive to construct than other CW types. Another advantage is that this wetland technology produces only small quantities of sludge [33]. Therefore, sludge treatment and disposal is not necessary. Furthermore, FWS-CWs stand out due to their possible multiple purpose use. FWS-CWs are flora and fauna habitats. Besides, they can be used as a park and for educational, aesthetical as well as for recreational purposes. Next to that the effluent might be reused e.g., for irrigation [25].

But, land requirements are high for FWS-CWs [25]. Therefore, this CW type is most cost effective in regions, where suitable areas are available for reasonable prices [33]. Next to that, starting time for FWS-CWs is long before they operate at full capacity. The created wetland area may promote mosquito breeding [31], and faecal coliforms are introduced into the area by birds and other wildlife. In contrast to renewable removal of biodegradable contaminants, pollutants like phosphorus and metals are bound in the wetland sediments and accumulate over time [33].

10.3.2.2 Horizontal Subsurface Flow CWs

A horizontal subsurface flow CW (HSSF-CW) consists of a filter bed, filled with gravel, sand or soil, and is planted with wetland vegetation (Fig. 10.3). From the inlet point, the water to be treated flows horizontally beneath the surface of the bed media passing plant roots and rhizomes to the outlet point [14]. This subsurface flow avoids mosquito problems as they might be found in FWS-CWs [7]. Furthermore, the risk of human or animal exposure to pathogenic organisms is minimized [14]. As the porous filter medium provides a greater contact surface for treatment processes, HSSF-CWs require less land compared to FWS-CWs [31].

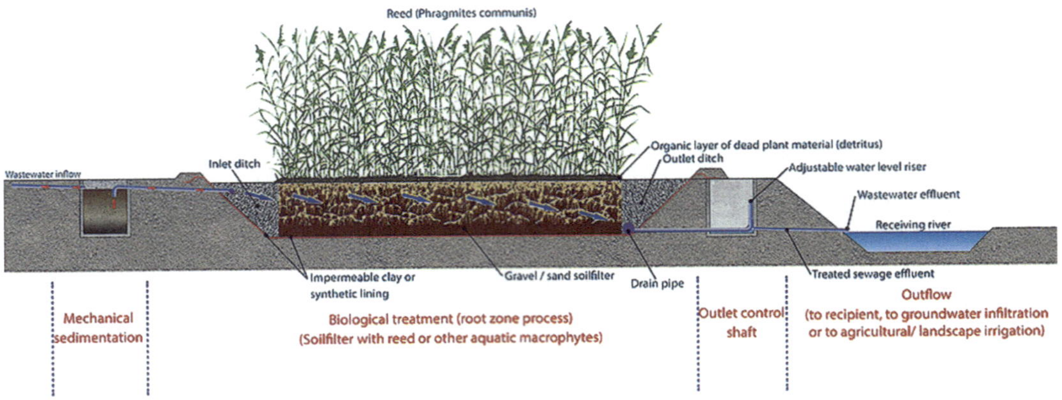

Fig. 10.3 Cross section of a horizontal subsurface flow constructed wetland planted with reed [34]

10.3.2.3 Vertical Subsurface Flow CWs

Vertical subsurface flow constructed Wetlands (VSSF-CWs) are typically sand or gravel beds, planted with wetland vegetation (Fig. 10.4) [14]. The water to be treated is intermittently applied across the surface of the filter medium and then percolates vertically through the root zone of the plants towards a drainage system at the bottom [35]. Due to the intermittently loading, the flow in VSSF-CWs is unsaturated, which leads to higher oxygen transfer to the filter medium compared to HSSF-CWs [32]. This set up results into smaller area requirement but higher construction and operation costs. VSSF-CWs start running quickly and operate better than FWS-CWs under cold weather conditions [36].

10.3.3 Constructed Wetlands Design

Before starting a calculation for CW design, the following general information should be determined according to Kadlec and Wallace [14]:

- inlet concentrations and flows,
- target concentrations (regulatory limits and allowable exceedance factors),
- allowable inflow and seepage rates,
- rain, evapotranspiration and temperature ranges for the project site,
- wetland type (FWS or SSF).

The specific input data necessary for the calculative step of the CW design is shown in Table 10.1. The presented values are

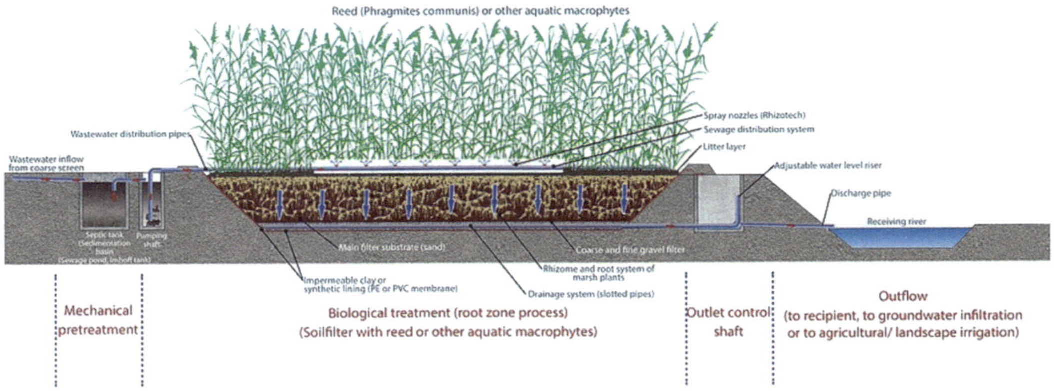

Fig. 10.4 Cross section of a vertical subsurface flow constructed wetland planted with reed [34]

Table 10.1 Summary of the required input data for CW design (suggested literature values and necessary site-specific values). The site-specific values need to be measured on site

	Unit	Value	Source
Literature values			
Depth	m	0.75	[37]
Porosity	–	0.65	[37]
Max. BOD_5 loading	kg/(ha * d)	80	[38–40]
Area safety factor	–	1.3	[37]
Max. HLR* (FWS-CW)	L/(m²* d)	100	[25, 40–42]
Max. HLR* (VSSF-CW)	L/(m²* d)	200	[7]
Basin geometry (aspect ratio)	–	1:4	[23, 37]
Site-specific values			
Inflow	m³/d	Must determined	
Inlet BOD_5 concentration	mg/L	Must determined	
Inlet BOD_5 load	kg/d	Must determined	

* HLR: Hydraulic loading rate

differentiated between suggested literature values and site-specific values.

The required area for the wetland system can be calculated in two ways. One option is to identify the area according to the maximum BOD_5 loading rate (see Table 10.1). To do so, the BOD_5 load of the influent needs to be put in relation to the maximum BOD_5 loading rate receiving the minimum CW area required for water treatment:

$$A_{CW} = \frac{B_{BOD_5,\,in}}{B_{BOD_5,\,max}}$$

where:

A_{CW} minimum CW area [ha]
$B_{BOD_5,\,in}$ BOD_5 load of influent [kg/d]
$B_{BOD_5,\,max}$ maximum BOD_5 loading rate [kg/(ha * d)]

According to EPA [37], a safety factor should be applied for buffers and setbacks. Consequently, the resulting area is calculated by multiplication of the determined CW area with the safety factor.

In an additional step, the corresponding hydraulic parameters should be calculated in order to match them with suggested hydraulic parameters given in the literature. To do so, the hydraulic residence time (HRT), (Eq. 10.1), and the hydraulic loading rate (HLR), (Eq. 10.2) should be determined for the resulting area according to the following formulas [14]:

$$HRT = \frac{A_{CW,s} * \varepsilon * h}{Q_{in}} \tag{10.1}$$

where:

HRT hydraulic residence time [d]
$A_{CW,s}$ CW area including safety factor [m²]
ε porosity [−]
h water depth [m]
Q_{in} daily inflow [m³/d]

$$HLR = \frac{Q_{in}}{A_{CW,s}} \tag{10.2}$$

where:

HLR hydraulic loading rate [m³/(ha * d)]
Q_{in} daily inflow [m³/d]
$A_{CW,s}$ CW area including safety factor [ha]

Alternatively, the area can be conducted according to the hydraulic loading rate e.g., in cases when the hydraulic parameters do not match with literature values after CW design according to BOD_5 loading rate. To do so, an HLR suggested in the literature can be applied.

Afterwards, the safety factor according to EPA [37] should be applied with reference to the minimum CW area (Eq. 10.3).

$$A_{CW} = \frac{1}{HLR} * Q_{in} \quad (10.3)$$

where:

A_{CW} minimum CW area [ha]
Q_{in} daily inflow [m^3/d]

Depending on the inlet water quality, a pre-treatment might be considered. The given calculations refer to the design of a single reactor. For better treatment performance, the "sequential model" approach should be used considering three different zones. Furthermore, when conducting the CW design more detailed, attention should be paid to implementation of the system in multiple parallel trains. In the best case, it should consist of three parallel trains in order to have the possibility to remove any single zone of one train from service and transfer its inflow to the same zone kind of a parallel train [37].

10.3.4 CWs for Wastewater and Leachate Treatment

Nowadays, CWs for municipal wastewater treatment are most often restricted to small communities due to their high land area requirements. Nevertheless, they provide a nature-based solution for wastewater treatment and are often used as "polishing" step after wastewater treatment in a conventional treatment plant. Generally, a pre-treatment with conventional processes is advised in order to avoid potential solids or oxygen demand overload. However, CWs may be used to perform all the functions of secondary treatment and higher [14].

CW systems might also be used for the treatment of industrial wastewater. Especially industries that produce wastewater, which is high in biodegradable organic and nitrogen content, such as potato, wine, olive oil, sugar, starch, alcohol, and meat processing industries, are potential users of CWs [14].

Using sanitary landfills for solid waste disposal entailsx the treatment and disposal of liquid leachates. Modern landfills are lined and thus enable the collection of leachate. The collected water can differ widely in chemical composition due to the various natures of solid waste, age differences as well as differences in decomposition and taken place reactions within the landfill [14]. However, studies have shown sufficient removal efficiencies of leachates with the help of CW systems [43].

10.3.5 CWs for River Water Treatment

CWs offer a secure and sustainable opportunity to treat raw domestic, agricultural and industrial wastewater. Next to that, they are used for effluent polishing as well as for treatment of runoff and river waters [2, 22]. A literature review on CWs used for river water treatment was conducted (Table 10.2). It revealed that mainly the free water surface type or hybrid systems are used. However, also horizontal subsurface flow CWs as well as vertical subsurface flow CWs were applied. Generally, satisfying removal efficiencies of standard parameters were reported, and many authors stated that the river water quality could be improved with feasible use of a CW system. Next to that, Zheng et al. [44] reached removal rates of approx. 20% for four phthalic acid esters (PAEs).

In CWs used for treatment of river water, free floating, emergent as well as submerged plants have been used. *Phragmites australis* and *Typha latifolia* were used most often [44–53]. In addition, *Typha spp., Typha orientalis, Phragmites communis, Sparganium erectum, Juncus effuses, Carex elata, Chrysopogon, Lythrum salicaria L., Iris pseudacorus L., Salix integra* and *Dactylis glomerata* were utilized in some of the studies. Planting densities differ from less than 5% of surface coverage [54] up to 90% surface coverage [55].

Table 10.2 Collection of relevant parameters of CW systems used for treatment of river water (Free water surface = FWS, Horizontal subsurface flow = HSSF, vertical subsurface flow = VSSF, subsurface flow = SSF)

CW type	Location	Construction and operation	Removal rates	Source
FWS-CW	Jinan, North China	• Media: washed river sand (particle size < 2 mm, mainly Si_2O_3, Al_2O_3, and Fe_2O_3) • 5 wetland units with inflow of 20 L, respectively	TN removal per mechanism: 8.4–34.3% (plant uptake) 20.5–34.4% (sediment storage) 0.6–1.9% (N_2O emission) ca. 2.0–23.5% (nitrification and denitrification)	[51]
FWS-CW	North-east Italy	• In-stream • Floating CW	Median: COD: 66% BOD_5: 52% TP: 65%	[48, 50]
FWS-CW	Ho-Bou-Yu Drainage, Taiwan	• 1.55 ha • Monthly average influent flow rate: 3,672–5,352 m^3/d • 5 zones in series	Average: BOD_5: 36.9% NH_4-N: 47.1% TSS: 71.8%	[55]
SSF-CW	Jinhe River, Tianjin, China	• Inflow: 0.16 L/min • 7 × 0.196 m^2 in surface area • Down-flow wetland unit + up-flow wetland unit • Media: coarse gravel, shale, PHPB	Mean: COD: 35% NO_3-N: 50.42% NH_4-N: 71.25% TN: 64.85% SRP*: 67.34% TP: 61.24%	[56]
HSSF-CW	Hai River, Tianjin, China	• 55 chambers in parallel (each 31 m × 20 m × 0.8 m) • Daily treatment capacity: 170 m^3 • Hydraulic retention time: 24 h • Vegetation: *Phragmites australis*	DMP*: 19.32% DEP*: 19.18% DBP*: 19.40% DEHP*: 48.56% PAEs* in water: 35.38–64.92% PAEs in soil: 1.02–31.33% PAEs in plant: 0.85–36.54% PAEs in air and biological transformation: 2.72–33.21%	[44]

(continued)

Table 10.2 (continued)

CW type	Location	Construction and operation	Removal rates	Source
VSSF-CW	Longdao River, Beijing, China	• Inflow: 200 m^3/d (1,25% of river runoff) • Off-stream • 602 m^2 • Media: nature soil	Annual mean: BOD_5: 87.2% COD: 81.8% TSS: 85.1% TP: 98.8% NH_3-N: 77.4%	[45]
VSSF-CW	Old Canal of Zhenjiang, China	• Hydraulic loading: approx. 750 mm/d	COD: 50–60% TN: 40–50% TP: 30–40%	[57]
Hybrid: FWS-CW + SSF-CW	Close to confluence of an urban stream to a larger river in Xi'an, China	• Average surface loading: 0.053 $m^3/(m^2*day)$ • Media: local gravel, sand or slag • One year operation	COD: 72.7% ± 4.5% BOD_5: 93.4% ± 2.1% NO_3-N: 54.0% ± 6.3% TN: 53.9% ± 6.0% TP: 69.4% ± 4.6%	[46]
Hybrid: Filter beds + FWS-CW	Pingtung, Taiwan	Filter bed: • 6 × 29 m^2 • 6 × 1,000 m^3/d FWS CW: 3 × 3 ha	Mean: BOD_5: 83% ± 15% SS: 81% ± 25% NH_4^+: 61% ± 28%	[58]
Hybrid: Floating- bed CW + HSSF-CW + FWS-CW	Yitong River, Changchun, China	• Sequential series systems • Designed capacity: 100 m^3/d • Hydraulic load: 0.10 m^3/m^2d • Hydraulic retention time: approx. 72 h	Average: COD: 74.79% NH_4^+-N: 80.90% TN: 71.12% TP: 78.44% SS: 91.90%	[59]

SRP: soluble reactive phosphorus, DMP: dimethyl phthalate, DEP: diethyl phthalate, DBP: di-n-butyl phthalate, DEHP: bis(2-ethylhexyl) phthalate, PAEs: Phthalate acid esters

10.3.6 Wetland Operation, Maintenance and Management

One advantage of CW systems is that there is generally only infrequent operational control or maintenance required due to their design and construction. However, there are some activities that should be executed daily, such as the monitoring and adjustment of flows, water levels, inflow and outflow water quality, and biological parameters [14].

Especially for FWS-CWs, skilled operator needs can be minimized, since the systems experience minimal ecological changes due to the conservative design, passive treatment and simple mechanical controls [14, 33]. In contrast, maintenance of HSSF-CWs and VSSF-CWs is more problematic, since HSSF-CWs require regular bed maintenance throughout the lifetime of the system, and VSSF-CWs depend on loading-and-resting regimes to maintain hydraulic conductivity, unless lightly loaded. Operations and maintenance activities that need to be conducted less frequently are, e.g., the repair of pumps and water control structures, vegetation management, pest control and removal of accumulated mineral solids.

To enable maintenance, CWs require access facilities such as dikes and berms, which need maintenance and mowed [14].

10.4 Case Study "Oued Hamdoun" River

In many countries, rivers are the main sources of pollutant transport to the sea. On the one hand, this is a serious threat to the river and marine ecosystem. On the other hand, it might be a concern for public health and tourism development, in case the sea discharge is located close to beaches, as it is the case at the Hamdoun River. The Hamdoun River is located on the east coast of central Tunisia at the Gulf of Hammamet and in the south of the City of Sousse. The river water quality is affected by the discharge of five wastewater treatment plants, industrial wastewater, untreated wastewater, surface runoff and illegal disposed wastes along river side. During summer months, more than half of the river discharge has its origin in the discharge of the wastewater treatment plants [60]. Because of polluted river water, the beaches near the river mouth had to be closed for bathing in several years resulting into serious economic constraints [61]. A case study was carried out to pre-design a FWS-CW in order to clean up the river water. Since the estimated calculated hydrograph of Hamdoun River shows clear fluctuation over the year, the 85% percentile value of 375,645 m^3/d was used according to ATV-DVWK [62] in analogy to the design of conventional WWTPs. Additional positive effects could be the possible reuse of the effluent water for irrigation and the establishment of a green area, which could also be used for educational purposes [63].

10.5 Conclusions

Because of their wide range of applications, constructed wetlands are very suitable for wastewater, river water and leachate treatment. They are also cost-effective and require hardly any energy. They show sufficient removal rates for standard pollution parameter as well as for emerging contaminants. Furthermore, in comparison with conventional wastewater treatment plants their little environmental impact, strong adaptability, low sludge generation and good self-purification capacity can be highlighted. In addition, CWs require low maintenance with generally mostly infrequent operational control requirements. One major disadvantage is the need for comparatively large land surfaces. The utilized vegetation in CWs is similar to the one of natural wetlands. Generally, pollutants get removed via several physical, chemical, and biological removal processes during their passage through the CW system. With regard to the removal of newly emerging pollutants, the oxygen input and thus the oxygen concentration is often a limiting factor.

CWs are often divided into different types according to their flow characteristics and can be combined to form so-called hybrid systems. FWS-CWs are characterised by their process stability and ability to tolerate fluctuating water levels and nutrient loads. This also leads to the fact that for FWS-CWs less maintenance is needed. But, FWS-CWs require more land. CWs that are used for sufficient river water treatment can be found frequently in China. Here, mostly FWS and hybrid systems are applied.

Acknowledgements The author would like to thank DAAD and Exceed Swindon project. Thanks also to Prof. Dr. Müfit Bahadir for his valuable support.

References

1. Yang Q, Wu Z, Liu L, Zhang F, Liang S. Treatment of oil wastewater and electricity generation by integrating constructed wetland with microbial fuel cell. Materials. 2016;9.
2. Gökalp Z, Karaman S, Taş I, Kirnak H. Constructed wetland technology to prevent water resources pollution. Curr Trends Nat Sci. 2016;5(9):125–32.
3. Liu S, Song H, Li X, Yang F. Power generation enhancement by utilizing plant photosynthate in microbial fuel cell coupled constructed wetland system. Int J Photoenergy. 2013.
4. Avellan CT, Ardakanian R, Gremillion P. The role of constructed wetlands for biomass production within the water-soil-waste nexus. Water Sci Technol. 2017;75(10):2237–45.
5. Small-Lorenz S. Wetlands do triple duty in a changing climate; 2014. [30.08.2019].
6. UNEP-DHI, UNEP-DTU, CTCN. Constructed wetlands for water treatment. Technology Compendium; 2017.
7. Hoffmann H, Platzer C, Winker M, Muench EV. Technology review of constructed wetlands—subsurface flow constructed wetlands for greywater and domestic wastewater treatment. Deutsche Gesellschaft für Int Zusammenarbeit (GIZ) GmbH; 2011.
8. Abrahams JC, Coupe SJ, Sañudo-Fontaneda LA, Schmutz U. The Brookside farm wetland ecosystem treatment (WET) system: a low-energy methodology for sewage purification, biomass production (Yield), flood resilience and biodiversity enhancement. Sustainability. 2017;9.
9. Meerburg BG, Vereijken PH, Visser WD, Verhagen J, Korevaar H, Querner EP, Blaeij ATD, Werf AVD. Surface water sanitation and biomass production in a large constructed wetland in the Netherlands. Wetlands Ecol Manage. 2010;18(4):463–70.
10. Qomariyah S, Ramelan AH, Setyono P, Sobriya. Linking climate change to water provision: greywater treatment by constructed wetlands. In: IOP conference series: earth and environmental science. 2018;129.
11. Gardner RC, Finlayson CM. Global wetland outlook: state of the world's wetlands and their services to people, ed. Dudley N. Gland, Switzerland: Ramsar Convention Secretariat; 2018.
12. Kanungo P, Kumawat DM, Billore SK. Carbon sequestration potential of constructed wetlands used for wastewater treatment. Int J Appl Pure Sci Agric. 2017;3(4):38–44.
13. Gokalp Z, Karaman S. Critical design parameters for constructed wetlands natural wastewater treatment systems. Curr Trends Nat Sci. 2017;6(12):156–64.
14. Kadlec RH, Wallace SD. Treatment wetlands. 2nd ed. Taylor & Francis Group; 2009.
15. Panrare A, Sohsalam P, Tondee T. Constructed wetland for sewage treatment and thermal transfer reduction. Energy Proc. 2015;79:567–75.
16. Crites RW, Gunther DC, Kruzic AP, Pelz JD, Tchobanoglous G. Design manual—constructed wetlands and aquatic plant systems for municipal water treatment. United States Environmental Protection Agency, Hydrik Wetlands Consultants; 1998.
17. Ainsworth D, Hedlund J. Wetlands and ecosystem services, in world wetlands day—wetlands for our future, C.o.B. Diversity, Editor; 2015.
18. EPA. Why are wetlands important? 2018 [19.08.2019]. https://www.epa.gov/wetlands/why-are-wetlands-important.
19. EPA. What is a wetland? 2018 [19.08.2019]. https://www.epa.gov/wetlands/what-wetland.
20. UNEP-DHI, UNEP-DTU, CTCN. Natural wetlands. In Climate change adaptation technologies for water—a practitioner's guide to adaptation technologies for increased water sector resilience; 2017.
21. EPA. Guiding principles for constructed treatment wetlands: providing for water quality and wildlife habitat, ed. O.a.W. Office of Wetlands. United States Environmental Protection Agency; 2000.
22. Vymazal J. Constructed wetlands for wastewater treatment. Water. 2010;2:530–49.
23. Tousignant E, Fankhauser O, Hurd S. Guidance manual for the design, construction and operations of constructed wetlands for rural applications in Ontario; 1999.
24. Hayes TD, Isaacson HR, Reddy KR, Chynoweth DP, Biljetina R. Water Hyacinth systems for water treatment. Aquatic plants for water treatment and resource recovery, ed. Reddy KR, Smith WH; 1987.
25. Gearheart R, Finney B, Lang M, Anderson J, Lagacé S. Free water surface wetlands for wastewater treatment: a technology assessment. Environmental Resources Engineering Department, CH2M-Hill, Wetland Management Services; 1999.

26. Healy MG, Newell J, Rodgers M. Harvesting effects on biomass and nutrient retention in Phragmites australis in a free-water surface constructed wetland in western Ireland. Biol Environ Proc R Irish Acad. 2007;107B(3):139–45.
27. Yang Z, Wang Q, Zhang J, Xie H, Feng S. Effect of plant harvesting on the performance of constructed wetlands during summer. Water. 2016;8(24).
28. Ding H, Ding Y, Bai S. Emerging organic contaminant removal in constructed wetlands. In 6th international conference on energy and environmental protection. Advances in engineering research, vol 143, pp 451–454. Atlantis Press; 2017.
29. Matamoros V, Bayona JM. Behavior of emerging pollutants in constructed wetlands. In: Hutzinger O, Barceló D, Kostianoy A (eds.) The handbook of environmental chemistry. Volume 5—Water pollution part s/2: emerging contaminants from industrial and municipal waste—removal technologies. Barceló D, Petrovic M (Volume Editors). Springer, Berlin, Heidelberg; 2008. pp. 199–217.
30. Masi F, Conte G, Lepri L, Martellini T, Bubba MD. Endocrine disrupting chemicals (EDCs) and pathogens removal in an hybrid CW system for a tourist facility wastewater treatment and reuse. ResearchGate; 2004.
31. EAWAG, B. Stauffer, Free-water surface CW, s.i.g. Swiss Federal Institute of Aquatic Science and Technology, Editor. Sustainable Sanitation and Water Management Toolbox; 2019.
32. Zhang DQ, Jinadasa KBSN, Gersberg RM, Liu Y, Ng WJ, Tan SK. Application of constructed wetlands for wastewater treatment in developing countries—a review of recent developments (2000–2013). J Environ Manage. 2014;141:116–31.
33. EPA, Wastewater technology fact sheet—free water surface wetlands. United States Environmental Protection Agency; 2000.
34. Blumberg-engineers. Constructed wetlands for wastewater treatment. 2021. https://www.blumberg-engineers.com/en/ecotechnologies/constructed-wetlands.
35. Eawag B Stauffer D Spuhler. Vertical Flow CW. 2019 [30.08.2019]. https://sswm.info/factsheet/vertical-flow-cw.
36. Maupin AJ Constructed wetlands. In Idaho water reuse conference. Idaho; 2011.
37. EPA. Constructed wetlands treatment of municipal wastewaters. In: Manual. U.S. Environmental Protection Agency; 1999.
38. Bendoricchio G, Cin LD, Persson J. Guidelines for free water surface wetland design. EcoSys. 2000;8:51–91.
39. Economopoulou MA, Tsihrintzis VA. Design methodology of free water surface constructed wetlands. Water Resour Manage. 2004;18:541–65.
40. Li X, Ding A, Zheng L, Anderson BC, Kong L, Wu A, Xing L. Relationship between design parameters and removal efficiency for constructed wetlands in China. Ecol Eng. 2018;123:135–40.
41. Rühmland S. Technische Feuchtgebiete zur Nachreinigung von Abwasser—Stickstoff, Abwasserdesinfektion, Spurenstoffe, in Fakultät VI—Planen Bauen Umwelt. Technischen Universität Berlin; 2015.
42. Novotny V, Ahern J, Brown P. Water centric sustainable communities: planning, retrofitting, and building the next urban environment; 2010.
43. Bakhshoodeh R, Alavi N, Oldham C, Santos RM, Babaei AA, Vymazal J, Paydary P. Constructed wetlands for landfill leachate treatment: a review. Ecol Eng. 2020;146.
44. Zheng L, Liu T, Xie E, Liu M, Ding A, Zhang B-T, Li X, Zhang D. Partition and fate of phthalate acid esters (PAEs) in a full-scale horizontal subsurface flow constructed wetland treating polluted river water. Water. 2020;12(3):865.
45. Chen ZM, Chen B, Zhou JB, Li Z, Zhou Y, Xi XR, Lin C, Chen GQ. A vertical subsurface-flow constructed wetland in Beijing. Commun Nonlinear Sci Numer Simul. 2008;13:1986–97.
46. Zheng Y, Wang X, Xiong J, Liu Y, Zhao Y. Hybrid constructed wetlands for highly polluted river water treatment and comparison of surface- and subsurface-flow cells. J Environ Sci. 2014;26(4):749–56.
47. Saeed T, Paul B, Afrin R, Al-Muyeed A, Sun G. Floating constructed wetland for the treatment of polluted river water: a pilot scale study on seasonal variation and shock load. Chem Eng J. 2016;287:62–73.
48. Kasak K, Kill K, Pärn J, Mander Ü. Efficiency of a newly established in-stream constructed wetland treating diffuse agricultural pollution. Ecol Eng. 2018;119:1–7.
49. Kill K, Pärn J, Lust R, Mander Ü, Kasak K. Treatment efficiency of diffuse agricultural pollution in a constructed wetland impacted by groundwater seepage. Water. 2018;10(11).
50. Stefani GD, Tocchetto D, Salvato M, Borin M. Performance of a floating treatment wetland for in-stream water amelioration in NE Italy. Hydrobiologia. 2011;674(1):157–67.
51. Wu H, Zhang J, Wei R, Liang S, Li C, Xie H. Nitrogen transformations and balance in constructed wetlands for slightly polluted river water treatment using different macrophytes. Environ Sci Pollut Res. 2013;20:443–51.
52. Bass KL. Evaluation of a small in-stream constructed wetland in North Carolina's coastal plain. North Carolina State University; 2000.
53. Saeed T, Majed N, Khan T, Mallika H. Two-stage constructed wetland systems for polluted surface water treatment. J Environ Manage. 2019;249(109379).
54. Schulz R, Peall SKC. Effectiveness of a constructed wetland for retention of nonpoint-source pesticide pollution in the Lourens river catchment, South Africa. Environ Sci Technol. 2001;35(2):422–6.

55. Juang DF, Chen PC. Treatment of polluted river water by a new constructed wetland. Int J Environ Sci Technol. 2007;4(4):481–8.
56. Tang X, Huang S, Scholz M, Li J. Nutrient removal in pilot-scale constructed wetlands treating eutrophic river water: assessment of plants, intermittent artificial aeration and polyhedron hollow polypropylene Balls. Water Air Soil Pollut. 2009;197:61–73.
57. Zhao J, Zhu W, Lianfang Z. Efficiency and mechanism of treating the polluted river water with constructed wetland. J Lake Sci. 2007;19(1):32–8.
58. Huang Y-P, Kuo W-C, Lee C-H, Ting C-S, Hsieh H-H. River water quality improvement using a large-scale constructed wetland in southern Taiwan. Environ Eng Manag. 2007;17:277–81.
59. Bai X, Zhu X, Jiang H, Wang Z, He C, Sheng L, Zhuang J. Purification effect of sequential constructed wetland for the polluted water in urban river. Water. 2020;12(4):1054.
60. Domscheit E. Contribution to a pre-feasibility study of a constructed wetland system for the treatment of the Hamdoun river water, Tunisia. In Leichtweiß-Institut für Wasserbau, Technical University Braunschweig; 2019.
61. I2E. Inventaire et caracterisation des sources de pollution, in Etude de depollution de l'Oued Hamdoun (Gouvernorat de Sousse et de Monastir). Societe d'ingenerie de l'environnement et de l'énergie SA; Direction Générale de l'Environnement et de la Qaulité de la Vie. 2008. p. 148.
62. ATV-DVWK. Vereinheitlichung und Herleitung von Bemessungswerten für Abwasseranlagen. In ATV-DVWK-A 198. 2003.
63. Domscheit E. Contribution to a pre-feasibility study of a constructed wetland system for the treatment of the Hamdoun River water, Tunisia. In Water perspectives in emerging countries, water resources and climate change—impacts, mitigation and adaptation. Amman, Jordan: Cuvillier, Göttingen; 2019.

11 An Overview of Process and Technologies for Industrial Wastewater and Landfill Leachate Treatment

Marcelo A. Nolasco, Gabriela Ribeiro L. da Silva, and Vitor Cano

Abstract

Industrial development and high urbanization are responsible for several environmental problems, as the effluents generated cause disturbances to the ecosystems and risks to people's health due to the release of pollutants that are not properly treated. Based on the nature of wastewater, quantitative and qualitative aspects, different types of technologies or combinations of them are necessary and should be used before final disposal. Thus, to address the wastewater (industrial and landfill leachate) is fundamental to design a suitable treatment process. The combination of different processes and technologies in a general manner can provide advantages over a single technology or a single process itself. To ensure the safety, efficacy and quality of the treated wastewater, laboratory and pilot scale tests should be deeply explored in order to improve the performance of the process already applied at full scale or for the development of a new treatment system.

Keywords

Bioprocess engineering · Biological wastewater treatment · Physicochemical treatment

11.1 Introduction

Fast urbanization growth rates and the increase of resources demand for the global population needs are responsible for the most of environmental problems, causing ecosystems degradation and an increase of global risks [1]. The significant increase in the volume of industrial wastewaters and leachates from landfills puts the environment's self-cleaning capacity at risk. Due to the need to treat wastewater before disposal, various methods and technologies were developed, such as biological treatment, natural systems, and physicochemical process [2]. Since the 1980s, several physical, chemical, and biological technologies have been reported such as flotation, precipitation, oxidation, air stripping, adsorption, ion exchange, membrane filtration, electrochemically assisted process, biodegradation, and phytoremediation [3, 4].

So far, there is no direct answer regarding the best methods, because each treatment has its own

M. A. Nolasco (✉) · V. Cano
Laboratory of Sanitation and Environmental Technology, University of São Paulo, Sao Paulo, Brazil
e-mail: mnolasco@usp.br

V. Cano
e-mail: vitorc@usp.br

G. R. L. da Silva
Faculty of Architecture and Urbanism, Federal University of Rio de Janeiro, Rio de Janeiro, Brazil

M. Bahadir and A. Haarstrick (eds.), *Water and Wastewater Management*,
Water and Wastewater Management, https://doi.org/10.1007/978-3-030-95288-4_11

advantages and disadvantages not only in terms of financial aspects but also of efficiency and feasibility for different types of wastewater and their environmental impacts (energy and carbon footprint). In general, pollutant removal is carried out by physical, chemical and biological process, usually using a combination of different methods and techniques to achieve the desired effluent quality[5–8].

11.2 Leachates from Landfills

The prioritization of urban solid waste disposal in landfills leads to an increase in the generation of highly polluted liquid material called percolate or leachate. Globally, 37% percent of solid waste is disposed of in some form of landfills [9–11]. Even after shutting down the landfill, organic waste continues to degrade over time, gas emission and leachate percolation continues to occur, contaminating air, soil, surface and groundwater [12, 13].

Landfill leachate is characterized by having high concentrations of recalcitrant and xenobiotic compounds, organic matter including humic and fulvic acids, ammoniac nitrogen, heavy metals, organochlorines and inorganic salts [14]. Emerging and persistent compounds, such as endocrine disrupting chemicals, pharmaceuticals, and personal care products that show adverse effects on aquatic species in receiving waters and human health are also found in landfill leachates [15–17]. However, the characteristics of the leachate is subject to the following variables: characteristics and depth of landfilled waste, degree of decomposition, climate conditions, season of the year, age of the landfill, type of landfill operation, and landfill location. All these must be taken into account, when choosing a system for leachate treatment [18].

The variation of characteristics is mainly related to the landfill's age, which occurs due to the phenomena of waste digestion during the aerobic stages, followed by the anaerobic ones with the acetogenesis and methanogenesis, until reaching the stabilization of the compounds present in the leachate. Thus, both pH and high molecular weight recalcitrant compounds progressively increase, and consequently the concentrations biodegradable compounds decrease [18, 19]. Due to the leachate characteristics, removal of organic material and ammonia nitrogen along with other toxic compounds are the main treatment goals [20]. Knowledge about the biodegradability level of the leachate supports the choice of treatment technology and processes.

The biodegradability of landfill leachate is quantified through the BOD (biological oxygen demand), COD (chemical oxygen demand) and their ratio. A broad range of BOD/COD ratio between 0.05 and 0.7 is found for landfill leachates, while lower values indicate high proportion of recalcitrant compounds (typical for old landfills). Recalcitrant organic matter consists of compounds resistant to biodegradation, and they usually persist and accumulate in the environment, but they are not necessarily toxic to microorganisms, such as humic and fulvic acids [21–26]. Furthermore, the COD/TOC ratio (total organic carbon) indicates the average oxidation state of carbon in organic compounds. In landfill leachates, it tends to decrease as the age of the different landfills increase. Higher COD/TOC ratios around 4 are observed for young landfills, while ratios around 2.7 are found for old landfills, indicating higher oxidization state of the organic carbon, meaning less readily available energy for microbial growth [27–29].

It is well known that technologies based solely on biological treatment are not successful options for the efficient removal of recalcitrant organic matter from landfill leachates. [30]. Thus, specific treatment goals and conditions of the landfill leachate must be considered to avoid problems, such as high maintenance and operation costs and low efficiencies [12].

11.3 Industrial Wastewater

The use of water by industry can occur in different ways, such as (i) washing machines, pipes, and floors, (ii) water for cooling systems and steam generators, (iii) water used directly in the stages of the industrial process, and (iv) for sanitary services of employees. Except for the volumes of water incorporated into the products and the evaporation losses, the wastewater may offer risks to the environment due to contamination by residues from the industrial processes or due to high temperature. For these reasons, there is no need to emphasise that industrial wastewater must be treated in advance at all costs before it is discharged into a municipal wastewater treatment plant or even untreated into the environment in order to prevent irreparable damage to aquatic systems [1].

Water pollution can be defined as any physical, chemical or biological alteration of a water body, capable of exceeding the standards established for that water body [31]. Traditionally, the main groups to be considered are physical materials (solids in suspension) or forms of energy (calorific and radiation), chemical materials (dissolved substances or substances with potential solubilization) and biological material (flora and fauna) [2].

Pollution originates from losses of energy, products, and raw materials, and due to inefficiency of industrial processes. The use of end-of-pipe techniques is the worst ones of all solutions and, therefore, also no longer up-to-date. Industry is challenged to make industrial wastewater treatment more efficient and to expose it to modern processes. This is not least a competitive issue, where improving production efficiency is also a matter of survival on the market. Industrial efficiency is also the first step towards environmental efficiency. Pollution by industrial liquid effluents must be initially controlled by reducing losses in processes, including the use of more modern processes, optimized general arrangements, reduction of water consumption including washing of industrial equipment and floors, reduction of product losses and discharging these or raw materials into the collecting network. Maintenance is also critical to reducing losses from leaks and waste power. In addition to verifying the efficiency of the process, one must question whether this is the most modern, considering its technical and economic feasibility.

11.4 Treatment Technologies for Leachates and Industrial Wastewaters

Many technologies are available for the treatment of landfill leachates and industrial wastewaters, including conventional/consolidated ones, emerging technologies and hybrid processes. These technologies are based on biological processes, chemical processes, separation techniques and some hybrid processes, such as nature-based systems and bioelectrochemical technologies. The decision for the most suitable technology should consider, among other things, the relationship between technical and economical feasibilities, volume and composition of the wastewater and permits as well as regulation and operating capacity.

11.4.1 Biotechnological Processes

In the biological processes, biodegradation of compounds occurs due to the action of microorganisms that degrade organic compounds into carbon dioxide (CO_2) in aerobic processes. Under anaerobic conditions, biogas is generated, which is a mixture that mainly comprises methane (CH_4), CO_2 and hydrogen sulphide (H_2S) [32–35].

Biological processes show good efficiency in removing carbonaceous organic matter in wastewater, when BOD_5/COD ratio is greater than 0.5, for both leachate and industrial wastewater [14]. In addition, microorganisms can remove also nitrogen based on nitrification–denitrification, commonly processes in the elimination of nitrogenous compounds from wastewaters. However, biological treatments are vulnerable and hampered by toxic substances and/or the presence of refractory organic compounds [21].

11.4.1.1 Moving Bed Biofilm Reactor (MBBR)

The moving bed biofilm reactor (MBBR) is a biological technology aiming at the development of compact and highly efficient wastewater treatment plants (WWTP), combining the processes based on the use of attached and suspended biomass [36–38]. Its operation is characterized using suspended porous polymers with densities lower than water. They are kept in free and continuous movement within a tank, normally aerated, with biofilm growth on their surface [30, 39, 40]. MBBR can be deployed either as a new standalone WWTP or as adaptation on an activated sludge system already in operation [41].

The concept of the system is based on the use of the support medium to create a surface area for the development of adhered biomass and consequently increase in sludge retention time. The biomass increase provided by the support medium allows for increased decomposition of carbonaceous organic matter and transformation of nitrogenous compounds, reducing the volume needed for treatment [38, 41]. The agitation of the support medium increases the exposure and contact of the biofilm with the suspended liquid mass [41]. In the case of aerobic systems, the support material is kept in motion by the flow of air injected into the tank, while in the case of anaerobic/anoxic systems a stirrer is used [38, 42]. An important feature of the support medium is its surface available for biofilm growth. The movement of carriers inside the reactor causes frequent collisions, resulting in loss of biomass adhered to the external face [38, 42].

One of the main reference parameters of the carriers is the specific surface area, which is given by the ratio between the total surface area and the volume [42, 43]. The importance of this parameter stems from the relationship it provides between the amount of support medium inside the reactor and the potential amount of adhered biomass [41]. Thus, the rate of degradation of organic matter and nitrification can proportionally be associated with the specific surface of the support medium used [22, 25]. However, for very high specific surface values, the gains in system performance are less accentuated since it is limited by mass transfer [23].

The MBBR is versatile and can be applied in different situations, such as aerobic, anaerobic or anoxic treatment, and with the objective of oxidizing carbonaceous matter, nitrifying or denitrifying [42] in the treatment of various types of wastewaters, including industrial effluents and landfill leachates [44–48]. In summary, the main advantages of the MBBR systems are [14, 44, 49]:

- The entire useful volume of the reactor is efficiently used for microbial growth;
- High interfacial area between biofilm and substrates;
- High resistance to overload;
- Operational flexibility;
- The treatment plant requires less area;
- Sludge recycling is not necessary to maintain the high biomass concentration;
- There is no need for backwashing, as there is no clogging;
- Short settling period and less sludge generation;
- Less sensitivity to toxic compounds;
- Possible occurrence of denitrification in anoxic zones in the deep layers of the biofilm;
- Operational stability.

However, the main disadvantages are associated with operating costs. The higher concentration of biomass in the reactor demands a large amount of dissolved oxygen (DO), in addition to the need for aeration to move the media, increasing expenses with aeration of the system due to the energy cost. In addition, the media used in the system has a high cost, which is an important issue in the final decision for treatment technologies [49, 50].

Despite their similarity to activated sludge systems, MBBR has specific characteristics of attached biomass reactors. In this sense, it is common to associate the organic load applied to the reactor with the total surface area of the carriers. The control parameter that best applies in this case is the superficial organic loading (SOL), Eq. 11.1, (Table 11.1) [51, 52], expressed as:

Table 11.1 Superficial organic loadings applied to MBBR

References	Type of wastewater	SOL (g/m^{2}d)	
		BOD	COD
Welander et al. [23]	Landfill leachate	–	1.13–3.96
Gaul et al. [47]	Sludge digester centrate	–	2.66–10.65
Luostarinen et al. [48]	Dairy industry	–	0.38
Oliveira et al. [51]	Sewage	6.4–9.6	–
Oliveira et al. [53]	Cellulose industry	43.8	–
Vanzetto [52]	Cellulose industry	–	2–60.4

$$\text{SOL}\left(\text{g BOD or COD/m}^2\right) = \frac{\text{BOD or COD applied load}\left(\text{g/d}^1\right)}{\text{Total surface of the medium }\left(\text{m}^2\right)} \quad (11.1)$$

11.4.1.2 Constructed Wetlands

The constructed treatment wetland, a nature-based solution, is an alternative engineering system, designed and constructed to use the natural functions of microbial populations, soil and vegetation to treat contaminants in surface, groundwater or wastewater [54–59]. The importance of vegetation in the system is due to complex chemical, physical and biological activity in the rhizosphere, which is the region influenced by the plant root [60]. The rhizosphere is an area densely populated by diverse organisms, in which plants compete for space and nutrients with microorganisms, insects and invading root systems from other plants [61].

Another important feature is the interaction of vegetation with the rhizosphere through decomposition processes and exudates [62, 63]. The release of organic matter can occur due to plant death, senescence processes and environmental conditions [64]. Wu et al. [62] observed a greater variety of organic acids and their release rates in plants exposed to conditions of low phosphorus availability and/or competition with other species. Thus, the organic component release mechanism can be applied to improve denitrification in the treatment of effluents with low biodegradable organic load, as in the case of landfill leachates and some types of industrial effluents [58, 59, 65, 66].

11.5 Physical and Chemical Treatment Technologies and Processes

11.5.1 Physicochemical Processes

These processes enable the removal of suspended solids, colloidal particles, floating material, colour and toxic compounds, having a pre-treatment character, sometimes increasing the biodegradability of the effluent or as a polish, providing efficiencies that were otherwise not achieved with biological treatments through processes such as flotation, coagulation/flocculation, adsorption, chemical oxidation and counter current gas desorption (air stripping). Physical–chemical treatments for leachates are used in addition to conventional biological treatment or in order to treat a specific substance and usually to achieve greater efficiencies against both isolates [14, 67, 68].

11.5.1.1 Flotation

The flotation process makes the density of solid particles less than that of water by agglomerating the particles, forming gas bubbles, and fixing the solid particles inside the formed bubbles. In dissolved air flotation, bubbles are produced by reducing the pressure in a stream of water saturated with air. It has been used for the treatment of wastewater with considerable concentrations of oils and greases as it is a well efficient process to separate low density particles in the form of flocs [2].

Flotation systems have their efficiency enhanced, when a physicochemical process of coagulation and flotation is applied before entering the floater. This occurs due to the formation of flakes with greater capacity to accumulate suspended solids and separating water present in the effluent from solid particles more efficiently [69].

11.5.1.2 Coagulation and Flocculation

The coagulation process associated with flocculation in wastewater treatment provides the removal of organic and inorganic pollutants, insoluble material, heavy metals, non-biodegradable organic matter, suspended solids, and colour, among others. It has the property of conferring primary character, in many cases preceding biological purification treatment with the purpose of reducing the dimensions of effluent treatment plants [70].

Coagulants cause a reduction in the repulsion potential between the colloids present in the effluent, that is, they destabilize the colloidal particles resulting in the formation of microparticles, usually followed by flocculation of these unstable particles which, when colliding with each other, form larger structures called flocs, promoting removal of suspended solids and colloidal particles as they group more easily [67, 71]. The control of pH in the process is essential, since the coagulants react with the alkalinity of the effluent, forming hydroxides that destabilize colloids due to the reduction of the zeta potential, tending to zero, that is, the isoelectric point. Iron (III) salts used as coagulants dissolve in water. During the dissolution process, several soluble complexes with high positive charges are formed, which are adsorbed on the surface of negative colloids [71].

Coagulation and flocculation depend on several factors to generate and to maintain process efficiencies, namely the choice to use the process as a pre- or post-treatment step, the nature and dose of the coagulant used, the pH of coagulation, time and speed gradient of rapid mixing and flocculation, leachate characteristics, for example age and pH, among others [67, 72].

In relation to other coagulants, those which are based on aluminum sulphate or ferric chloride are more efficient in removing organic matter and have little sensitivity to temperature changes [71]. In the treatment of landfill leachates, the removal of organic substances is mainly based on the precipitation (coagulation) of organic compounds that are difficult to biodegrade (e.g., humic acids). The bivalent and trivalent cations such as Fe^{2+} and Fe^{3+} interact with the humic acids and form complexes that lose their dissolving capability, leading to precipitation of these compounds as their molecular weight increases [67].

11.5.1.3 Adsorption

In the adsorption process, the transfer of substances from the liquid or gas to the solid phase occurs. The adsorbed substance is the one to be removed from the liquid or gaseous phase, and the adsorbent is, where the adsorbed substances accumulate [2]. Adsorption treatment is commonly used after biological treatment for effluent polishing purposes to remove the non-biodegradable compounds that are recalcitrant, and residual inorganic compounds such heavy metals, and substances that cause odour [4].

11.5.1.4 Air Stripping

Air stripping consists of the removal of volatile compounds from the effluent through the passage of atmospheric air through the liquid. This phenomenon usually occurs at high pH, favouring the transformation of the dissociated ammonium ion (NH_4^+) into free ammonia (NH_3). It provides the transfer of phases from liquid to gas and finally the release into the atmosphere [73–75].

11.5.1.5 Membrane Processes

Membrane filtration processes are often used after biological and/or conventional chemical treatments. Membranes aims to remove fine particulate matter, pathogens, recalcitrant compounds from organic matter, nutrients and dissolved substances that are not removed by conventional treatment processes [2].

Regarding the operational concepts, such as during treatment, the generation of permeate occurs, a portion of the influent that passes through the membranes and is filtered, and of concentrate, which is the portion of the influent that is rejected in the process and returns to the beginning of the system. There is also a difference between the Cross-flow and Dead-end systems. In the first, the concentrate is generated as described above, and in the second, the influent is filtered without the generation of concentrate [76].

Another recurrent concept when it comes to membranes is flow, which is the unit that measures how much of the influent passes through the membrane per unit of time and area, thus expressing the rate of permeate production. The flow rate varies according to the characteristics of the effluent to be treated, the possible pre-treatment applied, the feed rate and the characteristics of the membrane system [77, 78].

Membrane treatment processes for wastewater treatment purposes are: Microfiltration (particle sizes 0.02–2.0 mm), Ultrafiltration (0.002–0.2 m), Nanofiltration (0.001–0.02 mm), and Reverse Osmosis (0.00001–0.005 mm) [2]. Among the recalcitrant compounds known to be present in some industrial effluents and landfill leachate, the ultrafiltration process can remove part of the humic acids, nanofiltration reaches a portion both fulvic and humic acids, and only reverse osmosis can remove both compounds that impart colour and recalcitrance.

The application of coagulation/flocculation process prior to the ultrafiltration membranes helps to reduce the fouling in the membranes and can increase the final efficiency, while the ultrafiltration process is able to remove some hazardous substances that are not completely removed by traditional physicochemical processes [18, 78]. According to Marañón [79], through the process of coagulation and flocculation, it is possible to remove recalcitrant organic matter without pH adjustment.

11.5.2 Chemical Oxidation

In this process, oxidizing chemical compounds that react with the substances present in the effluent are added into the process, promoting the degradation of synthetic toxic substances or those resulting from anthropogenic activities. Advanced oxidation processes cause the formation of hydroxyl radicals (HO^-), thus aiming at its reaction with the substances present in the effluent, resulting in a more powerful oxidation than conventional oxidizing chemical compounds [2].

Advanced oxidation processes are quite efficient and enable a high degradation rate. That is one of the main reasons, why they have become feasible processes in landfill leachate and industrial wastewater treatment. AOPs provide the ability to eliminate colour, to reduce organic load and to increase the biodegradability of recalcitrant compounds from those wastewaters [80].

11.5.3 Bio-Electrochemical Process

A bio-electrochemical system with enzymes and microorganisms acting as biocatalysts can convert—based on the synergy between microbial metabolism and a solid electron acceptor—the chemical energy of a biodegradable substrate into electrical energy [81–84]. One of the best-known examples of bio-electrochemical process is the microbial fuel cell (MFC), also called biofuel cell. In the MFC, biochemical reactions are catalyzed by bacteria on the surface of an electrode, called anode, under anaerobic condition, producing protons and electrons from the degradation of organic or inorganic substrates [85]. A typical MFC consists of an anaerobic anode, oxidizing organic matter, and an aerobic cathode, separated by a proton transfer system [86].

Electrons produced by the oxidation of the organic substrate migrate from the anode to the cathode through an external circuit, generating electrical current. The protons migrate from the anode to the cathode through a proton transfer system (usually a proton exchange membrane,

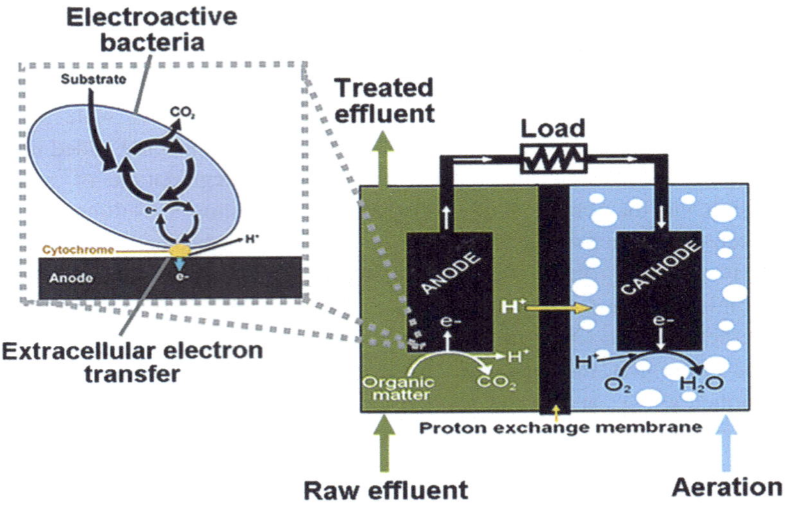

Fig. 11.1 Diagram of a typical double chamber MFC, including the exoelectrogen bacteria in the anode chamber and an aerated cathode chamber separated by a proton exchange membrane

PEM), where they are consumed by an electron acceptor (usually oxygen) in a reduction reaction, thereby closing the loop [87, 88]. The electron flow and the potential difference between the respiratory enzymes of anodic microorganisms and the oxygen reduction reaction (or another electron acceptor) at the cathode generate the current and voltage respectively [85]. Figure 11.1 schematically illustrates a typical MFC, consisting of an anodic chamber and a cathodic chamber, separated by a PEM.

Due to its operating mechanism, the MFC has important advantages over conventional technologies used for wastewater treatment and/or energy generation from organic matter [82, 85, 89]. Some MFC characteristics are (i) direct conversion of the organic substrate into the electrical energy without intermediate steps, avoiding losses and enabling greater efficiency, (ii) stable operation in different temperature ranges, (iii) does not demand biogas treatment, since under ideal conditions the final gaseous product is mainly composed of CO_2, (iv) there is no need for additional energy input with aeration for oxidation of organic matter, since a number of electron acceptors can be used, and (v) low sludge production rate, reducing the costs of its treatment and final disposal.

The energy generation in the MFC depends on microorganisms present in the anaerobic anodic compartment. During the metabolization of organic matter, they generate the electrons and protons used in the cathodic compartment. These microorganisms are called exoelectrogen (anodophilic) microorganisms due to their ability to transfer electrons outside their membranes to the surface of an electrode or a soluble or insoluble electron acceptor (minerals) [86].

The traditional electroactive microorganisms consume organic substrates and produce energy that is stored intracellularly in the form of NADH. The growth and maintenance of these microorganisms is possible by the energy recovered from the potential difference between electron donor and electron acceptor [82].

The final products of the biochemical reactions carried out by electrogens in an MFC are H_2O and CO_2, with no considerable contribution to global warming [82]. The half redox reactions that characterize the bio-electrochemical system are presented here, considering glucose as the organic substrate [90]:

$$C_6H_{12}O_6 + 6H_2O \rightarrow 6CO_2 + 24H^+ + 24e^-$$
$$E^\circ = -0.43\text{V vs. SHE at pH7}$$
$$O_2 + 4H^+ + 4e^- \rightarrow 2H_2O$$
$$E^\circ = 0.82\text{V vs. SHE at pH7}$$

Oxygen is the most widely used electron acceptor for cathodic reactions due to its easy availability, high redox potential, non-toxicity, and non-generation of greenhouse gases as a by-product of the reduction reaction [85, 90]. Oxygen reduction provides a standard cathodic potential of +0.82 V versus Standard Hydrogen Electrode (SHE) under typical operating conditions [86].

The anode must have a greater potential than the NADH to allow the transfer of electrons from the microorganism to the electrode. As the redox potential of NADH is −0.32 V versus SHE, a potential difference of less than 1.14 V (considering O_2 as an electron acceptor) is expected between the two MFC electrodes, regardless of the oxidized substrate [82].

The main group of bacteria belongs to the electrogens, which can use insoluble or solid-state electron acceptors with a mechanism to transport electrons out of the cell toward the electron acceptor, entitled extracellular electron transfer (EET) [86]. The EET can be carried out by two mechanisms: (i) direct and (ii) mediated electron transfer.

For the direct electron transfer, a direct contact between microbial cell and the solid-state electron acceptor is established by membrane redox proteins and/or cell appendages. It is reported that a large number of microorganisms are able to use the anode as an electron acceptor, but for most of them it remains unclear, how electrons are transported out of their cells in the direction of the electrode [91].

11.6 Conclusions

The aims of wastewater management are the protection of the environment, the public health and address the socio-economic demands. Based on the nature of wastewater in terms of quantitative and qualitative aspects, different types of technologies or combinations are feasible and should be used before the final disposal. Thus, to understand the nature of wastewater (leachates and industrial wastewater) is fundamental to design suitable treatment process, but also, local climate, topography, and sustainability criteria such as carbon footprint are important and recent elements for decision making. The combination of different process and technologies can provide advantages over a solo technology or a process itself. To ensure environmental security and acceptable quality of treated wastewater, laboratory and pilot scale tests should continue to be studied in depth in order to improve cleaning performance, but also to explore new efficient and safe treatment systems.

References

1. Ranade VV, Bhandari VM. Chapter 1—Industrial wastewater treatment, recycling, and reuse: an overview. In: Ranade VV, Bhandari VM (eds.) Industrial wastewater treatment, recycling and reuse. Oxford: Butterworth-Heinemann; 2014. p. 1–80. https://doi.org/10.1016/B978-0-08-099968-5.00001-5.
2. Metcalf & Eddy Inc an AC, Asano T, Burton F, Leverenz H. Water reuse: issues, technologies, and applications. New York: McGraw-Hill Education; 2007.
3. Rathoure AK, Dhatwalia VK. Toxicity and waste management using bioremediation. IGI Global; 2016.
4. Crini G, Lichtfouse E. Wastewater treatment: an overview. In: Crini G, Lichtfouse E (eds.) Green adsorbents for pollutant removal: fundamentals and design. Cham: Springer International Publishing; 2018. p. 1–21. https://doi.org/10.1007/978-3-319-92111-2_1.
5. Barbosa SA (2006) Avaliação de biofiltro aerado submerso no pós-tratamento de efluente de tanque séptico. Dissertação (Mestrado em Engenharia de Recursos Hídricos e Ambiental). Universidade Federal do Paraná
6. Buzzini AP, Nolasco MA, Springer AM, Pires EC (2006) Evaluation of aerobic and anaerobic treatment of Kraft pulp mill effluent for organochlorines removal. Water Practice Technol 1:1–8. https://doi.org/10.2166/wpt.2006.068
7. Ribeiro EN, de Sousa WC, de Julio M, Irrazabal WU, Nolasco MA (2013) Airports and environment: proposal of wastewater reclamation at Sao Paulo International Airport. Clean-Soil Air Water 41:627–34. https://doi.org/10.1002/clen.201100682
8. Campos F, Nolasco, MA (2021) Prospecção Científica e Tecnológica Aplicada ao Conceito de Estações de Tratamento de Esgoto Sustentáveis. Cadernos De Prospecção 14(3):964. https://doi.org/10.9771/cp.v14i3.37258
9. Clarke BO, Anumol T, Barlaz M, Snyder SA. Investigating landfill leachate as a source of trace organic pollutants. Chemosphere. 2015;127:269–75. https://doi.org/10.1016/j.chemosphere.2015.02.030.
10. Ilyas H, Masih I. The performance of the intensified constructed wetlands for organic matter and nitrogen removal: a review. J Environ Manage. 2017;198:372–83. https://doi.org/10.1016/j.jenvman.2017.04.098.

11. Lu M-C, Chen YY, Chiou M-R, Chen MY, Fan H-J. Occurrence and treatment efficiency of pharmaceuticals in landfill leachates. Waste Manage. 2016;55:257–64. https://doi.org/10.1016/j.wasman.2016.03.029.
12. Lebron YAR, Moreira VR, Brasil YL, Silva AFR, Santos LV de S, Lange LC et al. A survey on experiences in leachate treatment: common practices, differences worldwide and future perspectives. J Environ Manage. 2021;288:112475. https://doi.org/10.1016/j.jenvman.2021.112475.
13. Gouveia N, Prado RR do. Riscos à saúde em áreas próximas a aterros de resíduos sólidos urbanos. Revista de Saúde Pública. 2010;44:859–66. https://doi.org/10.1590/S0034-89102010005000029.
14. Renou S, Givaudan JG, Poulain S, Dirassouyan F, Moulin P. Landfill leachate treatment: review and opportunity. J Hazard Mater. 2008;150:468–93. https://doi.org/10.1016/j.jhazmat.2007.09.077.
15. Eggen T, Moeder M, Arukwe A. Municipal landfill leachates: a significant source for new and emerging pollutants. Sci Total Environ. 2010;408:5147–57. https://doi.org/10.1016/j.scitotenv.2010.07.049.
16. Tran NH, Reinhard M, Gin KY-H. Occurrence and fate of emerging contaminants in municipal wastewater treatment plants from different geographical regions—a review. Water Res. 2018;133:182–207. https://doi.org/10.1016/j.watres.2017.12.029.
17. Nivala J, Hoos MB, Cross C, Wallace S, Parkin G. Treatment of landfill leachate using an aerated, horizontal subsurface-flow constructed wetland. Sci Total Environ. 2007;380:19–27. https://doi.org/10.1016/j.scitotenv.2006.12.030.
18. Chang W-S, Chen S-S, Chang T-C, Nguyen N-T, Cheng H-H, Hsu H-T. Fouling potential and reclamation feasibility for a closed landfill leachate treated by various pretreatment processes on membrane system. Desalin Water Treat. 2015;55:3568–75. https://doi.org/10.1080/19443994.2014.946730.
19. Moreira CA, Braga AC de O. Anomalias de cargabilidade em aterro de resíduos sólidos domiciliares. Revista Brasileira de Geofísica. 2009;27:55–62. https://doi.org/10.1590/S0102-261X2009000100005.
20. Costa AM, Alfaia RG de SM, Campos JC. Landfill leachate treatment in Brazil—an overview. J Environ Manage. 2019;232:110–6. https://doi.org/10.1016/j.jenvman.2018.11.006.
21. Wiszniowski J, Robert D, Surmacz-Gorska J, Miksch K, Weber JV. Landfill leachate treatment methods: a review. Environ Chem Lett. 2006;4:51–61. https://doi.org/10.1007/s10311-005-0016-z.
22. Welander U, Henrysson T, Welander T. Nitrification of landfill leachate using suspended-carrier biofilm technology. Water Res. 1997;31:2351–5. https://doi.org/10.1016/S0043-1354(97)00080-8.
23. Welander U, Henrysson T, Welander T. Biological nitrogen removal from municipal landfill leachate in a pilot scale suspended carrier biofilm process. Water Res. 1998;32:1564–70. https://doi.org/10.1016/S0043-1354(97)00351-5.
24. Barr MJ, Robinson HD. Constructed wetlands for landfill leachate treatment. Waste Manage Res. 1999;17:498–504. https://doi.org/10.1034/j.1399-3070.1999.00075.x.
25. Loukidou MX, Zouboulis AI. Comparison of two biological treatment processes using attached- growth biomass for sanitary landfill leachate treatment. Environ Pollut. 2001;111:9.
26. Bulc TG. Long term performance of a constructed wetland for landfill leachate treatment. Ecol Eng. 2006;26:365–74. https://doi.org/10.1016/j.ecoleng.2006.01.003.
27. Öncü G, Reiser M, Kranert M. Aerobic in situ stabilization of Landfill Konstanz Dorfweiher: Leachate quality after 1 year of operation. Waste Manage. 2012;32:2374–84. https://doi.org/10.1016/j.wasman.2012.07.005.
28. Morello L, Raga R, Sgarbossa P, Rosson E, Cossu R. Storage potential and residual emissions from fresh and stabilized waste samples from a landfill simulation experiment. Waste Manage. 2018;75:372–83. https://doi.org/10.1016/j.wasman.2018.01.026.
29. Jiménez-Silva VA, Santoyo-Tepole F, Ruiz-Ordaz N, Galíndez-Mayer J. Study of the ibuprofen impact on wastewater treatment mini-plants with bioaugmented sludge. Process Saf Environ Prot. 2019;123:140–9. https://doi.org/10.1016/j.psep.2018.08.006.
30. Abbas AA, Jingsong G, Ping LZ, Ya PY, Al-Rekabi WS. Review on landfill leachate treatments. Am J Appl Sci. 2009;6:672–84. https://doi.org/10.3844/ajassp.2009.672.684.
31. (UNESCO) United Nations Educational S and CO, Programme (WWAP) World Water Assesment, Water UN, United Nations Educational, Scientific and Cultural Organization (UNESCO). Water in a Changing World (WWDR-3): the 3rd United Nations World Water Development Report. Geneva: UNESCO; 2009.
32. Carvalho T, Nolasco MA (2006) Créditos de carbono e geração de energia com uso de biodigestores no tratamento de dejetos suínos. Revista Acadêmica, Ciências Agrárias e Ambientais 3:23–32. https://doi.org/10.7213/cienciaanimal.v4i3.9405
33. Aisse MM, Nolasco MA, Andreoli FDN, Lobato MB, Savelli CS, Jurgensen D, Alem Sobrinho P (2000) Pós-tratamento de efluentes provenientes de reatores anaeróbios tipo UAS. In: Proceedings of the VI Latin-American Workshop and Seminar on Anaerobic Digestion, Recife, Brazil, pp 21–327
34. Nolasco MA, Baggio RB, Griebeler J (2005) Implicações ambientais e qualidade da água da produção animal intensiva. Revista Acadêmica Ciência Animal 3(2):19–26. https://doi.org/10.7213/cienciaanimal.v3i2.9081
35. Nolasco MA, Campos ALO, Springer AM, Pires EC (2002) Use of lysis and recycle to control excess sludge production inactivated sludge treatment: bench scale study and effect of chlorinated organic compounds. Water Sci Technol 10:55–61. https://doi.org/10.2166/wst.2002.0289

36. Ødegaard H, Rusten B, Westrum T. A new moving bed biofilm reactor—applications and results. Water Sci Technol. 1994;29:157–65. https://doi.org/10.2166/wst.1994.0757.
37. Ødegaard H. The moving bed biofilm reactor. Hokkaido Press; 1999.
38. Rusten B, Eikebrokk B, Ulgenes Y, Lygren E. Design and operations of the Kaldnes moving bed biofilm reactors. Aquacult Eng. 2006;34:322–31. https://doi.org/10.1016/j.aquaeng.2005.04.002.
39. Ciesielski S. Characterization of bacterial structures in two-stage moving-bed biofilm reactor (MBBR) during nitrification of the landfill leachate. J Microbiol Biotechnol. 2010;20:1140–51. https://doi.org/10.4014/jmb.1001.01015.
40. Oliveira ACDG, Blaich CI, Santana DDLSV, Prates K. NMP de bactérias nitrificantes e desnitrificantes e sua relação com os parâmetros físico-químicos em lodo ativado para remoção biológica de nitrogênio de lixiviado de aterro sanitário. Revista DAE. 2013;61:60–9. https://doi.org/10.4322/dae.2014.107.
41. Oliveira DVMD. Caracterização dos parâmetros de controle e avaliação de desempenho de um reator biológico com leito móvel (MBBR). Dissertação (Mestrado). Universidade Federal do Rio de Janeiro; 2008.
42. Rodgers M, Zhan X-M. Moving-medium biofilm reactors. Re/Views Environ Sci Bio/Technol. 2003;2:213–24. https://doi.org/10.1023/B:RESB.0000040467.78748.1e.
43. Aygun A, Nas B, Berktay A. Influence of high organic loading rates on COD removal and sludge production in moving bed biofilm reactor. Environ Eng Sci. 2008;25:1311–6. https://doi.org/10.1089/ees.2007.0071.
44. Chen S, Sun D, Chung J-S. Simultaneous removal of COD and ammonium from landfill leachate using an anaerobic–aerobic moving-bed biofilm reactor system. Waste Manage. 2008;28:339–46. https://doi.org/10.1016/j.wasman.2007.01.004.
45. Wang R-C, Wen X-H, Qian Y. Influence of carrier concentration on the performance and microbial characteristics of a suspended carrier biofilm reactor. Process Biochem. 2005;40:2992–3001. https://doi.org/10.1016/j.procbio.2005.02.024.
46. Jahren SJ, Rintala JA, Ødegaard H. Aerobic moving bed biofilm reactor treating thermomechanical pulping whitewater under thermophilic conditions. Water Res. 2002;36:1067–75. https://doi.org/10.1016/S0043-1354(01)00311-6.
47. Gaul T, Märker S, Kunst S. Start-up of moving bed biofilm reactors for deammonification: the role of hydraulic retention time, alkalinity and oxygen supply. Water Sci Technol. 2005;52:127–33. https://doi.org/10.2166/wst.2005.0191.
48. Luostarinen S, Luste S, Valentín L, Rintala J. Nitrogen removal from on-site treated anaerobic effluents using intermittently aerated moving bed biofilm reactors at low temperatures. Water Res. 2006;40:1607–15. https://doi.org/10.1016/j.watres.2006.02.022.
49. Nocko LM. Remoção de carbono e nitrogênio em reator de leito móvel submetido à aeração intermitente. Dissertação (Mestrado). Universidade de São Paulo; 2008. https://doi.org/10.11606/D.18.2008.tde-11022009-173925.
50. de Oliveira DVM, Volschan I, Piveli RP. Avaliação comparativa entre custos dos processos MBBR/IFAS e lodo ativado para o tratamento de esgoto sanitário. Revista DAE. 2013;61:46–55. https://doi.org/10.4322/dae.2014.110.
51. Minegatti de Oliveira DV, Volschan Junior I, Pacheco Jordão E. Comportamento e desempenho do processo reator biológico com leito móvel (MBBR) para a remoção da matéria orgânica e compostos nitrogenados. Revista AIDIS de Ingeniería y Ciencias Ambientales Investigación, Desarrollo y Práctica. 2011;4:12–26. https://doi.org/10.22201/iingen.0718378xe.2011.4.1.26008.
52. Vanzetto SC. Estudos de viabilidade de tratamento de efluente de indústria de celulose kraft por reator biológico com leito móvel (MBBR). Dissertação (Mestrado). Universidade Tecnológica Federal do Paraná; 2012.
53. Oliveira DVM de, Filho AC de O, Rabelo MD, Nariyosh YN. Avaliação de uma Planta Piloto de MBBR (Moving Bed Biofilm Reactor—Reator Biológico com Leito Móvel) para Tratamento de Efluente de uma Fábrica de Celulose e Papel. O Papel. 2012;73:75–80.
54. ITRC. Technical and regulatory guidance document for constructed treatment wetlands; 2003.
55. Valentim MAA. Desempenho de leitos cultivados ("constructed wetland") para tratamento de esgoto: contribuições para concepção e operação. Tese (doutorado). Universidade Estadual de Campinas; 2003.
56. Clarke E, Baldwin AH. Responses of wetland plants to ammonia and water level. Ecol Eng. 2002;18:257–64. https://doi.org/10.1016/S0925-8574(01)00080-5.
57. Akinbile CO, Yusoff MS, Ahmad Zuki AZ. Landfill leachate treatment using sub-surface flow constructed wetland by Cyperus haspan. Waste Manage. 2012;32:1387–93. https://doi.org/10.1016/j.wasman.2012.03.002.
58. Cano V, Vich DV, Rousseau DPL, Lens PNL, Nolasco MA. Influence of recirculation over COD and N-NH $_4$ removals from landfill leachate by horizontal flow constructed treatment wetland. Int J Phytorem. 2019;21:998–1004. https://doi.org/10.1080/15226514.2019.1594681.
59. Cano V, Vich DV, Andrade HHB, Salinas DTP, Nolasco MA. Nitrification in multistage horizontal flow treatment wetlands for landfill leachate treatment. Sci Total Environ. 2020;704:135376. https://doi.org/10.1016/j.scitotenv.2019.135376
60. Bais HP, Weir TL, Perry LG, Gilroy S, Vivanco JM. The role of root exudates in rhizosphere interactions with plants and other organisms. Annu Rev Plant

Biol. 2006;57:233–66. https://doi.org/10.1146/annurev.arplant.57.032905.105159.

61. Bais HP, Park S-W, Weir TL, Callaway RM, Vivanco JM. How plants communicate using the underground information superhighway. Trends Plant Sci. 2004;9:26–32. https://doi.org/10.1016/j.tplants.2003.11.008.
62. Wu FY, Chung AKC, Tam NFY, Wong MH. Root exudates of wetland plants influenced by nutrient status and types of plant cultivation. Int J Phytorem. 2012;14:543–53. https://doi.org/10.1080/15226514.2011.604691.
63. Zhu H, Yan B, Xu Y, Guan J, Liu S. Removal of nitrogen and COD in horizontal subsurface flow constructed wetlands under different influent C/N ratios. Ecol Eng. 2014;63:58–63. https://doi.org/10.1016/j.ecoleng.2013.12.018.
64. Koottatep T, Polprasert C. Role of plant uptake on nitrogen removal in constructed wetlands located in the tropics. Water Sci Technol. 1997;36:1–8. https://doi.org/10.1016/S0273-1223(97)00725-7.
65. Kozub DD, Liehr SK. Assessing denitrification rate limiting factors in a constructed wetland receiving landfill leachate. Water Sci Technol. 1999;40:75–82. https://doi.org/10.1016/S0273-1223(99)00459-X.
66. Mendonça AAJ (2016) Avaliação de um sistema descentralizado de tratamento de esgotos domésticos em escala real composto por tanque séptico e wetlands construída híbrida. Dissertação (Mestrado), Universidade de São Paulo
67. Mello VFB, Abreu JP da G, Ferreira JM, Jucá JFT, Motta Sobrinho MA da. Variáveis no processo de coagulação /floculação/decantação de lixiviados de aterros sanitários urbanos. Revista Ambiente & Água. 2012;7:88–100. https://doi.org/10.4136/ambi-agua.861.
68. Queiroz LM, Amaral MS, Morita DM, Yabroudi SC, Sobrinho PA. Aplicação de processos físico-químicos como alternativa de pré e pós-tratamento de lixiviados de aterros sanitários. Engenharia Sanitaria e Ambiental. 2011;16:403–10. https://doi.org/10.1590/S1413-41522011000400012.
69. Cecchet J, Gomes BM, Costanzi RN, Gomes SD. Tratamento de efluente de refinaria de óleo de soja por sistema de flotação por ar dissolvido. Revista Brasileira de Engenharia Agrícola e Ambiental. 2010;14:81–6. https://doi.org/10.1590/S1415-43662010000100011.
70. Nunes JA. Tratamento físicoquímico de águas residuárias industriais. Chiado Books; 2019.
71. Matilainen A, Vepsäläinen M, Sillanpää M. Natural organic matter removal by coagulation during drinking water treatment: a review. Adv Coll Interface Sci. 2010;159:189–97. https://doi.org/10.1016/j.cis.2010.06.007.
72. Felici EM, Kuroda EK, Yamashita F, da Silva SMCP. Remoção de carga orgânica recalcitrante de lixiviado de resíduos sólidos urbanos pré-tratado biologicamente por coagulação química-floculação-sedimentação. Engenharia Sanitária e Ambiental. 2013;18:177–84. https://doi.org/10.1590/S1413-41522013000200010.
73. Yabroudi Bayram SC. Remoção de matéria orgânica e nitrogênio de lixiviados de aterro sanitário: tratamento por nitritação/desnitritação biológica e processos físico-químicos. Doutorado em Engenharia Hidráulica. Universidade de São Paulo; 2012. https://doi.org/10.11606/T.3.2012.tde-29072013-161002.
74. Souto GD de B. Lixiviado de aterros sanitários brasileiros: estudo de remoção do nitrogênio amoniacal por processo de arraste com ar (stripping). Tese (doutorado). Universidade de São Paulo; 2009. https://doi.org/10.11606/T.18.2009.tde-19022009-121756.
75. Abdanur A (2005) Remediação de solo e água subterrânea contaminados por hidrocarbonetos de petróleo: estudo de caso da refinaria Duque de Caxias/RJ. Dissertação (Mestrado) Universidade Federal do Paraná
76. Youm KH, Fane AG, Wiley DE. Effects of natural convection instability on membrane performance in dead-end and cross-flow ultrafiltration. J Membr Sci. 1996;116:229–41. https://doi.org/10.1016/0376-7388(96)00047-6.
77. Federation WE. Membrane systems for wastewater treatment. 1st ed. New York: McGraw-Hill Education; 2005.
78. Pi KW, Li Z, Wan DJ, Gao LX. Pretreatment of municipal landfill leachate by a combined process. Process Saf Environ Prot. 2009;87:191–6. https://doi.org/10.1016/j.psep.2009.01.002.
79. Marañón E, Castrillón L, Fernández-Nava Y, Fernández-Méndez A, Fernández-Sánchez A. Coagulation-flocculation as a pretreatment process at a landfill leachate nitrification-denitrification plant. J Hazard Mater. 2008;156:538–44. https://doi.org/10.1016/j.jhazmat.2007.12.084.
80. Cortez S, Teixeira P, Oliveira R, Mota M. Evaluation of Fenton and ozone-based advanced oxidation processes as mature landfill leachate pre-treatments. J Environ Manage. 2011;92:749–55. https://doi.org/10.1016/j.jenvman.2010.10.035.
81. Pant D, Van Bogaert G, Diels L, Vanbroekhoven K. A review of the substrates used in microbial fuel cells (MFCs) for sustainable energy production. Biores Technol. 2010;101:1533–43. https://doi.org/10.1016/j.biortech.2009.10.017.
82. Sun H, Xu S, Zhuang G, Zhuang X. Performance and recent improvement in microbial fuel cells for simultaneous carbon and nitrogen removal: a review. J Environ Sci. 2016;39:242–8. https://doi.org/10.1016/j.jes.2015.12.006.
83. Tee P-F, Abdullah MO, Tan IAW, Mohamed Amin MA, Nolasco-Hipolito C, Bujang K. Performance evaluation of a hybrid system for efficient palm oil mill effluent treatment via an air-cathode, tubular upflow microbial fuel cell coupled with a granular activated carbon adsorption. Biores Technol. 2016;216:478–85. https://doi.org/10.1016/j.biortech.2016.05.112.

84. Cano V, Cano J, Nunes SC, Nolasco MA. Electricity generation influenced by nitrogen transformations in a microbial fuel cell: assessment of temperature and external resistance. Renew Sustain Energy Rev. 2021;139:110590. https://doi.org/10.1016/j.rser.2020.110590
85. Al-Mamun A, Baawain MS. Accumulation of intermediate denitrifying compounds inhibiting biological denitrification on cathode in microbial fuel cell. J Environ Health Sci Eng. 2015;13:81. https://doi.org/10.1186/s40201-015-0236-5.
86. Logan BE. Microbial fuel cells. Hoboken, N.J: Wiley-Interscience; 2008.
87. Zhang F, He Z. Integrated organic and nitrogen removal with electricity generation in a tubular dual-cathode microbial fuel cell. Process Biochem. 2012;47:2146–51. https://doi.org/10.1016/j.procbio.2012.08.002.
88. Sakdaronnarong CK, Thanosawan S, Chaithong S, Sinbuathong N, Jeraputra C. Electricity production from ethanol stillage in two-compartment MFC. Fuel. 2013;107:382–6. https://doi.org/10.1016/j.fuel.2012.10.030.
89. Rabaey K, Verstraete W. Microbial fuel cells: novel biotechnology for energy generation. Trends Biotechnol. 2005;23:291–8. https://doi.org/10.1016/j.tibtech.2005.04.008.
90. Logan BE, Rossi R, Ragab A, Saikaly PE. Electroactive microorganisms in bioelectrochemical systems. Nat Rev Microbiol. 2019;17:307–19. https://doi.org/10.1038/s41579-019-0173-x.
91. Philips J, Verbeeck K, Rabaey K, Arends J. Electron transfer mechanisms in biofilms. Microbial electrochemical and fuel cells: fundamentals and applications. Elsevier. Woodhead Publishing; 2015. pp. 67–113. https://doi.org/10.1016/B978-1-78242-375-1.00003-4.

Modelling and Control of Wastewater Treatment Processes: An Overview and Recent Trends

12

Victor Alcaraz-Gonzalez

Abstract

In this study, some modelling and process control approaches used in Wastewater Treatment Plants (WWTP) are recalled. Two principal kinds of WWTP are used as frameworks: continuous anaerobic and aerobic. The highly nonlinear nature of the system models representing such processes is highlighted. In addition, parametric uncertainties that characterize them, and disturbances to which they are subject and that affect their performance are also underlined. Thus, mention is made of how various modelling and process control techniques have been used to face such issues in different ways. Typical deterministic models proposed by the International Water Association (IWA) are recalled, but some useful simpler models and even knowledge-based ones, like neural networks and fuzzy approaches, are also mentioned for specific applications. Particular emphasis is placed on the main parameters and variables to be monitored and controlled in order to ensure the optimal performance of WWTP: In the case of anaerobic digestion, alkalinity, and in the case of aerobic processes oxygen transfer efficiency. Thus, unlike classical Proportional Integral Derivative (PID) controllers, two kinds of nonlinear control approaches, namely adaptive and predictive, which are robust against uncertainties, nonlinearities, and perturbations are cited as the most used in this kind of process. Finally, some implications are highlighted in terms of energy consumption and cost optimization, and how different control strategies in the frame of benchmarking are used to minimize their impact.

Keywords

Automatic control · Operating conditions · Operational difficulties · Process control · Process optimization · Wastewater treatment simulation

12.1 Introduction

Wastewater Treatment Plants (WWTP) commonly involve a series of biological transformation processes that are usually very complex in terms of biochemical and biokinetics understanding [1], and therefore, also in terms of everything related to its modelling, optimization, and automatic control [2, 3]. The type of waste treated is mostly organic, including urban sewage and agricultural effluents with a multi-variety of organic compounds like carbohydrates, lipids, and proteins. These compounds are typically

V. Alcaraz-Gonzalez (✉)
Universidad de Guadalajara-CUCEI, Guadalajara, Mexico
e-mail: victor.alcaraz@cucei.udg.mx

M. Bahadir and A. Haarstrick (eds.), *Water and Wastewater Management*,
Water and Wastewater Management, https://doi.org/10.1007/978-3-030-95288-4_12

expressed as Chemical Oxygen Demand (COD) and are the main substrate for a wide variety of microorganisms working in a symbiotic relationship. WWTP may be classified into two large groups: Aerobic (in the presence of oxygen), and Anaerobic (in the absence of oxygen), although Anoxic units [4] (anaerobic, but oxygen is available from nitrates, nitrites, or NO) are also included in industrial and municipal WWTP. The most representative aerobic WWTP include Activated Sludge (AS) [2] and Aerated Lagoons [5], while Anaerobic Digestion (like Upflow Anaerobic Sludge Blanket (UASB) [6]) and Dark Fermentation [7] are among the most common anaerobic WWTP [8]. Other WWTP configurations also combine anaerobic and aerobic units [9]. In these processes, organic matter is degraded and converted into microbial biomass, residual organic matter, and eventually in biogas in the case of Anaerobic Digestion (AD) processes [10, 11], being this last the only biological WWT process capable of treating directly wastewater with an organic load higher than 15–20 g_{COD}/L [9]. In the following in this contribution, in order not to disperse the information too much, efforts will be concentrated mainly on AS and AD continuous processes.

Whether in industry, in the countryside, or in cities, WWTP technologies are undoubtedly indispensable elements to ensure the conditions of industrialization, urbanization, and sustainability of modern society. Thus, their good enough functioning is necessary to meet all kinds of required standards. However, talking about proper operation involves more than just periodic monitoring and maintenance. In reality, WWTP are subject to several disturbances and uncertainties that make difficult to meet such objectives. Effluents that reach the WWTP have a varying and changing composition affecting their physical, chemical, and physicochemical characteristics. Variations in flow, temperature, pH, organic compounds concentration and even the presence of inhibitors affect the biological performance of microbial communities, which are very sensitive to all these parameters, and therefore, also impact their efficiency. Changes in these parameters represent serious disturbances of different nature. The most typical example is seasonal abrupt flow rate step changes (e.g., in the rainy season), but important steps in all these parameters can also occur weekly or even daily simply by changes in environmental conditions and even by the effect of human activity. Most of these disturbances are difficult to foresee or are of a relatively high monitoring cost.

Taking into account these issues, it is clear that current WWTP operate under important uncertainty conditions from a process control point of view, which may cause serious problems in their industrial, agro-industrial or urban implementation. These features, together with the ecological norms requirements, make evident the importance to implement Instrumentation, Control, and Automation (ICA) systems in WWTP [3].

In order to show the benefits of the application of modern control approaches to WWTP, this paper is intended to present different approaches that have proved their usefulness in the modelling, parameter estimation, and regulation of WWTP. These approaches certainly have demonstrated to be robust against process input disturbances, and uncertainties, and have a strong potential for optimizing the performance of WWTP. Actual application examples and successful cases are also shown and demonstrate that they are a useful and efficient alternative that may be implemented at a relatively low cost.

12.2 Modelling

Modelling WWTP can become a difficult task mainly because their growth kinetics exhibit a highly nonlinear behavior [12], which in turn is due to the involved microbial consortia being among the most diverse ecosystems in nature [13]. An additional disadvantage is that the high complexity in the microbial populations leads also to high uncertainty in the kinetic parameters of current WWTP models.

Certainly, the attempt to adapt microbial communities to different operating conditions, and to the abovementioned disturbances makes them very changeable. The constant distribution of the different species in the consortium causes that their kinetic parameters are also uncertain and change over time. For these reasons, the first step to control and to optimize WWTP is to have models as closely as possible to reality in such a way that the sources of uncertainty are reduced, but at the same time, a relative simplicity is preserved to facilitate the application of control techniques.

Nowadays, one of the most complete and ambitious AD model is the ADM1 (Anaerobic Digestion Model) [14], while concerning AS, the most used models are the series ASM (Activated Sludge Model) [2]. Both models have been developed by the International Water Association (IWA). Several updates and reviews have been developed until now and still remain an active research field. For instance, in [15, 16] seven variants of the ASM series are exhaustively analyzed, verified and evaluated concerning aspects like stoichiometric and kinetic rate expressions, fermentation features, autotrophic and heterotrophic microbial growth, phosphorus removal, nitrification and denitrification, alkalinity, and others. Model parameter estimation and recalibration are also carried out frequently as in [17–19], where stochastic parameter estimation approaches together with Monte Carlo simulations and sensitivity analysis are used to fit ASM models to industrial actual data. Theoretical advances have been also further developed recently in [20].

Regarding AD, an additional issue in comparison to aerobic WWTP is its high sensibility to pH variations, and more exactly to alkalinity. Certainly, inlet concentration disturbances may affect partially or even completely the so-called System Operational Stability (SOS) [21, 22]. Such disturbances may compromise the physicochemical equilibrium that is directly related to the SOS preservation [23]. Furthermore, it is known that the presence of strong ions provokes a physicochemical unbalancing that affects directly microbial activity [24, 25]. Even when ADM1 consider anion-cation balancing, pH is not explicit in their equations [14, 26]. In this concern, some efforts have been developed for taking into account a more physicochemical framework [27, 28]. However, even with these modifications, the ADM1 model remains very large to be useful for control purposes, but it is still used rather for state variables estimation [29], experimental validation [30], and monitoring. Thus, in order to obtain greater applicability, reduced order AD models have been also proposed [31, 32]. Particularly, the so-called AM2 Model [33] has taken on great importance not only for being easier to handle from a mathematical point of view, but for taking into account specifically the strong ions concentration, which at pH 7 is very close to alkalinity [34, 35]. Without demerit of any other model described in the literature, and only for illustrating one of them in this contribution, AM2 model is depicted as follows [33]:

$$\begin{aligned}
\dot{x}_1 &= (\mu_1 - \alpha D)x_1 \\
\dot{x}_2 &= (\mu_2 - \alpha D)x_2 \\
\dot{S}_1 &= D(S_1^{in} - S_1) - k_1\mu_1 x_1 \\
\dot{S}_2 &= D(S_2^{in} - S_2) + k_2\mu_1 x_1 - k_3\mu_2 x_2 \\
\dot{Z} &= D(Z_1^{in} - Z) \\
\dot{C} &= D(C^{in} - C) - k_L \\
&\quad (C + S_2 - Z - k_H P_{CO_2}) + k_4\mu_1 x_1 - k_5\mu_2 x_2
\end{aligned} \tag{12.1}$$

with

$$\mu_1 = \frac{\mu_{max1} S_1}{k_{S1} + S_1},\ \mu_2 = \frac{\mu_{max2} S_2}{k_{S2} + S_2 + \left(\frac{S_2}{k_I}\right)^2} \tag{12.2}$$

where x_1 (g/L) and x_2 (g/L) represent the acidogenic and methanogenic biomass concentrations respectively, S_1 (g/L) is the organic matter concentration expressed as COD, and S_2 (mmol/L) is the volatile fatty acids concentration. C (mmol/L) is the total inorganic carbon and Z (mEq/L) is the strong ions concentration. The superscript *in* represents "input concentration" for these variables. k_L (1/d) is the liquid–gas transfer coefficient, P_{CO_2} is the CO_2 partial pressure, while k_H is the Henry's constant. μ_1 (1/d) (Monod type) and μ_2 (1/d) (Haldane type) are the microbial specific growth rates for acidogenic bacteria and methanogenic archea, respectively, where μ_{max1} and μ_{max2} (1/d) are the maximum growth rates, k_{S1} (g/L), and k_{S2} (mmol/L) represent the half saturation constants, while k_I [(mmol/L)]$^{1/2}$ represents the inhibition constant. The parameters k_1 to k_5 are yield coefficients in the corresponding units. D (1/d) is the dilution rate, which usually is used as manipulable control input in several control approaches. The parameter α (dimensionless) denotes the biomass fraction that is retained for the reactor bed, i.e., $\alpha = 0$ stands for an ideal fixed-bed reactor, while $\alpha = 1$ stands for an ideal continuous stirred reactor tank [33].

This model has been also used for optimization, state and parameter estimation, and process control purposes (see for instance [36–39]), and for modeling the alkalinity spatial distribution in an up-flow fixed bed anaerobic digester [40].

12.3 Control and Optimisation

It is important to remark that the application of modern control approaches (i.e., those developed from the 60s) to WWTP is relatively recent. Control algorithms used in the past in the field of WWTP were basically on/off type, Proportional Integral (PI) or Proportional Integral Derivative (PID) [41]. However, this situation is changing very quickly. Even when classical control methods may offer satisfactory results in relatively simple systems, more recent approaches nowadays do take into account specifically the nonlinear features of biological WWT systems. For instance, adaptive control or predictive control, which are better adapted to face nonlinear behavior, are finding more and more applications in this field. Furthermore, as it has been mentioned before, the main "SOS" lack indicator in AD is alkalinity, because it involves all the physical–chemical equilibria, and therefore, several studies have been developed for regulating it [42].

One of the most typical control variables in WWTP is the COD, since its reduction is certainly the ultimate goal in these processes. However, due to the uncertainties and disturbances, to which treatment plants are exposed in terms of process inputs, it has become necessary to include control approaches being robust against such disturbances and lack of knowledge [22]. In [43], for example, interval observers have been included to estimate ranges of values in biomass concentrations, as these cannot be measured, and then they were combined with an adaptive control approach for regulating COD, also within a preset control interval. But, uncertainties in the growth kinetics of microbial populations in WWTP often lead researchers to propose different nonlinear type control techniques, see for instance [44, 45]. Moreover, some types of nonlinear tracking controls have been proposed. For instance in [46], an adaptive nonlinear tracking control was proposed for regulating the concentration of biomass in an AS-type WWTP based on the input flow, and thus, proportional to it, using a simplified model.

Furthermore, in particular in relation to aerobic processes such as AS-type WWTP, it is well known that in order to maintain adequate oxygen levels, the main parameter to be controlled is oxygen transfer efficiency, and therefore, the main unit operation is aeration. This demands high levels of energy consumption and other costs that need to be optimized [47–49]. Different adaptive and predictive control strategies have been proposed in the past for this purpose [50, 51]. In addition to COD reduction, and not less important, the removal of nitrogen and phosphorus from wastewater is also one of the main targets in WWTP [52]. Several classical and advanced control techniques proposed in this sense have been used and applied successfully [53]. However, as the efficiency of the operation is closely related to the aeration energy, which is a costly process, control objectives are compromised at the same time with concentration setpoints and minimal use of energy, and thus, with cost reduction. For instance in [54], adaptive control techniques were used to improve the removal of these nutrients, while reduced the aeration energy required in three WWTP in Switzerland. Nevertheless, Model Predictive Control and other optimization approaches seem to be preferred [55, 56]. However, it is important to note that a common feature in many of the strategies used today, both for AS and AD WWTP is the use of benchmarking [57–59]. In such strategies, accumulated knowledge in monitoring, modeling, and control is used to provide a framework for assessing both well-known techniques to be industrially implemented and new approaches to evaluate its correct performance. Certainly, it is in the systematic current form more efficient to simulate WWTP and to take profit from models representing them.

Many modern applications include artificial networks as well lagged regression models, among other methods, that have been used for preventing indirectly the failure of the system due to alkalinity mismatches [60, 61]. These approaches has been also used in cases, when bioreactors have been stopped and then launched after a long time of starvation for bacterial communities in order to recover their correct functioning [62]. Fuzzy logic approaches have also seen a growing interest in the last three decades. For instance in [63], it is depicted how fuzzy logic based modules are used to assess the state of an up-flow fixed bed AD bioreactor, while in [64] similar approaches together with expert systems are proposed and experimentally validated as a methodology for assessing and improving the efficiency of agricultural biogas plants. Neural network and fuzzy approaches have been also used for optimizing energy consumption [65] and regulating the dissolved oxygen concentration [66] in AS processes. Finally, it is important to mention an important current trend in nonlinear control of WWTP, consisting of Sliding Mode approaches that have been used together with models both classical deterministic type [30], and those described by neural networks and fuzzy logic as well [67–69]. Finally, it is important to underline that modern control and optimization approaches are currently possible thanks to the existence of Supervisory Control and Data Acquisition (SCADA) systems that play a preponderant role in real-time ICA approaches implementation.

12.4 Conclusions

The use of Instrumentation Control and Automation (ICA) approaches in WWTP has allowed a greater development of these technologies for wide-scale implementation. The modeling, which is complemented and fed back with monitoring, as well as parameter and state estimation techniques have allowed for a greater understanding of the chemical, physicochemical and biological processes that are carried out in WWTP. The most widely disseminated models, e.g., the ADM1 models and the ASM series of the International Water Association (IWA), are of deterministic type, but models developed from other approaches such as neural networks and diffuse logic have also had extensive development.

As these models are highly nonlinear, the great majority of researchers have developed robust and nonlinear predictive adaptive control approaches. Such approaches allow capturing this nonlinear nature and admitting variations in kinetics and other parameters as well as in operating conditions, which make them robust in the face of these variations as well as in the face of uncertainties and disturbances. This paper has underlined the most important variables to consider in anaerobic and aerobic processes, being alkalinity and oxygen transfer efficiency, respectively, and examples of the main approaches developed for modeling and controlling such variables have been cited.

Advances made in these approaches and technologies over the last 3 decades have enabled good practice reference frameworks in modeling, monitoring, and control of WWTP, which have been validated in different scenarios and have given way to the development of benchmarking strategies that are widely used nowadays. Thus, the use of these techniques has allowed both biological technologies to remove contaminants increasingly efficiently as well as to achieve continuous optimization and improvement of WWTP.

Acknowledgements This publication is an output of the global collaborative project “EXCEED—Swindon—Sustainable Water Management in Developing Countries”. The author highly acknowledge the support of German Academic Exchange Service DAAD for taking part.

References

1. Sari T, Wade MJ. Generalised approach to modelling a three-tiered microbial food-web. Math Biosci. Epub ahead of print 2017. https://doi.org/10.1016/j.mbs.2017.07.005.
2. Henze M, Gujer W, Mino T, et al. Activated sludge models ASM1, ASM2, ASM2d and ASM3. 1st ed. London: IWA Publishing; 2000.
3. Jimenez J, Latrille E, Harmand J, et al. Instrumentation and control of anaerobic digestion processes: a review and some research challenges. Rev Environ Sci Biotechnol. 2015;14:615–48.
4. Jetten MSM, Strous M, Van De Pas-Schoonen KT et al. The anaerobic oxidation of ammonium. FEMS Microbiol Rev. Epub ahead of print 1998. https://doi.org/10.1016/S0168-6445(98)00023-0.
5. Marais GR, Ekama GA, Wentzel MC. Application of the activated sludge model to aerated lagoons. Water SA. 2017;43:238–57.
6. Chong S, Sen TK, Kayaalp A, et al. The performance enhancements of upflow anaerobic sludge blanket (UASB) reactors for domestic sludge treatment—a state-of-the-art review. Water Res. 2012;46:3434–70.
7. Guo XM, Trably E, Latrille E et al Hydrogen production from agricultural waste by dark fermentation: a review. Int J Hydrogen Energy. Epub ahead of print 2010. https://doi.org/10.1016/j.ijhydene.2010.03.008.
8. Zhang R, El-Mashad HM, Hartman K et al. Characterization of food waste as feedstock for anaerobic digestion. Bioresour Technol. Epub ahead of print 2007. https://doi.org/10.1016/j.biortech.2006.02.039.
9. Chan YJ, Chong MF, Law CL et al. A review on anaerobic-aerobic treatment of industrial and municipal wastewater. Chem Eng J. 2009;1–18.
10. Holm-Nielsen JB, Al Seadi T, Oleskowicz-Popiel P. The future of anaerobic digestion and biogas utilization. Bioresour Technol. 2009;100:5478–84.
11. Mao C, Feng Y, Wang X, et al. Review on research achievements of biogas from anaerobic digestion. Renew Sustain Energy Rev. Epub ahead of print 2015. https://doi.org/10.1016/j.rser.2015.02.032.
12. Donoso-Bravo A, Mailier J, Martin C, et al. Model selection, identification and validation in anaerobic digestion: a review. Water Res. 2011;45:5347–64.
13. Delbès C, Moletta R, Godon JJ. Bacterial and archaeal 16S rDNA and 16S rRNA dynamics during an acetate crisis in an anaerobic digestor ecosystem. FEMS Microbiol Ecol. 2001;35:19–26.
14. Batstone DJ, Keller J, Angelidaki I, et al. The IWA anaerobic digestion model no 1 (ADM1). Water Sci Technol. 2002;45:65–73.
15. Hauduc H, Rieger L, Takács I, et al. A systematic approach for model verification: application on seven published activated sludge models. Water Sci Technol. 2010;61:825–39.
16. Hauduc H, Rieger L, Oehmen A, et al. Critical review of activated sludge modeling: State of process knowledge, modeling concepts, and limitations. Biotechnol Bioeng. 2013;110:24–46.
17. Sin G, De Pauw DJW, Weijers S, et al. An efficient approach to automate the manual trial and error calibration of activated sludge models. Biotechnol Bioeng. 2008;100:516–28.
18. Keskitalo J, Leiviskä K. Application of evolutionary optimisers in data-based calibration of activated sludge models. Expert Syst Appl. 2012;39:6609–17.
19. Alikhani J, Takacs I, Al-Omari A, et al. Evaluation of the information content of long-term wastewater characteristics data in relation to activated sludge model parameters. Water Sci Technol. 2017;75:1370–89.
20. Fortela DLB, Farmer K, Zappi A et al. A methodology for global sensitivity analysis of activated sludge models: case study with activated sludge model no. 3 (ASM3). Water Environ Res. 2019;91:865–876.
21. Angelidaki I, Boe K, Ellegaard L. Effect of operating conditions and reactor configuration on efficiency of full-scale biogas plants. Water Sci Technol. 2005;52:189–94.
22. Méndez-Acosta HO, Palacios-Ruiz B, Alcaraz-González V, et al. A robust control scheme to improve the stability of anaerobic digestion processes. J Process Control. 2010;20:375–83.
23. Batstone DJ, Amerlinck Y, Ekama G, et al. Towards a generalized physicochemical framework. Water Sci Technol. 2012;66:1147–61.
24. Chen Y, Cheng JJ, Creamer KS. Inhibition of anaerobic digestion process: a review. Bioresour Technol. 2008;99:4044–64.
25. Chen S, Zhang J, Wang X. Effects of alkalinity sources on the stability of anaerobic digestion from food waste. Waste Manag Res. 2015;33:1033–40.
26. Patón M, González-Cabaleiro R, Rodríguez J. Activity corrections are required for accurate anaerobic digestion modelling. Water Sci Technol. 2018;77:2057–67.
27. Zhang Y, Piccard S, Zhou W. Improved ADM1 model for anaerobic digestion process considering physico-chemical reactions. Bioresour Technol. 2015;196:279–89.
28. Shi E, Li J, Leu SY, et al. Modeling the dynamic volatile fatty acids profiles with pH and hydraulic retention time in an anaerobic baffled reactor during the startup period. Bioresour Technol. 2016;222:49–58.
29. Montiel-Escobar JL, Alcaraz-González V, Méndez-Acosta HO et al. ADM1-based robust interval

observer for anaerobic digestion processes. Clean Soil Air Water. 40. Epub ahead of print 2012. https://doi.org/10.1002/clen.201100718.
30. Torres Zúñiga I, Villa-Leyva A, Vargas A et al. Experimental validation of online monitoring and optimization strategies applied to a biohydrogen production dark fermenter. Chem Eng Sci. Epub ahead of print 2018. https://doi.org/10.1016/j.ces.2018.05.039.
31. Xue L, Li D, Xi Y. Nonlinear model predictive control of anaerobic digestion process based on reduced ADM1. In: 2015 10th Asian control conference emerging control techniques for a sustainable world, ASCC 2015. https://doi.org/10.1109/ASCC.2015.7244539.
32. Hassam S, Ficara E, Leva A et al. A generic and systematic procedure to derive a simplified model from the anaerobic digestion model no. 1 (ADM1). Biochem Eng J. 2015;99. https://doi.org/10.1016/j.bej.2015.03.007.
33. Bernard O, Hadj-Sadok Z, Dochain D, et al. Dynamical model development and parameter identification for an anaerobic wastewater treatment process. Biotechnol Bioeng. 2001;75:424–38.
34. Hassam S, Ficara E, Leva A et al. A generic and systematic procedure to derive a simplified model from the anaerobic digestion model no. 1 (ADM1). Biochem Eng J. 2015;99:193–203.
35. Attar S, Haugen FA. Model-based optimal recovery of methane production in an anaerobic digestion reactor. Model Identif Control. 2020;41:121–8.
36. Bernard O, Chachuat B, Hélias A, et al. An integrated system to remote monitor and control anaerobic wastewater treatment plants through the internet. Water Sci Technol. 2005;52:457–64.
37. Didi I, Dib H, Cherki B. A Luenberger-type observer for the AM2 model. J Process Control. 2015;32:117–26.
38. Rodríguez A, Quiroz G, Femat R, et al. An adaptive observer for operation monitoring of anaerobic digestion wastewater treatment. Chem Eng J. 2015;269:186–93.
39. Vargas A, Sepúlveda-Gálvez A, Barrios-Pérez JD. A fast extremum-seeking approach for the methanisation of organic waste in an anaerobic bioreactor. IFAC-PapersOnLine. 2019;52:269–74.
40. Hmissi M, Harmand J, Alcaraz-Gonzalez V et al. Evaluation of alkalinity spatial distribution in an upflow fixed bed anaerobic digester. Water Sci Technol. 77. Epub ahead of print 2018. https://doi.org/10.2166/wst.2017.612.
41. García-Sandoval JP, Méndez-Acosta HO, González-Alvarez V et al. VFA robust control of an anaerobic digestion pilot plant: experimental implementation. IFAC-PapersOnLine. 49. Epub ahead of print 2016. https://doi.org/10.1016/j.ifacol.2016.07.328.
42. Alcaraz-González V, Fregoso-Sanchez FA, Mendez-Acosta HO, et al. Robust regulation of alkalinity in highly uncertain continuous anaerobic digestion processes. Clean: Soil, Air, Water. 2013;41:1157–64.
43. Alcaraz-González V, Harmand J, Rapaport A, et al. Robust interval-based regulation for anaerobic digestion processes. Water Sci Technol. 2005;52:449–56.
44. Méndez-Acosta HO, Palacios-Ruiz B, Alcaraz-González V, et al. Robust control of volatile fatty acids in anaerobic digestion processes. Ind Eng Chem Res. 2008;47:7715–20.
45. Flores-Estrella RA, Alcaraz-González V, García-Sandoval JP et al. Robust output disturbance rejection control for anaerobic digestion processes. J Process Control. 75. Epub ahead of print 2019. https://doi.org/10.1016/j.jprocont.2018.12.012.
46. Georgieva P, Ilchmann A. Adaptive λ-tracking control of activated sludge processes. Int J Control. 2001;74:1247–59.
47. Leu S-Y, Rosso D, Larson LE, et al. Real-time aeration efficiency monitoring in the activated sludge process and methods to reduce energy consumption and operating costs. Water Environ Res. 2009;81:2471–81.
48. Åmand L, Carlsson B. Optimal aeration control in a nitrifying activated sludge process. Water Res. 2012;46:2101–10.
49. Füreder K, Svardal K, Frey W, et al. Energy consumption of agitators in activated sludge tanks—actual state and optimization potential. Water Sci Technol. 2018;77:800–8.
50. Shen W, Chen X, Corriou JP. Application of model predictive control to the BSM1 benchmark of wastewater treatment process. Comput Chem Eng. 2008;32:2849–56.
51. Belchior CAC, Araújo RAM, Landeck JAC. Dissolved oxygen control of the activated sludge wastewater treatment process using stable adaptive fuzzy control. Comput Chem Eng. 2012;37:152–62.
52. Sedlak RI. Phosphorus and nitrogen removal from municipal wastewater. Principles and practice, 2nd ed. CRC Press, Taylor & Francis Group; 1991.
53. Stare A, Vrečko D, Hvala N, et al. Control of nutrient removing activated sludge system. IFAC Proc. 2007;40:61–6.
54. Rieger L, Takács I, Siegrist H. Improving nutrient removal while reducing energy use at three Swiss WWTPs using advanced control. Water Environ Res. 2012;84:170–88.
55. Steffens MA, Lant PA. Multivariable control of nutrient-removing activated sludge systems. Water Res. 1999;33:2864–78.
56. Zhang M, Peng Y, Wang C, et al. Optimization denitrifying phosphorus removal at different hydraulic retention times in a novel anaerobic anoxic oxic-biological contact oxidation process. Biochem Eng J. 2016;106:26–36.
57. Samuelsson P, Halvarsson B, Carlsson B. Cost-efficient operation of a denitrifying activated sludge process. Water Res. 2007;41:2325–32.
58. Chen W, Yao C, Lu X. Optimal design activated sludge process by means of multi-objective optimization: case study in benchmark simulation model 1 (BSM1). Water Sci Technol. 2014;69:2052–8.

59. Gernaey KV, Jeppsson U, Vanrolleghem PA et al. Benchmarking of control strategies for wastewater treatment plants IWA task group on benchmarking of control strategies for wastewater treatment plants.
60. Wang X, Bai X, Li Z, et al. Evaluation of artificial neural network models for online monitoring of alkalinity in anaerobic co-digestion system. Biochem Eng J. 2018;140:85–92.
61. Barampouti EMP, Mai ST, Vlyssides AG. Dynamic modeling of the ratio volatile fatty acids: bicarbonate alkalinity in a UASB reactor for potato processing wastewater treatment. Environ Monit Assess. 2005;110:121–8.
62. Holubar P, Zani L, Hager M, et al. Advanced controlling of anaerobic digestion by means of hierarchical neural networks. Water Res. 2002;36:2582–8.
63. Lardon L, Punal A, Steyer JP. On-line diagnosis and uncertainty management using evidence theory—experimental illustration to anaerobic digestion processes. J Process Control. 2004;14:747–63.
64. Djatkov D, Effenberger M, Martinov M. Method for assessing and improving the efficiency of agricultural biogas plants based on fuzzy logic and expert systems. Appl Energy. 2014;134:163–75.
65. Kusiak A, Wei X. Optimization of the activated sludge process. J Energy Eng. 2013;139:12–7.
66. Du X, Wang J, Jegatheesan V et al. Dissolved oxygen control in activated sludge process using a neural network-based adaptive PID algorithm. Appl Sci (Switzerland). 8. Epub ahead of print 2018. https://doi.org/10.3390/app8020261.
67. Jaramillo-Morán MA, Peguero-chamizo JC. Sliding mode control of a wastewater plant with. 2007;120–129.
68. Han H, Wu X, Qiao J. A self-organizing sliding-mode controller for wastewater treatment processes. IEEE Trans Control Syst Technol. 2019;27:1480–91.
69. Zlateva P. Sliding mode control of wastewater treatment process with activated sludge under extreme weather events. IOP Conf Ser Earth Environ Sci. 2021;776:012001.

Part V

Wastewater Management (Pollutants)

Fingerprint of Persistent Organic Pollutants (POPs) in the Environment: Ecological Assessment and Human Health Effects

13

Fatma Beduk, Senar Aydin, Arzu Ulvi, and Mehmet Emin Aydin

Abstract

Persistent organic pollutants (POPs) are considered as emerging contaminants due to their bio-accumulative, persistent and toxic natures. These chemicals are present everywhere in the environment including human tissues. These priority organic pollutants are released to the environment during their intended use and/or their production unintentionally. The POPs regulation aims to prohibit production and use of POPs, and to ensure safe management of POPs' contaminated wastes. POPs containing effluents of Wastewater Treatment Plants pose risk for receiving water bodies. In this study, it is aimed to review recent literature in order to give an overall picture about continuing risks of POPs.

Keywords

Ecological assessment · Endocrine disrupting chemicals · Human exposure pathways · POPs elimination network · POPs fingerprint

F. Beduk (✉) · S. Aydin · A. Ulvi
Engineering Faculty, Environmental Engineering Department, Necmettin Erbakan University, Konya, Turkey
e-mail: fabeduk@erbakan.edu.tr

M. E. Aydin
Engineering Faculty, Civil Engineering Department, Necmettin Erbakan University, Konya, Turkey

13.1 Introduction

Negative effects of environmental pollutants on human health and ecological balance have reached terrifying levels. Babies are born with environmental pollutants in their tissues and further continue to accumulate these chemicals during their lifetimes [1, 2]. A new type of pollutant is on the agenda of researchers every day. Persistent organic pollutants (POPs) are among most emerging priority pollutants, created international public opinion. They are persistent to chemical or biological degradation, and exist in every compartment of the environment for a long time [3]. These pollutants are transported to long distances and can be determined in poles, where never been produced [4].

POPs are halogenated organic compounds, including aldrin, chlordane, DDT, dieldrin, endrin, heptachlor, mirex, and toxaphene from the group of pesticides; polychlorinated biphenyls (PCBs) and hexachlorobenzene (HCB) from the group of industrial chemicals; polychlorinated dibenzo-p-dioxins and dibenzofurans (PCDDs/PCDFs) produced by combustion of chlorine containing materials as well as in oxychlorination reactions in the chemical industry as unintended by-products, e.g., while copper catalyzed synthesis of vinyl chloride for PVC polymerization. The preliminary list, namely "dirty dozen" was announced in the Stockholm Convention in 2001. New POPs, including

M. Bahadir and A. Haarstrick (eds.), *Water and Wastewater Management*,
Water and Wastewater Management, https://doi.org/10.1007/978-3-030-95288-4_13

polybrominated flame retardants, and some other chemicals entered the list in the course of time. Alternative chemicals substituted POPs to meet the needs in production processes. For example, organophosphate esters (OPEs) were introduced as flame retardants to replace polybrominated diphenyl ethers (PBDEs). Organophosphorus pesticides (OPPs) were also alternatives to organochlorine pesticides (OCPs). It is not known, what will be the result of bulk consumption of these substituted chemicals, but it is known that POPs list will cover more chemicals in the course of time.

POPs are characterized through their persistency, lipophilicity, bioavailability, and toxicity [1, 5–7]. They resist to physical, chemical, and biological degradation processes in the environment. Therefore, monitoring POPs are important in order to reduce their occurrence in the environment. Studies about POPs generally focus on resource, exposure and adverse effects of mother compounds, however, a variety of more toxic by-products have potential adverse effects as well [8].

While water demand is increasing for domestic, industrial and agricultural use in parallel to population growth, clean water sources are decreasing as a result of climate change, improper wastewater management, agricultural activities, excessive abstraction of surface and ground water, etc. Contamination of limited fresh water sources by toxic chemicals is a major environmental concern. Shortage of clean water sources pushes the countries to use treated wastewater as an alternative water source. Utilizing treated wastewater for irrigation purposes is a spreading solution of water scarcity. It seems sure that clean water shortage is going to be a common problem of more and more countries in the world as a result of climate change.

Regulations for discharge or reuse of reclaimed water mainly include conventional parameters, such as chemical oxygen demand (COD), biological oxygen demand (BOD), total nitrogen (TN), total phosphorus (TP), total suspended solids (TSS), bacteria etc. While secondary effluents of Wastewater Treatment Plants (WWTPs) can generally meet most of the criteria, effluents of tertiary treatment can successfully meet all criteria set for conventional parameters. However, a significant number of emerging pollutants are not commonly monitored in wastewater, since their presence in the environment is not regulated by legislations. While pesticides, PAHs, PCBs, nutrients and heavy metals are regulated contaminants, pharmaceuticals and microplastics are candidates for future regulations. Fates of emerging pollutants in the environment or their human health effects are not well characterized. Lack of good understanding of potential risks to human health results in conflicting opinions about regulatory frameworks and measures. Besides, there is also a need for highly qualified analytical laboratories for determination POPs at low detection limits. Sampling and analysis restricts are important factors affecting regulatory frameworks and measures of emerging pollutants.

POPs mostly pass through WWTPs unchanged or unaffected, since conventional wastewater treatment methods are insufficient to remove POPs from wastewater [9]. Urbaniak et al. [10] evaluated effluents of 14 conventional WWTPs in central Poland in the means of PCDDs/PCDFs concentrations. 2.99–177 pg/L PCDDs and 6.05–51.3 pg/L PCDFs were determined in samples taken during stable conditions of WWTPs. Research studies show that some advanced treatment methods are also not sufficient to produce ecologically safe reclaimed water [11]. Özcan et al. [12] reported on PAHs and PCBs in sewage sludge samples taken from two WWTPs in Konya (Turkey). These sludges from WWTP were not suitable for using in agriculture as soil conditioners due to their high PAH contents ($\sum$PAHs: 1,077–17,509 μg/kg dry matter). The origin of PAHs in sewage sample from WWTP-1 is petroleum combustion, while the possible source of PAHs in sludge samples from WWTP-2 was combustion of kerosene, grass, coal, and wood. The ecotoxicological tests examined on *Vibrio fischeri* (luminous bacteria) and *Lepidium sativum* (garden cress) showed that these sludge samples had acute toxic properties. Effluents of WWTPs are generally discharged to surface water bodies. POPs containing effluents pose risk for vegetation and biota in receiving ecosystems [13]. Aydin et al. [14] found 0.015–0.065 μg/L ΣHCHs and n.d.–0.047 μg/L ΣDDTs in samples taken from 9 monitoring stations in Konya Closed Basin in the central part of Turkey. DDE isomer was reported to be in the highest concentration ranged from not detected (n.d.) to 0.037 μg/L. This finding shows not new, but old use of DDT as a result of restrictions.

Although the production of POPs has declined as a result of elimination and restriction consensus between the parties, exposure risk for some of the recently phased out POPs has not started to decline. Hence, in this study it is aimed to review the fingerprint of POPs in the environment from source to receiver, and to draw attention to their adverse health effects.

13.2 POPs and New POPs

The Stockholm Convention is a global agreement that entered into force on 17 May 2004. The Convention on POPs prohibits and restricts the use of chemicals that adversely affect the environment and human health. Preparing a national implementation plan for POPs and updating this plan periodically, taking measures to reduce or eliminating the stocks and releases of these chemicals, keeping a record of the special exemptions granted under the contract and regularly reporting the data to the contract secretariat are among the obligations to be fulfilled by 179 parties consented to be bound by the Stockholm Convention.

POPs are classified in 3 categories in Stockholm Convention.

Annex A: Subjected to elimination: Aldrin, chlordane, dieldrin, endrin, heptachlor, hexachlorobenzene, mirex, toxaphene, polychlorinated biphenyls (PCBs), chlordecone, lindane, hexa- and penta-bromodiphenyl ethers, tetra- and penta-bromodiphenyl ethers, α- and β-hexachlorocyclohexane, technical endosulfan and its related isomers, pentachlorobenzene, hexabromobiphenyl.

Annex B: Subjected to restriction: DDT (1,1,1-trichloro-2,2-bis (4-chlorophenyl)ethane), Perfluorooctane sulfonic acid (PFOS), its salts and perfluorooctane sulfonyl fluoride (PFOSF).

Annex C: Unintentionally produced and subjected to reduction: Polychlorinated dibenzo-p-dioxins and dibenzofurans (PCDDs/PCDFs), hexachlorobenzene (HCB), pentachlorobenzene (PeCB), polychlorinated biphenyls (PCBs).

In recent years, halogenated organic compounds have received increased attention, including polybrominated diphenyl ethers (PBDEs) used as flame retardants and perfluorinated compounds (PFCs) used as surfactants, and oil and water repellents for consumer products. Chemicals under evaluation are chlorinated naphthalene, hexabromocyclododecane (HBCD), short-chained chlorinated paraffin, hexachlorobutadiene, and pentachlorophenol.

13.3 Fingerprints of POPs

There is a global concern about long term chronic exposure to POPs that result in possible bioaccumulation in lipid rich tissues of biota and human [15]. POPs can be detected in serum, adipose tissue and breast milk samples taken from human [2, 16]. Sources of POPs are pesticide applications in agricultural activities, industrial wastes and by-products of industrial production [17]. People are exposed to POPs by consuming contaminated food and drinking water, inhalation, and dermal contact. Besides, infants are exposed to POPs during intrauterine development and breastfeeding [18].

POPs adhere to aerosols in air and to sediment in water, and can be transported over very long distances, where they are released from to the environment. Wind, water currents, and migratory animals help for this transportation [19]. These slowly metabolized contaminants biomagnificate in food web and bioaccumulate in adipose tissues. Majority of POPs have high lipid solubility and semi-volatility. High lipid solubility of POPs enables them to pass through placenta. Unlikely, PFCs accumulate in the blood and liver.

Even POPs are not in use any more, they can be determined in every compartment of the environment; in wastewater, sediments, air, animal tissues, etc. Since POPs have hydrophobic character, they bind to the particle fraction in water media and accumulate in sediment. Remobilization of POPs from sediment to water media makes sediment a second source for water contamination. Kilunga et al. [20] evaluated POPs contamination levels of surface sediments in three rivers from Democratic Republic of the Congo. 169 μg/kg total PCBs and 270.6 μg/kg total OCPs were determined in Makelele River, reported as the most polluted river in the region. Sakan et al. [21] reported DDT and its byproducts as the most abundant OCPs in sediment samples taken from rivers and lakes in Serbia. Highest total DDT and its by-products (DDE and *p,p'*-DDD,) was determined as 295 μg/kg. DDT/(DDE + DDD) ratio is generally being used to evaluate recent or historical usage of DDT. In the study, DDT/(DDE + DDD) ratio over 0.1 is attributed to recent applications. These studies reveal the still continuing usage of DDT either because of stocks or illegal productions.

Sediments are defined as contamination sources of water living organisms. Organic fraction and texture (silt/clay fraction) of the sediment also affect accumulation and transportation of POPs in water media. Baran et al. [22] determined the levels of polychlorinated dibenzo-p-dioxin and polychlorinated dibenzofuran (PCDD/F) in sediment samples taken from Rybnic Reservoir in Poland. PCDD/F concentration levels in sediment samples were 2–38 folds higher than sediment quality guidelines in Poland. PCDD/F concentrations were highly correlated with organic matter, especially humic substances in sediment samples. PCDD/F movement potential was reported to be positively correlated with silt/clay fraction and negatively correlated with sand fraction.

Defining source of contaminant exposure for biota is generally conflicting for multiple contaminant sources. Obtaining satisfactory information is especially harder for migrating animals. O'Neill et al. [19] reported polybrominated diphenyl ether (PBDEs) flame retardants and polychlorinated biphenyls (PCBs) exposure of natural and hatchery origin Juvenile Chinook salmon in US Pacific Northwest. This Pacific Salmon is exposed to water contaminants from multiple sources on its migration route. Collected samples were analyzed for PBDEs, PCBs, nitrogen (^{15} N/^{14} N ratio, δ^{15}N), carbon (^{13}C/^{12}C ratio, δ^{13}C), and sulfur isotope (^{34}S/^{32}S ratio, δ^{34}S) ratios. Higher PBDEs were determined for natural origin type, while PCBs were equal for natural and hatchery origin salmon. A correlation between depletion of stable nitrogen isotopes (δ^{15}N) of the salmon and higher PBDE concentrations revealed common wastewater source for both nitrogen and PBDEs. Tracking stable isotopes of carbon (δ^{13}C), nitrogen (δ^{15}N), and sulfur (δ^{34}S) besides POPs help to define ecological process, migratory patterns, diet, trophic level, exposure to sewage, and wastewater etc. [23].

While bioaccumulation of POPs in aquatic species is well documented, there are limited studies about bioaccumulation potential of POPs in terrestrial species [24]. Aydin et al. [25] reported traceable amounts of OCPs, PCBs, and PBDEs in raw and ultra-high temperature (UHT) cow's milk samples from Turkey. Food and feed crops are also defined as important OCPs sources studied [26]. Aydin and Ulvi [27] reported residue levels of 227 pesticides in 13 different types of dried nuts and fruits. Prohibited pesticides were detected in nuts, and among them, cypermethrin, chlorpyrifos and chlorpyrifos-methyl were determined as to pose an acute and chronic risk for children and adults.

Bioaccumulation of POPs is related with body size, lipid content, and trophic level of biota, while trophic magnification factors of POPs is highly related with logarithmic n-octanol/water (log P_{ow}) and n-octanol/air (log P_{oa}) partition coefficients. Degradation rates and bioaccumulation of POPs in tissues of the organism depends not only on chemical structure of POPs, but also on biological structure of the organism itself. Pesticides accumulation was reported to be in male gonads, eggs, liver, and muscles of Pacific salmon species (ordered from the highest to lowest concentrations) [28]. Species in upper trophic levels bioaccumulate POPs in higher concentrations. When invertebrates and fishes in aquatic media are evaluated in the means of POPs accumulation, concentrations of POPs were 2–22 times greater in fish samples than in invertebrate samples [29]. *Cetacea* (marine mammals like whales, dolphins, etc.) defined as high trophic level species store contaminants in their lipid-rich tissues. POPs level in these organisms varied according to species, sex, age, geographic region, and temporal scale [30]. Long life span of these top predators result in binding the contaminants to fatty acids in their blubber, and female dolphins make gestational and lactational offload of their pollutants by lipid rich milk [31].

There is a similar transfer from mother to infant through gestation and breastfeeding. 16 different OCPs were determined in breast milk samples collected from eastern and central cities in Saudi Arabia [32]. Average concentrations of $\sum$HCHs, $\sum$DDTs, $\sum$PCBs and $\sum$PBDEs in 45 individual human milk samples collected from Konya City, Turkey, were reported to be 22.6, 37.1, 105 and 67.3 ng/g lipid wt., respectively [33]. Among urine samples taken from 111 infants, metabolites of organophosphate esters (OPEs) were determined for >70% of samples [18].

POPs can be transferred over long distances from its source. Many studies revealed the arctic marine ecosystem as a global sink of POPs, even there are nearly no POPs production sources in these regions of the world [4]. Low volatility and low degradation of POPs in arctic media was attributed to the very low temperatures. Besides, arctic organisms have slower metabolism and higher lipid storage, resulting in higher levels of POPs bioaccumulation [23, 34]. A 25 years temporal air monitoring study conducted by Arctic Monitoring and Assessment Program (AMAP) in eight arctic stations gave promising results that POPs in the arctic air had a declining trend [35]. Similar POPs declining trend was reported also for biota. Samples taken from male bowhead whales from Alaska were analyzed for POPs. Declining concentrations were reported for the 2006–2015 period [36]. Conducted analyses and obtained results confirmed the effectiveness of control measures taken by the Stockholm Convention.

13.4 Health Effects

POPs are responsible for many lethal diseases, including diabetes, neurodevelopmental disorders, endocrine disturbance, cancer, reproductive and cardiovascular problems, etc. [6, 37, 38]. Adverse effects are more severe in case of fetal and childhood exposure than on adults [39, 40]. Halogenated POPs are more resistant to environmental degradation processes. Among them, chlorinated POPs show the most consistent results in human health studies [41].

PBDEs in flame retardants were proved to have neurodevelopmental disorders on children. Lam et al. [39] evaluated intelligence and attention related deficits of children exposed to PBDEs during their perinatal period and childhood. Results of the work showed that PBDEs exposure caused decrements on IQ (3.7 points in tenfold). Sex is a factor affecting neurodevelopmental disorders caused by PBDEs. Azar et al. [1] conducted a research work to evaluate sex difference on cognitive ability of 592 children exposed to PBDEs. Maternal plasma samples taken in early pregnancy were analyzed for PBDEs. IQ scores of children were evaluated in three years old. Prenatal exposure to PBDEs was found to decrease IQ scores of boys, but not of girls.

It is known that the synergistic effects of chemicals reach a much different dimension than their effects alone [38]. Humans account for continuous combined exposures to POPs. Davidsen et al. [42] studied on 29 different POPs mixtures to evaluate neurodevelopmental disorders in human. Results showed that mixtures of POPs, especially those including brominated and chlorinated compounds had adverse effects on human neural stem cells. POPs mixtures altered synaptogenesis and neurite outgrowth in human. These neurodevelopmental disorders result in impairment of memory and learning in children.

POPs have defined as potential endocrine disrupting chemicals (EDCs) [5]. Many epidemiological and experimental studies well documented endocrine disrupting effects of POPs, not only for livestock, but also for humankind [43, 44]. Exposure to POPs adversely affects reproductive functions of both male and female [45, 46]. POPs bind to steroid receptors and disrupt biosynthesis or metabolism of steroid hormones [47]. Hormonal disruptions make disorders for genital development, sperm production and quality, etc. Warembourg et al. [40] reported disruption of hormonal levels in cord blood as a result of POPs exposure in prenatal period. Highest adverse effects were attributed to phthalates, bisphenol-A and PCBs; especially pesticides such as DDT, PCB153, HCB, PBDEs, and PFCs [48, 49]. Studies in males have shown decreased testosterone production as a result of PCBs exposure [50]. PCBs congeners have different endocrine disrupting effects. Pliskova et al. [51] reported estrogenic effects for lower chlorinated congeners of PCBs, while higher chlorinated congeners of PCBs, such as 138, 153, 170, etc. behaved as anti-estrogens.

There are studies that reveal a correlation between increased cancer cases and exposure to POPs [52, 53]. Park et al. [54] determined an association between serum concentration of chlordane and PCBs, and lung cancer risk in the general population. A dose-dependent effect was reported for PCBs, regardless of their chlorination degree. POPs accumulated in adipose tissues can influence tumor type and stimulate metastasis [55].

POPs cause chronic inflammation in adipose tissues, affecting key mechanisms of obesity and increasing the risk for type 2 diabetes [56]. The increase in the incidence of obesity in recent years has increased concerns about public health. Obesity is also the main risk factor for cardiovascular diseases. Both epidemiological and experimental studies give substantial evidence about multi-dimensional relation between POPs and obesity [57, 58].

13.5 Conclusions

POPs are still in traceable amounts in every compartment of the environment. Although mitigative and preventive measures resulted in declining of production and use of these chemicals, they can still be found even in human tissues. There are new candidate chemicals to extend the POPs list. Hence, it is understood that these chemicals will continue to threaten human health. Although it is known that many health problems are caused by these chemicals, there are not enough awareness about food and drinking water contamination. There is also a need to develop effective and economic wastewater treatment technologies so as to protect clean water sources.

References

1. Azar N, Booij L, Muckle G, Arbuckle TE, Seguin JR, Asztalos E, Fraser WD, Lanphear BP, Bouchard MF. Prenatal exposure to polybrominated diphenyl ethers (PBDEs) and cognitive ability in early childhood. Environ Int. 2021; 146:106296
2. Arrebola JP, Martin-Olmedo P, Fernandez MF, Sanchez-Cantalejo E, Jimenez- Rios JA, Torne P, Porta M, Olea N. Predictors of concentrations of hexachlorobenzene in human adipose tissue: a multivariate analysis by gender in Southern Spain. Environ Int. 2009;35:27–32.
3. Batinau C, Gavrilescu M. Migration and fate of persistent organic pollutants in the atmosphere—a modelling approach. Environ Eng Manag J. 2008;7:743–61.
4. Mckınney MA, Pedro S, Dıetz R, Sonne C, Fısk AT, Roy D, Jenssen BM, Letcher RJ. A review of ecological impacts of global climate change on persistent organic pollutant and mercury pathways and exposures in arctic marine ecosystems. Current Zoology. 2015;61(4):617–28.
5. Kumar M, Sarma DK, Shubham S, Kumawat M, Verma V, Prakash A, Tiwari R. Environmental endocrine-disrupting chemical exposure: role in non-communicable diseases. Front Public Health. 2020;24(8):553850.

6. Alharbi OML, Basheer AA, Khattab RA, Ali I. Health and environmental effects of persistent organic pollutants. J Mol Liq. 2018;263:442–53.
7. United Nations Environment Program (UNEP). Final act of the conference of plenipotentiaries on The Stockholm 732 convention on persistent organic pollutants, Stockholm, Sweden, 22 to 23 May 2001, Geneva, 733 Switzerland.
8. D'eon JC, Mabury SA. Is indirect exposure a significant contributor to the burden of perfluorinated acids observed in humans? Environ Sci Technol. 2011;45:7974–84.
9. Szulzyk-Cieplak J. Removal of hardly biodegradabale organic compounds from wastewater by means of reagentless methods. J Ecol Eng. 2017;18(5):63–71.
10. Urbaniak M, Kiedrzyńska E, Grochowalski A. The variability of PCDD/F concentrations in the effluent of wastewater treatment plants with regard to their hydrological environment. Environ Monit Assess. 2017;189:90.
11. Lin X, Xu J, Keller AA, He L, Gu Y, Zheng W, Sun D, Lu Z, Huang J, Huang X., Li G. Occurrence and risk assessment of emerging contaminants in a water reclamation and ecological reuse Project. Sci Total Environ. 2020;744:140977.
12. Özcan S, Tor A, Aydin ME. Investigation on the levels of heavy metals, polycyclic aromatic hydrocarbons, and polychlorinated biphenyls in sewage sludge samples and ecotoxicological testing. CLEAN-Soil Air Water. 2013;41(4):411–8.
13. Jin P, Jin X, Wang XC, Shi X. An analysis of the chemical safety of secondary effluent for reuse purposes and the requirement for advanced treatment. Chemosphere. 2013;91(4):558–62.
14. Aydin ME, Ozcan S, Beduk F, Tor A. Levels of organochlorine pesticides and heavy metals in surface waters of Konya closed basin, Turkey. Sci World J. 2013;849716:6.
15. Fång J, Nyberg E, Winnberg U, Bignert A, Bergman Å. Spatial and temporal trends of the Stockholm Convention POPs in mothers' milk—a global review. Environ Sci Pollut Res. 2015;22:8989–9041.
16. Aydin S, Aydin ME, Beduk F. Organochlorine pesticides in human milk samples. Int J Ecosyst Ecol Sci. 2016;6(3):323–8.
17. Venier M, Salamova A, Hites RA. How to distinguish urban vs. agricultural sources of persistent organic pollutants? Curr Opin Environ Sci Health. 2019;8:23–8.
18. Hammel SC, Zhang S, Lorenzo AM, Eichner B, Stapleton H., Hoffman K. Young infants' exposure to organophosphate esters: breast milk as a potential source of exposure. Environ Int. 2020;143:106009.
19. O'Neill SM, Carey AJ, Harding LB, West JE, Ylitalo GM, Chamberlin JW. Chemical tracers guide identification of the location and source of persistent organic pollutants in juvenile Chinook salmon (Oncorhynchus tshawytscha), migrating seaward through an estuary with multiple contaminant inputs. Sci Total Environ. 2020;712:135516.
20. Kilunga PI, Sivalingam P, Laffite A, Grandjean D, Mulaji CK, de Alencastro LF, Mpiana PT, Pote J. Accumulation of toxic metals and organic micro-pollutants in sediments from tropical urban rivers, Kinshasa, Democratic Republic of the Congo. Chemosphere. 2017;179:37–48.
21. Sakan S, Ostojić B, Đorđević D. Persistent organic pollutants (POPs) in sediments from river and artificial lakes in Serbia. J Geochem Explor. 2017;180:91–100.
22. Baran A, Mierzwa-Hersztek M, Urbaniak M, Gondek K, Tarnawski M, Szara M, Zieliński M. An assessment of the concentrations of PCDDs/Fs in contaminated bottom sediments and their sources and ecological risk. J Soils Sediments. 2020;20:2588–97.
23. Ko FC, Pan WL, Cheng JO, Chen TH, Kuo FW, Kao SJ, Chang CW, Ho HC, Wang WH, Fang LS. Persistent organic pollutants in Antarctic notothenioid fish and invertebrates associated with trophic levels. PLoS ONE. 2018;13(4):0194147.
24. Fremlin KM, Elliott JE, Green DJ, Drouillard KG, Harner T, Eng A, Gobas FAPC. Trophic magnification of legacy persistent organic pollutants in an urban terrestrial food web. Sci Total Environ. 2020;714:136746.
25. Aydin S, Aydin ME, Beduk F, Ulvi A. Organohalogenated pollutants in raw and UHT cow's milk from Turkey: a risk assessment of dietary intake. Environ Sci Pollut Res. 2019;26:12788–97.
26. Beduk F, Aydin ME, Aydin S, Tekinay A, Bahadir M. Pesticide pollution in soil and wheat: risk assessment of contaminated food consumption. Fresenius Environ Bull. 2017;26(3):2330–9.
27. Aydın S, Ulvi M. Residue levels of pesticides in nuts and risk assessment for consumers. Quality Assur Saf Crops Foods. 2019;11(6):539–48.
28. Lukyanowa ON, Tsygankov VN, Boyarova MD, Khristoforova NK. Bioaccumulation of HCHs and DDTs in organs of Pacific salmon (genus Oncorhynchus) from the Sea of Okhotsk and the Bering Sea. Chemosphere. 2016;157:174–80.
29. Windsor FM, Pereira MG, Tyler CR, Ormerod SJ. River organisms as indicators of the distribution and sources of persistent organic pollutants in contrasting catchments. Environ Pollut. 2019;255:113144.
30. Balmer B, Ylitalo G, Watwood S, Quigley B, Bolton J, Mullin K, Rosel P, Rowles T, Speakman T, Wilcox L, Zolman E, Schwacke L. Comparison of persistent organic pollutants (POPs) between small cetaceans in coastal and estuarine waters of the northern Gulf of Mexico. Mar Pollut Bull. 2019;145:239–47.
31. Murphy S, Law RJ, Deaville R, Barnett J, Perkins MW, Brownlow ., Penrose R, Davison NJ, Barber JL, Jepson PD. Organochlorine contaminants and reproductive implication in cetaceans: a case study of the common dolphin, organochlorines and

reproductive implication in Cetaceans Chapter 1, Marine mammal ecotoxicology, impacts of multiple stressors on population health; 2018. pp. 3–38.
32. EL-Saeid MH, Hassanin AS, Bazeyad AY. Levels of pesticide residues in breast milk and the associated risk assessment. Saudi J Biol Sci. 2021. Article in press.
33. Özcan S, Tor A, Aydın ME. Levels of organohalogenated pollutants in human milk samples from Konya City, Turkey. CLEAN-Soil Air Water. 2011;39(10):978–83.
34. Riget F, Bignert A, Braune B, Da M, Dietz R, Evans M, Green N, Gunnlaugsdottir H, Hoydal KS, Kucklick J, Letcher R, Muir D, Schuur S, Sonne C, Stern G, Tomy G, Vorkamp K, Wilson S. Temporal trends of persistent organic pollutants in Arctic marine and freshwater biota. Sci Total Environ. 2019;649:99–110.
35. Wong F, Hung H, Dryfhout-Clark H, Aas W, Bohlin-Nizzetto P, Breivik K, Mastromonaco MN, Lundén EB, Ólafsdóttir K, Sigurðsson A, Vorkamp K, Bossi R, Skov H, Hakola H, Barresi E, Sverko E, Fellin P, Li H, Vlasenko A, Zapevalov M, Samsonov D, Wilson S. Time trends of persistent organic pollutants (POPs) and chemicals of emerging arctic concern (CEAC) in Arctic air from 25 years ofmonitoring. Sci Total Environ. 2021;775:145109.
36. Bolton JL, Ylitalo GM, Chittaro P, George JC, Suydam P, Person BT, Gates JB, Baugh KA, Sformo T, Stimmelmayr R. Multi-year assessment (2006–2015) of persistent organic pollutant concentrations in blubber and muscle from Western Arctic bowhead whales (Balaena mysticetus), North Slope, Alaska. Marine Pollut Bull. 2020;151:110857.
37. Myhre O, Zimmer K., Hudecova AM, Hansen KEA, Khezri A, Berntsen HF, Berg V, Lyche JL, Mandal S, Duale N, Ropstad E. Maternal exposure to a human based mixture of persistent organic pollutants (POPs) affect gene expression related to brain function in mice offspring hippocampus. Chemosphere. 2021;276:130123.
38. Berntsen HF, Duale N, Bjørklund CG, Rangel-Huerta OD, Dyrberg K, Hofer T, Rakkestad KE, Østby G, Halsne R, Boge G, Paulsen RE, Myhre O, Ropstad E. Effects of a human-based mixture of persistent organic pollutants on the in vivo exposed cerebellum and cerebellar neuronal cultures exposed in vitro. Environ Int. 2021;146:106240.
39. Lam J, Lanphear BP, Bellinger D, Axelrad DA, McPartland J, Sutton P, Davidson L, Daniels N, Sen S, Woodruff TJ. Developmental PBDE exposure and IQ/ADHD in childhood: a systematic review and meta-analysis. Environ Health Perspect. 2021;125 (8):086001.
40. Warembourg C, Debost-Legrand A, Bonvallot N, Massart C, Garlante´zec R, Monfort C, Gaudreau E, Chevrier C, Cordier S. Exposure of pregnant women to persistent organic pollutants and cord sex hormone levels. Hum Reprod. 2016;31(1):190–198.
41. Lind PM, Lind L. Endocrine-disrupting chemicals and risk of diabetes: an evidence-based review. Diabetologia. 2018;61:1495–502.
42. Davidsen N, Lauvås AJ, Myhre O, Ropstad E, Carpi D, Gyves EM, Berntsen HF, Dirven H, Paulsen RE, Price A, Pistollato F. Exposure to human relevant mixtures of halogenated persistent organic pollutants (POPs) alters neurodevelopmental processes in human neural stem cells undergoing differentiation. Reprod Toxicol. 2021;100:17–34.
43. Elobeid MA, Padilla MA, Brock DW, Ruden DM, Allison DB. Endocrine disruptors and obesity, an examination of selected persistent organic pollutants in the NHANES 1999–2002 data. Int J Environ Res Public Health. 2010;7:2988–3005.
44. Rasinger JD, Carroll TS, Maranghi F, Tassinari R, Moracci G, Altieri I, Mantovani A, Lundebye AK, Hogstrand C. Low dose exposure to HBCD, CB-153 or TCDD induces histopathological and hormonal effects and changes in brain protein and gene expression in juvenile female BALB/c mice. Reprod Toxicol. 2018;80:105–16.
45. Toft G, Hagmar L, Giwercman A, Bonde JP. Epidemiological evidence on reproductive effects of persistent organochlorines in humans. Reprod Toxicol. 2004;19:5–26.
46. Gregoraszczuk EL, Ptak A. Endocrine-disrupting chemicals: some actions of POPs on female reproduction. Int J Endocrinol. 2013;828532:9.
47. WHO, UNEP. State of the science of endocrine disrupting chemicals. World Health Organization; 2012.
48. Vested A, Giwercman A, Bonde JP, Toft G. Persistent organic pollutants and male reproductive health. Asian J Androl. 2014;16:71–80.
49. Emeville E, Giton F, Giusti A, Oliva A, Fiet J, Thomé JP, Blanchet P, Multigner L. Persistent organochlorine pollutants with endocrine activity and blood steroid hormone levels in middle-aged men. PLoS ONE. 2013;8(6):66460.
50. Fergusona KK, Hauserb R, Altshulb L, Meekera JD. Serum concentrations of P, P'-DDE, HCB, PCBs and reproductive hormones among men of reproductive age. Reprod Toxicol. 2012;34(3):429–35.
51. Plísková M, Vondrácek J, Canton RF, Nera J, Kocan A, Petrík J, Trnovec T, Sanderson T, van den Berg M, Machala M. Impact of Polychlorinated Biphenyls Contamination On Estrogenic Activity In Human Male Serum. Environ Health Perspect. 2005;113(10):1277–84.
52. Galban-Velazquez S, Esteban J, Çakmak G, Artacho-Cordon F, Leone J, Barril J, Vela-Soria F, Martin-Olmedo P, Fernandez MF, Pellín MC, Arrebola JP. Associations of persistent organic pollutants in human adipose tissue with retinoid levels and their relevance to the redox microenvironment. Environ Res. 2021;195:110764
53. Koukoulakis KG, Kanellopoulos PG, Chrysochou E, Costopoulou D, Vassiliadou I, Leondiadis L, Bakeas E. Atmospheric concentrations and health

implications of PAHs, PCBs and PCDD/Fs in the Vicinity of a heavily industrialized site in Greece. Appl Sci. 2020;10:9023.
54. Park EY, Park E, Kim J, Oh JK, Kim B, Hong YC, Lim MK. Impact of environmental exposure to persistent organic pollutants on lung cancer risk. Environ Int. 2020;143:105925.
55. Kouala M, Cano-Sanchod G, Batsa AS, Tomkiewicza C, Kaddouch-Amara Y, Douay-Hausera N, Ngo, Bonsang H, Deloménie M, Lecurua F, Le Bizecd B, Marchand P, Botton J, Barouki R, Antignac J, Coumoul X. Associations between persistent organic pollutants and risk of breast cancer metastasis. Environ Int. 2019;132:105028.
56. Lee YM, Jacobs DR Jr, Lee D. Persistent organic pollutants and type 2 diabetes: a critical review of review articles. Front Endocrinol. 2018;9:712.
57. Arrebola JP, Pumarega J, Gasull M, Fernandez MF, Martin-Olmedo P, Molina- Molina JM, Fernandez-Rodriguez M, Porta M, Olea N. Adipose tissue concentrations of persistent organic pollutants and prevalence of type 2 diabetes in adults from Southern Spain. Environ Res. 2013;122:31–7.
58. Arrebola JP, Ocana-Riola R, Arrebola-Moreno AL, Fernandez-Rodriguez M, Martin-Olmedo P, Fernandez MF, Olea N. Associations of accumulated exposure to persistent organic pollutants with serum lipids and obesity in an adult cohort from Southern Spain. Environ Pollut. 2014;195:9–15.

14 Wastewater-Based Epidemiology (WBE) Studies for Monitoring of Covid-19 Spread

Bilge Alpaslan Kocamemi, Halil Kurt, Esra Erken, and Ahmet Mete Saatçi

Abstract

Severe acute respiratory syndrome coronavirus 2 (SARS-CoV-2) started in Wuhan, China, in December 2019 and was announced as a worldwide pandemic spread by WHO on March 11, 2020. The countries have started to monitor surveillance of SARS-CoV-2 through medical tests. However, people with no or very light symptoms are usually not medically tested or never hospitalized, and their test results missed. In March 2020, it was recognized that the urine and feces of all infected people contain SARS-CoV-2. After that, wastewater-based epidemiology studies have gained significant importance around the world. This chapter aims to describe the basics of wastewater-based epidemiology (WBE) studies, the current situation in the world together with the post-Covid-19 approaches. Additionally, the major challenges in Covid-19 WBE studies are discussed.

Keywords

Covid-19 · Pandemic · SARS-CoV-2 · Wastewater Based Epidemiology (WBE)

14.1 Introduction

Covid-19, which is an infectious disease caused by a newly discovered corona virus, became a global pandemic in about two months after outbreak in Wuhan, China, in December 2019. The Covid-19 pandemic differs from previous pandemics in terms of its impact and prevalence across countries and sectors. Due to easy transportation around the world and the broad communication network, no pandemic has spread so fast or led to such extensive information-sharing.

As of 15th of July 2021, there have been 188,128,952 confirmed cases infected by SARS-CoV-2 and 4,059,339 people lost their lives globally [1]. Diagnosis of Covid-19 cases has been performed through polymerase chain reaction (PCR) test of samples from nasopharyngeal and throat swabs. But there are numerous studies that detected SARS-CoV-2 in feces of both symptomatic and asymptomatic patients [2–6].

B. A. Kocamemi (✉) · E. Erken
Environmental Engineering Department, Marmara University, Göztepe, Turkey
e-mail: bilge.alpaslan@marmara.edu.tr

H. Kurt
International Faculty of Medicine, Department of Medical Biology, Saglik Bilimleri University, Istanbul, Turkey

A. M. Saatçi
Turkish Water Institute (SUEN), Üsküdar, Turkey

M. Bahadir and A. Haarstrick (eds.), *Water and Wastewater Management*,
Water and Wastewater Management, https://doi.org/10.1007/978-3-030-95288-4_14

Moreover, some studies suggested the possibility of extended duration of viral shedding in feces after the patients' respiratory samples were negative [7–12]. In this respect, SARS-CoV-2 wastewater-based epidemiology (WBE), i.e., wastewater surveillance, aiming to estimate the distribution of asymptomatic and symptomatic individuals in a specific region has received worldwide attention. Various research groups worldwide have started SARS-CoV-2 detection in wastewater, since WBE provides tracking the whole population by testing a small number of wastewater samples in a specific region and can predict SARS-CoV-2 RNA in human feces a few days to a week before the onset of symptoms [13–53]. This makes WBE a quite economic tool for continual tracking of decreasing or increasing trend of the Covid-19 in a particular region. Many countries have been defining hot spots, e.g., dormitories, metro stations, and airports to develop an early-warning tool for determining the presence of Covid-19 in a community [54, 55]. However, wastewater surveillance studies have not been properly integrated into the Covid-19 management strategy by decision-makers yet. There are only few countries in the world (e.g., Turkey, The Netherlands, and Canada) applying WBE studies nationwide. A major challenge of WBE studies for tracking Covid-19 is the duration and recovery efficiency of SARS-CoV-2 RNA in wastewater prior to real time quantitative polymerase chain reaction (RT-qPCR) measurements. The recovery of SARS-CoV-2 from wastewater samples has been accomplished through recovery methods (e.g., polyethylene glycol (PEG), Ultrafiltration (UF), skimmy milk), which are in use for other viruses. However, all these methods have some deficiencies in the application of WBE studies for routine Covid-19 tracking. Additionally, new primers are under development for the mutant variants of SARS-CoV-2 including B.1.1.7., B.1.351, P1 and B.1.617.2 that have been identified up to now.

This chapter provides a brief overview of the WBE studies applied to track Covid-19 by discussing the major benefits, current activities in the world, molecular methods applied and post-covid approaches.

In view of the recent Covid-19 pandemic, this chapter will also highlight future perspectives for the management of water and wastewater systems under conditions of public health crises.

14.2 Corona Viruses and SARS-CoV-2

Viruses are microscopic pathogenic agents that are abundant in water and wastewater [56]. Transmission of viruses to humans generates several diseases and epidemics that affect all age groups, being fatal for children in some cases (e.g., gastroenteritis leading to mortality during the first five years of life) [57]. The most common groups of viruses generating disease and epidemics for human being are Picornaviridae (e.g., polioviruses, echoviruses, and hepatitis A), Reoviridae (e.g., reoviruses, rotaviruses), Adenoviridae (e.g., adenovirus A.), Coronaviridae (e.g., corona viruses), Caliciviridae (e.g., caliciviruses), small round viruses (e.g., astroviruses, Norwalk), and bacteriophages [56, 58–63]. Among these, corona viruses cause respiratory and intestinal infections in animals and humans [64].

Corona viruses, which are important pathogens for human and vertebrates, belong to the subfamily of Coronavirinae. According to ICTV 2020 classification (https://talk.ictvonline.org), Coronavirinae subfamily consist of four genera: Alphacoronavirus, Betacoronavirus, Gammacoronavirus, and Deltacoronavirus. Virions are spherical shape of 120–160 nm diameter. The particles are classically covered with large, club- or petal-like surface spike proteins. Under electron microscopy, spherical particles generate an impression similar of the solar corona. Thus, entire family of this virus are called coronaviruses. Nucleocapsids of virus are helical and can be removed by detergent treatment.

So far, six types of coronaviruses have been identified that can infect humans. Among them, HCoV-229E and HCoV-NL63 belong to Alphacoronavirus, while HCoV-OC43, HCoV-

HKU1, SARS-CoV, MERS-CoV belong to Betacoronavirus (Fig. 14.1). They are considered to be highly pathogenic to humans with the outbreak of severe acute respiratory syndrome (SARS) in 2002 and 2003 [65]. Ten years after SARS, another highly pathogenic coronavirus, Middle East respiratory syndrome coronavirus (MERS-CoV) emerged in Middle Eastern countries [66]. Emerging human pathogens of SARS, MERS-CoV and SARS-CoV-2 viruses are members of Betacoronavirus.

SARS-CoV-2 is an enveloped, non-segmented, positive-sense RNA virus, which is broadly distributed in humans and other mammals. Its diameter is about 65–125 nm, containing single strands of RNA and provided with crown-like spikes on the outer surface [67]. Structurally, SARS-CoV-2 has four main structural proteins including spike (S) glycoprotein, small envelope (E) glycoprotein, membrane (M) glycoprotein, and nucleocapsid (N) protein, and also several accessory proteins [68]. SARS-CoV genome structure and functional domains of S protein are shown in Fig. 14.2.

14.3 Covid-19 Pandemic

In December 2019, a local outbreak of pneumonia caused by severe acute respiratory syndrome coronavirus 2 (SARS-CoV-2) was detected in Wuhan (Hubei, China) [61].The outbreak has spread so quickly that on March 11, 2020, the World Health Organization (WHO) announced the Coronavirus Disease 2019 (Covid-19) outbreak as a global pandemic. John Hopkins Coronavirus Resource Center interactive dashboard has been reporting global confirmed cases and deaths [70].

Different tools (Fig. 14.3) are used to monitor Covid -19, (a) RT-qPCR for SARS-CoV-2 RNA in rhinopharyngeal swabs, (b) serological surveillance through testing the blood of individuals for antibodies against SARS-CoV-2, and (c) wastewater surveillance for detection of SARS-CoV-2 RNA. It is noteworthy to mention that, while swab test and serological surveillance are used to diagnose Covid-19, wastewater surveillance helps to monitor the spread of the disease. RT–qPCR tests can detect the RNA

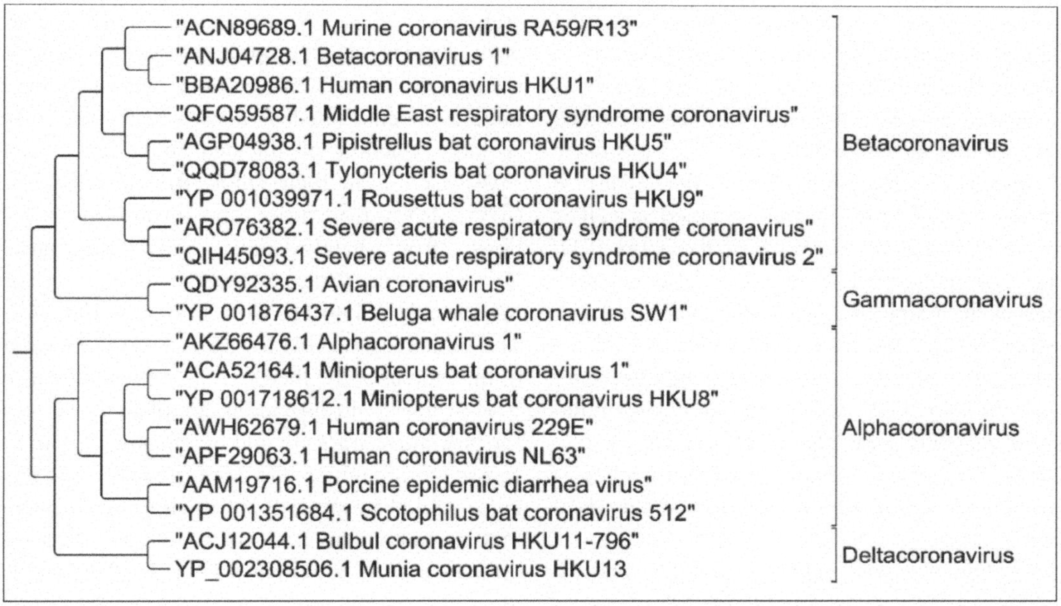

Fig. 14.1 Phylogenetic relationships among the members of the subfamily Coronavirinae. A rooted Neighbor-joining tree was generated from amino acid sequence alignments of the spike protein. The tree reveals four main monophyletic clusters corresponding to genera Alpha-, Beta-, Gamma- and Deltacoronavirus [66]

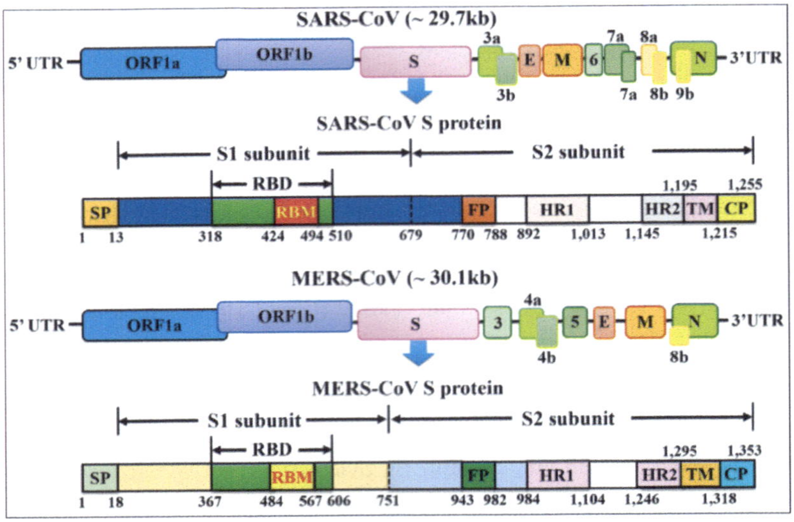

Fig. 14.2 SARS-CoV and MERS-CoV genome structure and functional domains of S protein. 16 non-structural proteins (nsp1–nsp16) are encoded by the ORF1a and ORF1b genes. Spike (S), envelope (E), membrane (M), and nucleocapsid (N) genes are encode structural proteins. Spike protein has two subunits called S1 and S2. CP, cytoplasm domain; FP, fusion peptide; HR, heptad repeat; RBD, receptor-binding domain; RBM, receptor-binding motif; SP, signal peptide; TM, transmembrane domain [69]. (*This picture is licensed under the Creative Commons Attribution 4.0 International license.*)

genome of SARS-CoV-2 by nucleic acid assays, however, the detection of the analyte requires that a sample contains sufficient viral genome copies or levels of viral proteins that exceed the limit of detection for a given assay. Moreover, the abundance of these viral analytes varies significantly between different anatomical locations and different stages of infection. The sensitivity of tests for detecting SARS-CoV-2 can be dependent on the time and site sampled as well as on the technical performance of the assay [71]. One systematic review reported false negative rates in the range of 2–33% in repeated sample testing [72]. Moreover, technical problems including contamination during sampling, contamination by PCR amplicons, and contamination of reagents could also result in false positive results. Therefore, a reliable detection method is required for sustainable monitoring of Covid-19 spread.

Serological tests help to monitor and to understand, how many individuals in a population have been infected with SARS-CoV-2. However, serosurveillance is a retrospective tool, as the development of an immune response and detectable antibody levels in blood require several weeks [73]. Wastewater surveillance has been proposed as a complementary approach for tracing the virus transmission within human population connected to a wastewater network. A large fraction of infected individuals shed SARS-CoV-2 in their stool [8]. Also, SARS-CoV-2 RNA from urine and respiratory secretions (from handwashing, showering, nasal lavages, tissues) may contribute to the load of SARS-CoV-2 into the sewer system [29, 73]. The virus may survive for up to several days out of the human body, therefore, its measurement in wastewater could provide a complementary tool for preventive tracking and diagnosing Covid-19 across communities [74]. Wastewater surveillance could be especially informative considering that asymptomatic patients cannot be detected during clinical surveillance [75]. Further details regarding wastewater surveillance are described under Sect. 14.4.

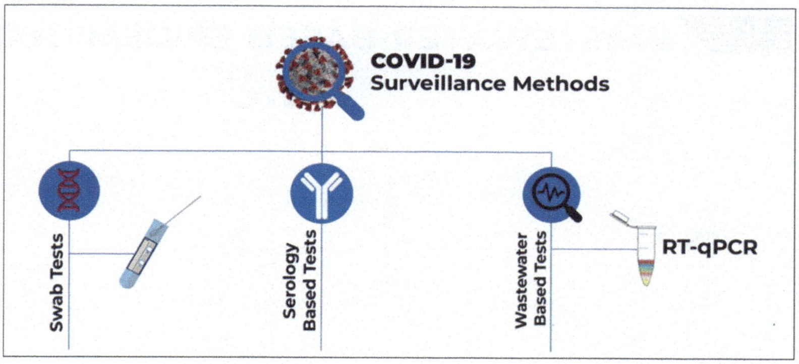

Fig. 14.3 Various Covid-19 surveillance methods

14.4 Wastewater-Based Epidemiology and Its Use for Covid-19 Track

Wastewater is all the water from homes and urban public facilities (hospitals, schools, etc.) as well as from certain industries (if it does not require specific treatment). This water is conveyed, via the "sewer-age system" to treatment plants, where it is treated and then discharged into the environment. WBE is an approach based on the chemical analysis of pollutants and biomarkers in raw wastewater in order to obtain qualitative and quantitative data on the activity of inhabitants within a given wastewater catchment [76]. Figure 14.4 summarizes the major applications of WBE. At the moment, majority of the studies on WBE are focused on the assessment of the presence and consumption of illicit (e.g., cocaine, heroin) and licit (e.g., caffeine, nicotine, alcohol) drugs in different areas [76]. However, community-wide exposure to pharmaceuticals and personal care products (PPCPs), toxicants, contaminants and carcinogens (e.g., nicotine, plasticizers, pesticides, PAFS, phthalates) are already in their incipient stages. It is also known that the health status of the population (e.g., diabetes, oxidative stress, cancer) may be evaluated by monitoring a number of biomarkers. Epidemiology of some pathogens (e.g., cryptosporidiosis, giardiasis), antibiotic resistance genes and viruses is also possible. WBE monitoring is a tool to guide epidemiological surveillance and mitigation efforts for infectious diseases, such as the Global Polio Eradication Initiative [77]. Figure 14.5 presents distinct milestones in the use of WBE studies for public health.

In WBE studies, representative wastewater samples (24-h composite) are collected at the influent of a wastewater treatment plant (WWTP). These samples are considered a complex matrix with relatively high concentrations of other compounds that might interfere with the determination of target biomarkers present at trace or ultratrace levels (ng/L range). The sample preparation processes commonly include sample filtering or centrifugation and preconcentration steps, usually by solid-phase extraction (SPE). Recently, there has been an increase in none target analysis studies to identify new substances susceptible to be used as biomarkers in WBE [8]. While extracts are mainly analyzed by liquid chromatography and mass spectrometry (LC–MS), RT-qPCR analyses are performed to detect the presence of specific genetic material in any pathogen, including viruses.

WBE has been shown to have an excellent potential for early detection of the outbreak of diseases, e.g., of Covid-19. Considering the spread rate and high population in various parts of the world, surveillance of disease prevalence during pandemic such as Covid-19 is a crucial task. The massive testing of the population to monitor the spread of the virus is a challenge. Moreover, as majority of the infected individuals are asymptomatic, it is not possible to obtain reliable data just with testing approach. Asymptomatic and symptomatic infections result in

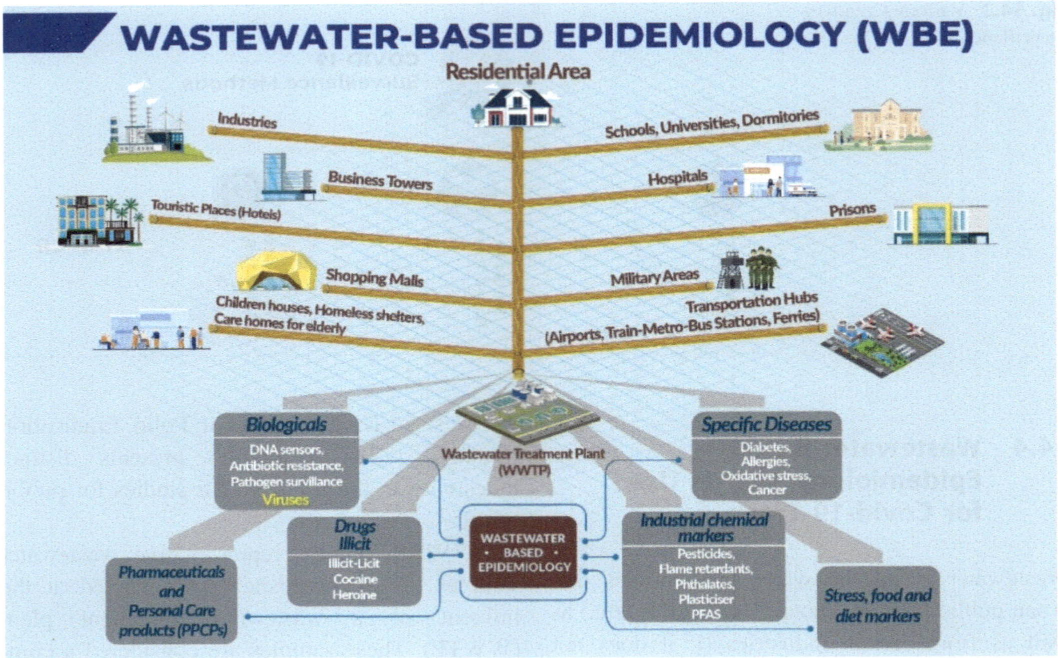

Fig. 14.4 Major applications of WBE studies

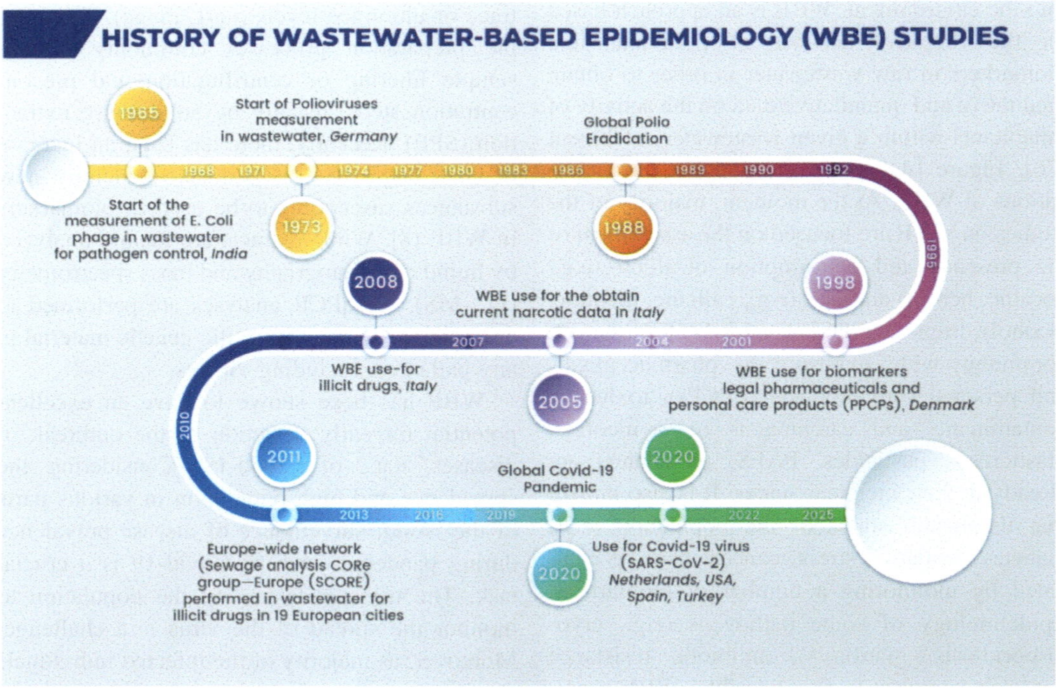

Fig. 14.5 Distinct milestones in WBE studies

significant uncertainty in the estimated extent of SARS-CoV-2 infection [64]. A high proportion of cases had persistently positive viral tests from rectal swabs even after nasopharyngeal swab results became negative, suggesting that the duration of virus shedding in the gastrointestinal tract could be longer than in the respiratory tract [78]. Viral shedding by asymptomatic or mild cases of Covid-19 are common [79]. In particular, asymptomatic transmission route of SARS-CoV-2 challenges contact tracing efforts [80]. Regarding the testing capacity and costs as well as the limitations in availability of these tests equally throughout the world countries, there is a need for alternative strategies to assess the spread of the disease. WBE provides a response to emerging infections [81–83]. Furthermore, wastewater surveillance can be used to determine the burden of undiagnosed infections at the population level. Circulation of SARS-CoV-2 can be monitored within communities, both large communities such as inhabitants of a large city or smaller communities, such as nursing homes, university campuses, prisons, touristic locations and airports. It is a very cost-efficient tool for population surveillance. Also, it does not require repeated sampling of individuals (such as nursing home inhabitants) to survey for (absence of) SARS-CoV-2 circulation in that community. Sewage surveillance has been suggested for areas with limited access to health or testing facilities provided that adequate sewage collection system is available [73]. Few studies demonstrated that virus RNA could be detected in the wastewater before the cases were clinically reported [34, 49, 73, 84]. These studies prove the power of WBE as an early warning system of increased transmission potential of the disease and identification of mutants. Based on WBE studies, decision makers may implement more significant measures to protect public.

14.5 Covid-19 Wastewater-Based Epidemiology Studies in the World

Detection of RNA fragments of SARS-CoV-2 in untreated wastewater has been reported in a number of studies including The Netherlands [13–15], USA [7, 16–21], Australia [22], France [23–25], China [26], Spain [27–31], Italy [32–34], Israel [35], Turkey [36–39], Germany [40], Japan [41, 42], India [43, 44], Pakistan [45], Brazil [46, 47], Chile [48], Denmark, France, Belgium [49], Ecuador [50], Sweden [51], Czech Republik [52] and Bangladesh [53]. However, majority of these studies were conducted either in only a couple of treatment plants or in a confined area. Nationwide studies are limited to few countries like Turkey and The Netherlands [15, 38, 39]. An important point to consider is that in most of the studies influent samples are screened for SARS-CoV-2, whereas effluent and sludge samples also need to be monitored for a comprehensive understanding of the fate of the virus [21, 24, 25, 37, 42–44, 48, 50, 84, 85]. Efforts are ongoing to establish a correlation between wastewater SARS-CoV-2 RNA concentrations and Covid-19 clinical case reports. As of now, very few countries have been using the information from surveillance studies to support decision making about whether to adjust public health and social measure, and few of them have been sharing their data with dashboards.

Although, wastewater based epidemiological surveillance can be used as an efficient early warning tool for outbreaks such as Covid-19, there are some challenges such as the sampling strategy, measurement methodology and availability of required facilities and reagents. Selection of the sampling methods is a longstanding challenge. For wastewater and sludge, grab samples represent a single time and are highly influenced by daily fluctuations in wastewater flow and composition. Composite samples are collected by pooling multiple grab samples at a specified frequency over a set time period—typically 24 h for wastewater surveillance. Moore swab method, which is another surveillance method, has been suggested to be more sensitive than the grab sampling method and can easily be applied at any building level, like schools, nursing homes, prisons, and public/private buildings [78]. Dilution of wastewater due to precipitation or larger flows in the morning and stability of virus and genome fragments in sewage may lead to inconsistent results [86]. Also, conflicting results have been obtained between manual sampling and automated sampling approaches [16]. Another challenge is the requirement of a large sample volume in low infected regions or time periods, when the virus concentration is low. In such cases, sampling location within a treatment plant (influent, effluent or sludge) gains more importance. Sludge is an important compartment to monitor SARS-CoV-2, particularly, during surge declining phases. Turkey is the first country that demonstrated SARS-CoV-2 surveillance with sewage sludge samples [37]. Absence of a standard or optimized protocol for SARS-CoV-2 detection or quantification in sewage is also an important challenge [9]. Besides difficulties associated with isolating culturable viruses from feces to the different protocols have ended up with contradictory results [87].

Various virus concentration methods (ultrafiltration, filtering through electronegative membranes, PEG precipitation, ultracentrifugation, aluminum-driven flocculation) have been used to improve the maximum recovery efficiency of SARS-CoV-2, however, issues such as cost, availability of reagents and access to instruments as well as scalability have been major challenges in this context. In addition, previous studies have shown that some standard virus concentration methods are inefficient to recover enveloped viruses from environmental water samples, since most of these studies have established the methods for non-enveloped enteric viruses [8, 88].

In fact, the RT-qPCR assays used in wastewater based epidemiology have been developed for clinical testing, and thus not meant for a heterogeneous sample matrix like wastewater [8]. The sample complexity is a challenge, as it can lead to PCR inhibition due to impurities or irrelevant genes. Access to laboratories with advanced containment capabilities, meaning those with a biosafety level (BSL) of 2 or higher to conduct RT-qPCR test for SARS-CoV-2 is also a challenge.

14.6 Quantification of SARS-CoV-2 and Mutant Variants

Effective and reliable extraction of SARS-CoV-2 virus from wastewater is the most important step in wastewater-based epidemiological studies. Thus, different concentration methods such as ultracentrifugation, PEG 8000 adsorption, ultrafiltration, aluminum-driven flocculation, skimmed-milk flocculation and electronegative membrane have been used [22, 23, 37, 89–91]. Using rapid, cost-effective, and sensitive methods for concentrating SARS-CoV-2 RNA in wastewater are vital for SARS-CoV-2 studies. There is no standard or optimal methods for concentrating SARS-CoV-2 RNA in wastewater. Comparisons between different concentration methods are very difficult as each study uses different methods and different surrogate coronavirus. Methodological studies were carried out to eliminate this deficiency and to compare different methods [92–95]. LaTurner et al. tested electronegative filtration (HA filtration) with bead beating, electronegative filtration with elution, ultrafiltration, PEG precipitation, ultracentrifugation and direct extraction [92]. They found HA filtration with bead beating was best in terms of sensitivity and cost. Barril et al. tested specific recovery of SARS-CoV-2 with 11 concentration techniques [93]. They found PEG precipitation and aluminum polychloride flocculation have high efficiency with 62.2% and 45.0%, respectively. Ahmed et al. compared seven concentration methods [94]. They showed adsorption-extraction methods could provide rapid recovery of SARS-CoV-2 from wastewater. Best recovery was observed in electronegative membrane with the addition of $MgCl_2$. Philo et al. also compared different methods for virus concentration [95]. They concluded that skimmed milk flocculation was best due to its detection consistency and simplicity. Despite all the work done, unfortunately, consensus among the methods could not be achieved. Skimmed milk flocculation, PEG 8000 and electronegative filtration stand out, because they are easy to use and cheap.

Quantitative virus analyses are performed using primers, which are targeting genes encoding N, orf1ab and S protein by RT-QPCR approach. Currently, Kitajama et al. summarized available primers and probes for SARS-CoV-2 [8]. The performance of seven primer sets targeting RdRp, E, and N genes were evaluated in clinical samples by van Kasteren et al. [96]. The authors found that RT-qPCR assays targeting N and E gene performed best and showed the highest sensitivity. Due to the low amount of virus in wastewater and the presence of various inhibitors in the wastewater matrix, care should be taken when evaluating the results. Mutations in the regions targeted by the RT-qPCR primers reduce the effectiveness of the primers used. Because of all these difficulties, it is very important to use molecular process controls while experimenting. Because of all these difficulties, it is very important to use synthetic DNA fragments of target with known copy number as a control for RT-qPCR experiment.

The main purposes of WBE studies are the quantification of SARS-CoV-2 in wastewater as well as the detection of new and potential future virus variants. For this purpose, the detection of variants with new generation sequencing technologies such as illumina amplicon sequencing is becoming widespread along with RT-qPCR with special primers to detect variants [97–99].

Although, the RT-qPCR method is considered the gold standard for the quantification of viruses in wastewater, using electrochemical immunosensors methods has the potential to provide faster results and is more practical than using the conventional PCR-based approach [100].

SARS-CoV-2 viral RNA synthesis is executed by the nsp12 RNA- dependent RNA polymerase (RdRP). Coronaviruses has unique RNA proof-reading activity. Nsp14 is responsible for the 3′–5′ exonuclease activity. Because of error-correcting nature of Nsp14 protein, replication error rates are more than tenfold lower than that of other RNA viruses [101]. Dilucca et al. showed different selective pressure over viral proteins [102]. For example, M and E integral proteins are under low natural selection. However, the S protein is under high selective pressure due to host response [102]. Simultaneous emergence of SARS-CoV-2 variants on different continents creates a great concern about latest pandemic. The SARS-CoV-2 variant with a single D614G mutation in the S protein was detected in March 2020 and became the global dominant variant within the next few months due to increased transmissibility [103, 104]. The emergence of three SARS-CoV-2 variants [B.1.1.7 (UK), B.1.351 (SA), and P.1 (B.1.1.28.1) (Brazil)] since the last months of 2020 has raised further concerns [105]. Lastly, B.1.617.2 variants are emerged in India. Those are recent Variants of Concern due to increased transmissibility and virulence of Covid-19 epidemiology. World Health Organization (WHO) start using Greek alphabet to assign variants. Alpha, beta, gamma and delta are assigned to B.1.1.7, B.1.351, P.1 and B.1.617.2, respectively. Along with Variants of Concern, Variants of interest were also listed in the WHO variant web site (https://www.who.int/en/activities/tracking-SARS-CoV-2-variants/).
These variants have become dominant variants due to their advantages in virus replication, transmissibility and ability to escape from the immune system. For this reason, excessive monitoring should be performed for SARS-CoV-2 variants in wastewater.

14.7 Post-Covid-19 Approaches

Decision makers are the most critical players in protection of the population from the Covid-19 disease. Though reliable information on the virus and its transmission mechanisms are limited, the rational policy decision should be built on the best available scientific evidence such as dashboards based on wastewater based surveillance studies along with the direction of scientific committees including public health specialists. Turkey is the first country presenting the results of nationwide Covid-19 wastewater surveillance study results with dashboards on a governmental web page [106].

Biosensors provide a great advantage to detect the minimal level of viruses in complex matrices, such as wastewater, provided that appropriate concentration methods are applied. Furthermore, biosensors can potentially provide rapid and real-time data for government agencies to monitor wave trends and to establish an effective early warning system to prevent community-based disease outbreaks [107].

The SARS-CoV-2 is a serious concern for the irrigation uses, particularly for the wastewater stakeholders of the world. In developing countries, the unregulated use of wastewater for irrigation demands has to be considered for the WBE. In this case, methods for irrigation with wastewater without proper treatment may be advised to operate with caution in order to avoid contact with SARS-CoV-2. Currently, there is no evidence confirming the survival of SARS-CoV-2 virus after the disinfection process for both drinking water treatment plant and centralized WWTP. However, medical wastewater may concentrate the SARS-CoV-2 RNA, hence, disinfection of treated effluent in medical WWTP should follow specific guidelines that include the concentration of SARS-CoV-2 RNA copy number in such water matrix.

WWTPs have been identified as an important reservoir of antibiotic resistance genes (ARG), which are often expressed in antibiotic resistance bacteria (ARB) and thus posing significant risks to human health and ecosystem. Thus, the extensive experience gained in monitoring and quantification of the SARS-CoV-2 virus in wastewater should be adapted to detect ARGs as well.

Interdisciplinary studies that include epidemiological, environmental, social, and economic interaction as well as transformative innovation are necessary to create strategies to new challenges of possible future pandemics.

References

1. WHO Coronavirus (COVID-19) Dashboard. https://covid19.who.int/. Accessed on 29 Mar 2021.
2. Jones DL, Baluja MQ, Graham DW, Corbishley A, McDonald JE, Malham SK, Hillary LS, Connor TR, Gaze WH, Moura IB, Wilcox MH, Farkas K. Shedding of SARS-CoV-2 in feces and urine and its potential role in person-to-person transmission and the environment-based spread of COVID-19. Sci Total Environ. 2020;749. https://doi.org/10.1016/j.scitotenv.2020.141364.
3. van Doorn AS, Meijer B, Frampton CM, Barclay ML, de Boer NK. Systematic review with meta-analysis: SARS-CoV-2 stool testing and the potential for faecal-oral transmission. Aliment Pharmacol Ther. 2020.
4. Jiang X, Luo M, Zou Z, Wang X, Chen C, Qiu J. Asymptomatic SARS-CoV-2 infected case with viral detection positive in stool but negative in nasopharyngeal samples lasts for 42 days. J Med Virol. 2020;92(10):1807–9. https://doi.org/10.1002/jmv.25941.
5. Wölfel R, Corman VM, Guggemos W, Seilmaier M, Zange S, Mueller MA, Niemeyer D, Vollmar P, Rothe C, Hoelscher M, Bleicker T, Bruenink S, Schneider J, Ehmann R, Zwirglmaier K, Drosten C, Wendtner C. Virological assessment of hospitalized cases of coronavirus disease 2019. Nature. 2020. https://doi.org/10.1038/s41586-020-2196-x.
6. Tang A, Tong ZD, Wang HL, Dai YX, Li KF, Liu JN, Wu WJ, Yuan C, Yu ML, Li P, Yan JB. Detection of novel coronavirus by RT-PCR in stool specimen from asymptomatic child, China. Emerg Infect Dis. 2020;26(2020):1–5. https://doi.org/10.3201/eid2606.200301.
7. Wu Y, Guo C, Tang L, Hong Z, Zhou J, Dong X, Yin H, Xiao Q, Tang Y, Qu X, Kuang L, Fang X, Mishra N, Lu J, Shan H, Jiang G, Huang X. Prolonged presence of SARS-CoV-2 viral RNA in faecal samples. Lancet Gastroenterol Hepatol. 2020;5(5):434–5. https://doi.org/10.1016/S2468-1253(20)30083-2.
8. Kitajima M, Ahmed W, Bibby K, Carducci A, Gerba CP, Hamilton KA, Haramoto E, Joan B, Rose JB. SARS-CoV-2 in wastewater: State of the knowledge and research needs. Sci Total Environ. 2020;739. https://doi.org/10.1016/j.scitotenv.2020.139076.
9. Wang W, Xu Y, Gao R, Lu R, Han K, Wu G, Tan W. Detection of SARS-CoV-2 in different types of clinical specimens. JAMA. 2020;323(18):1843–4. https://doi.org/10.1001/jama.2020.3786.

10. Ling Y, Xu SB, Lin YX, Tian D, Zhu ZQ, Dai FH, Wu F, Song Gang Z, Huang W, Chen J, Hu BJ, Wang S, Mao EQ, Zhu L, Zhang WH, Lu HZ. Persistence and clearance of viral RNA in 2019 novel coronavirus disease rehabilitation patients. Chin Med J. 2020;133(9):1039–1043.
11. Xiao F, Tang M, Zheng X, Liu Y, Li X, Shan H. Evidence for gastrointestinal infection of SARS-CoV-2. Gastroenterology. 2020;158(6):1831-1833.e3. https://doi.org/10.1053/j.gastro.2020.02.055.
12. Wang X, Zheng J, Guo L, Yao H, Wang L, Xia X, Zhang W. Fecal viral shedding in COVID-19 patients: clinical significance, viral load dynamics and survival analysis. Virus Res. 2020;289:198147. https://doi.org/10.1016/j.virusres.2020.198147.
13. Medema G, Heijnen L, Elsinga G, Italiaander R, Brouwer A. Presence of SARS-Coronavirus-2 in sewage. medRxiv: 2020.2003.2029.20045880; 2020.
14. Lodder W, de Roda Husman AM. SARS-CoV-2 in wastewater: potential health risk, but also data source. Lancet Gastroenterol Hepatol. 2020. https://doi.org/10.1016/S2468-1253(20)30087-X.
15. Medema G, Heijnen L, Elsinga G, Italiaander R, Brouwer A. Presence of SARS-Coronavirus-2 RNA in sewage and correlation with reported COVID-19 prevalence in the early stage of the epidemic in The Netherlands. Environ Sci Technol Lett. 2020;7(7):511–6. https://doi.org/10.1021/acs.estlett.0c00357.
16. Nemudryi A, Nemudraia A, Surya K, Wiegand T, Buyukyoruk M, Wilkinson R, Wiedenheft B. Temporal detection and phylogenetic assessment of SARS-CoV-2 in municipal wastewater. 2020. MedRxiv. 2020.04.15.20066746. https://doi.org/10.1101/2020.04.15.20066746.
17. Green H, Wilder M, Middleton FA, Collins M, Fenty A, Gentile K, Kmush B, Zeng T, Larsen DA. Quantification of SARS-CoV-2 and cross-assembly phage (crAssphage) from wastewater to monitor coronavirus transmission within communities. 2020. https://doi.org/10.1101/2020.05.21.20109181.
18. Wu F, Xiao A, Zhang J, Moniz K, Endo N, Armas F, Bonneau R, Brown MA, Bushman M, Chai PR, Duvallet C, Erickson TB, Foppe K, Ghaeli N, Gu X, Hanage WP, Huang KH, Lee WL, Matus M, McElroy KA, Nagler J, Rhode SF, Santillana M, Tucker JA, Wuertz S, Zhao S, Thompson J, Alm EJ. SARS-CoV-2 titers in wastewater foreshadow dynamics and clinical presentation of new COVID-19 cases. 2020. https://doi.org/10.1101/2020.06.15.20117747.
19. Curtis K, Keeling D, Yetka K, Larson A, Gonzalez R. Wastewater SARS-CoV-2 concentration and loading variability from grab and 24-hour composite samples. 2020. https://doi.org/10.1101/2020.07.10.20150607.
20. Weidhaas J, Aanderud Z, Roper D, Vanderslice J, Gaddis E, Ostermiller J, Hoffman K, Jamal R, Heck P, Zhang Y, Torgersen K, Laan JV, Lacross N. Correlation of SARS-CoV-2 RNA in wastewater with COVID-19 disease burden in sewersheds. 2020. https://doi.org/10.21203/rs.3.rs-40452/v1.
21. Sherchan SP, Shahin S, Ward LM, Tandukar S, Aw TG, Schmitz B, Ahmed W, Kitajima M. First detection of SARS-CoV-2 RNA in wastewater in North America: a study in Louisiana, USA. Sci Total Environ. 2020;743:140621. https://doi.org/10.1016/j.scitotenv.2020.140621.
22. Ahmed W, Angel N, Edson J, Bibby K, Bivins A, O'Brien JW, Choi PM, Kitajima M, Simpson SL, Li J, Tscharke B, Verhagen R, Smith WJM, Zaugg J, Dierens L, Hugenholtz P, Thomas KV, Mueller JF. First confirmed detection of SARS-CoV-2 in untreated wastewater in Australia: a proof of concept for the wastewater surveillance of COVID-19 in the community. Sci Total Environ. 2020;138764. https://doi.org/10.1016/J.SCITOTENV.2020.138764.
23. Wurtzer S, Marechal V, Mouchel JM, Moulin L. Time course quantitative detection of SARS-CoV-2 in Parisian wastewaters correlates with COVID-19 confirmed cases. 2020. MedRxiv. 2020.04.12.20062679. https://doi.org/10.1101/2020.04.12.20062679.
24. Wurtzer S, Marechal V, Mouchel JM, Maday Y, Teyssou R, Richard E, Almayrac JL, Moulin L. Evaluation of lockdown impact on SARS-CoV-2 dynamics through viral genome quantification in Paris wastewaters. 2020. https://doi.org/10.1101/2020.04.12.20062679.
25. Trottier J, Darques R, Mouheb NA, Partiot E, Bakhache W, Deffieu MS, Gaudin R. Post-lockdown detection of SARS-CoV-2 RNA in the wastewater of Montpellier. France. 2020. https://doi.org/10.1101/2020.07.08.20148882.
26. Wang J, Feng H, Zhang S, Ni Z, Ni L, Chen Y, Zhuo L, Zhong Z, Qu T. SARS-CoV-2 RNA detection of hospital isolation wards hygiene monitoring during the Coronavirus Disease 2019 outbreak in a Chinese hospital. Int J Infect Dis. 2020;94:103–6. https://doi.org/10.1016/j.ijid.2020.04.024.
27. Randazzo W, Truchado P, Ferrando EC, Simon P, Allende A, Sanchez G. SARS-CoV-2 RNA titers in wastewater anticipated COVID-19 occurrence in a low prevalence area. 2020. MedRxiv. 2020.04.22.20075200. https://doi.org/10.1101/2020.04.22.20075200.
28. Randazzo W, Ferrando EC, Sanjuan R, Calap PD, Sanchez G. Metropolitan wastewater analysis for COVID-19 epidemiological surveillance. 2020. https://doi.org/10.1101/2020.04.23.20076679.
29. Mera F, Río R, Fuente J, Sancho M, Hervas D, Moreno I, Dominguez M, Domínguez L, Gortázar C. COVID-19 in a rural community: outbreak dynamics, contact tracing and environmental RNA. 2020. Preprints 2020, 2020050450. https://www.preprints.org/manuscript/202005.0450/v1.

30. Miró GC, Estrada EA, Guix S, Paraira M, Galofré B, Sáanchez G, Pintó R, Bosch A. Sentinel surveillance of SARS-CoV-2 in wastewater anticipates the occurrence of COVID-19 cases. 2020. https://doi.org/10.1101/2020.06.13.20129627.
31. Vallejo JA, Feal SR, Perez KC, Oriona AL, Tarrio J, Reif R, Ladra S, Janeiro BKR, Nasser M, Cid A, Veiga MC, Acevedo A, Lamora C, Bou G, Cao R, Poza M. Highly predictive regression model of active cases of COVID-19 in a population by screening wastewater viral load. 2020. https://doi.org/10.1101/2020.07.02.20144865.
32. La Rosa G, Iaconelli M, Mancini P, Ferraro GB, Veneri C, Bonadonna L, Lucentini L, Suffredini E. First detection of Sars-Cov-2 in untreated wastewaters in Italy. 2020. MedRxiv 2020.04.25.20079830. https://doi.org/10.1101/2020.04.25.20079830.
33. Rimoldi SG, Stefani F, Gigantiello A, Polesello S, Comandatore F, Mileto D, Maresca M, Longobardi C, Mancon A, Romeri F, Pagani C, Moja L, Gismondo MR, Salerno F. Presence and vitality of SARS-CoV-2 virus in wastewaters and rivers. 2020. https://doi.org/10.1101/2020.05.01.20086009.
34. La Rosa G, Mancini P, Ferraro GB, Veneri C, Iaconelli M, Bonadonna L, Lucentini L, Suffredini E. SARS-CoV-2 has been circulating in northern Italy since December 2019: evidence from environmental monitoring. 2020. https://doi.org/10.1101/2020.06.25.20140061.
35. Or IB, Yaniv K, Shagan M, Ozer E, Erster O, Mendelson E, Mannasse B, Shirazi R, Winter EK, Nir O, Ali HA, Ronen Z, Rinott E, Lewis YE, Friedler EF, Paitan Y, Bitkover E, Berchenko Y, KushmaroA. Regressing SARS-CoV-2 sewage measurements onto COVID-19 burden in the population: a proof-of-concept for quantitative environmental surveillance. 2020. medRxiv 2020.04.26.20073569. https://doi.org/10.1101/2020.04.26.20073569.
36. Alpaslan Kocamemi B, Kurt H, Hacıoglu S, Yaralı C, Saatci AM, Pakdemirli B. First data-set on SARS-CoV-2 detection for Istanbul wastewaters in Turkey. MedRxiv. 2020. https://doi.org/10.1101/2020.05.03.20089417.
37. Alpaslan Kocamemi B, Kurt H, Sait A, Sarac F, Saatci AM, Pakdemirli B. SARS-CoV-2 detection in Istanbul wastewater treatment plant sludges. 2020. medRxiv. 2020.2005.2012.20099358.
38. Alpaslan Kocamemi B, Kurt H, Sait A, Kadi H, Sarac F, Aydin I, Saatci AM, Pakdemirli B. Nationwide SARS-CoV-2 surveillance study for sewage and sludges of wastewater treatment plants in Turkey. 2020. https://doi.org/10.1101/2020.11.29.20240549.
39. Alpaslan Kocamemi B, Kurt H, Sait A, Kadi H, Sarac F, Aydın I, Saatci AM, Pakdemirli B. Routine SARS-CoV-2 wastewater surveillance results in Turkey to follow Covid-19 outbreak. 2020. https://doi.org/10.1101/2020.12.21.20248586.
40. Döhla M, Wilbring G, Schulte B, Kümmerer BM, Diegmann C, Sib E, Richter E, Haag A, Engelhart S, Eis-Hübinger AM, Exner M, Streeck H, Schmithausen RM. SARS-CoV-2 in environmental samples of quarantined households. 2020. https://doi.org/10.1101/2020.05.28.20114041.
41. Hata A, Honda R, Hara-Yamamura H, Meuchi Y. Detection of SARS-CoV-2 in wastewater in Japan by multiple molecular assays-implication for wastewater-based epidemiology (WBE). 2020. https://doi.org/10.1101/2020.06.09.20126417.
42. Haramoto F, Malla B, Thakali O, Kitajima M. First environmental surveillance for the presence of SARS-CoV-2 RNA in wastewater and river water in Japan. Sci Total Environ. 2020;737:40405. https://doi.org/10.1016/j.scitotenv.2020.140405.
43. Kumar M, Patel AK, Shah AV, Raval J, Rajpara N, Joshi M, Joshi CG. First proof of the capability of wastewater surveillance for COVID-19 in India through detection of genetic material of SARS-CoV-2. Sci Total Environ. 2020;746. https://doi.org/10.1016/j.scitotenv.2020.141326.
44. Arora S, Nag A, Sethi J, Rajvanshi J, Saxena S, Shrivastava SK, Gupta AB. Sewage surveillance for the presence of SARS-CoV-2 genome as a useful wastewater-based epidemiology (WBE) tracking tool in India. 2020. https://doi.org/10.1101/2020.06.18.20135277.
45. Sharif S, Ikram A, Khurshid A, Salman M, Mehmood N, Arshad Y, Ahmad J, Angez M, Alam MM, Rehman L, Mujtaba G, Hussain J, Ali J, Akthar R, Malik MW, Baig ZI, Rana MS, Usman M, Ali MQ, Ahad A, Badar N, Umair M, Tamim S, Ashraf A, Tahir F, Ali N. Detection of SARS-Coronavirus-2 in wastewater, using the existing environmental surveillance network: an epidemiological gateway to an early warning for COVID-19 in communities. 2020. https://doi.org/10.1101/2020.06.03.20121426.
46. Fongaro G, Stoco PH, Souza DSM, Grisard EC, Magri ME, Rogovski P, Schorner MA, Barazzetti FH, Christoff AP, Valter de Oliveira LP, Bazzo ML, Wagner G, Hernandez M, Lazaro DR. SARS-CoV-2 in human sewage in Santa Catalina, Brazil, November 2019. 2020. https://doi.org/10.1101/2020.06.26.20140731.
47. Prado T, Fumian TM, Mannarino CF, Maranhão AG, Siqueira MM, Miagostovich MP. Preliminary results of SARS-CoV-2 detection in sewerage system in Niterói municipality. Rio de Janeiro, Brazil. Memórias do Instituto Oswaldo Cruz, 115, e200196. 2020. Epub July 27.
48. Ampuero M, Valenzuela S, Echeverria FV, Rifo RS, Barriga GP, Chnaiderman J, Rojas C, Leiva SG, Diez B, Gaggero A. SARS-CoV-2 detection in sewage in Santiago, Chile—preliminary results. https://doi.org/10.1101/2020.07.02.20145177.
49. Jorgensen AU, Gamst J, Hansen LV, Knudsen IIH, Jensen SKS. Eurofins Covid-19 sentinel TM wastewater test provide early warning of a potential

COVID-19 outbreak. 2020. https://doi.org/10.1101/2020.07.10.20150573.
50. Latorre LG, Ballesteros I, Granda IV, Granda MG, Paspuel BF, Touma BR. SARS-CoV-2 in river water: implications in low sanitation countries. Sci Total Environ. 2020;743:140832. https://doi.org/10.1016/j.scitotenv.2020.140832.
51. Jafferali MH, Khatami K, Atasoy M, Birgersson M, Williams C, Cetecioglu Z. Benchmarking virus concentration methods for quantification of SARS-CoV-2 in raw wastewater. Sci Total Environ. 2020. https://doi.org/10.1016/j.scitotenv.2020.142939.
52. Mlejnkova H, Sovova K, Vasickova P, Ocenaskova V, Jasikova L, Juranova E. Preliminary study of Sars-Cov-2 occurrence in wastewater in the Czech Republic. Int J Environ Res Public Health. 2020;17(15):5508. https://doi.org/10.3390/ijerph17155508.
53. Ahmed F, Islam A, Kumar M, Hossain M, Bhattacharya P, Islam T, Hossen F, Hossain S, Islam S, Uddin M, Islam N, Bahadur NM, Alam D, Reza HM, Jakariya M. First detection of SARS-CoV-2 genetic material in the vicinity of COVID-19 isolation centre through wastewater surveillance in Bangladesh. 2020. https://doi.org/10.1101/2020.09.14.20194696.
54. Betancourt WQ, Schmitz BW, Innes GK, Prasek SM, Pogreba Brown KM, Stark ER, Foster AR, Sprissler RS, Harris DT, Sherchan SP, Gerba CP, Pepper IL. COVID-19 containment on a college campus via wastewater-based epidemiology, targeted clinical testing, and an intervention. Sci Total Environ. 2021;779:146408.
55. Colosi LM, Barry KE, Kotay SM, Porter MD, Poulter MD, Ratliff C, Simmons W, Steinberg LI, Wilson DD, Morse R, Zmick P, Mathers AJ. Development of wastewater pooled surveillance of SARS-CoV-2 from congregate living settings. 2020. medRxiv: 2020.2010.2010.20210484.
56. Gantzer C, Senouci S, Maul A, Levi Y, Schwartzbrod L. Enterovirus genomes in wastewater: concentration on glass wool and glass powder and detection by RT-PCR. J Virol Methods. 1997;65(2):265–71.
57. Parashar UD, Nelson EA, Kang G. Diagnosis, management, and prevention of rotavirus gastroenteritis in children. Br Med J. (Clinical Research Ed.). 2013;347:f7204.
58. Girard M, Hirth L. Virologie moleculaire Doin Cd. Paris. Appl Enriron Microbial. 1989;53.
59. Havelaar AH. Bacteriophages as model viruses in water quality control.Water Res (Oxford). 1991;25(5):529–541.
60. Grabow WOK, Neubrech TE, Holtzhausen CS, Jofre J. Bacteroides fragilis and Escherichia coli bacteriophages: excretion by humans and animals. Water Sci Technol. 1995;31(5–6):223–30.
61. Havelaar AH, Furuse K, Hogeboom WM. Bacteriophages and indicator bacteria in human and animal faeces. J Appl Microbiol. 1986;60(3):255–62.
62. Havelaar AH, Pot-Hogeboom WM, Furuse K, Pot R, Hormann MP. F-specific RNA bacteriophages and sensitive host strains in faeces and wastewater of human and animal origin. J Appl Microbiol. 1990;69(1):30–7.
63. Joffre J. Les bacteriophages dans les milieux hydriques. In: Schwartzbrod L(ed.) Virologie des milieux hydriques. Tee Dot-Lavoisier, Paris; 1991. pp. 253–274.
64. Masters PS, Perlman S. Coronaviridae. In: Knipe DM, Howley PM (eds) Fields virology, 6th edn, vol 2; 2013. pp. 825–858.
65. Zhong NS, Zheng BJ, Li YM, Poon Xie ZH, Chan KH, Li PH, Tan SY, Chang Q, Xie JP, Liu XQ, Xu J, Li DX, Yuen KY, Peiris, Guan Y. Epidemiology and cause of severe acute respiratory syndrome (SARS) in Guangdong, People's Republic of China, in February 2003. Lancet (London, England). 2003;362(9393):1353–1358. https://doi.org/10.1016/s0140-6736(03)14630-2.b
66. Zaki AM, van Boheemen S, Bestebroer TM, Osterhaus ADME, Fouchier RAM. Isolation of a novel coronavirus from a man with pneumonia in Saudi Arabia. N Engl J Med. 2012;367(19):1814–20.
67. Astuti I, Ysrafil,. Severe Acute Respiratory Syndrome Coronavirus 2 (SARS-CoV-2): an overview of viral structure and host response. Diab Metab Syndr. 2020;14(4):407–12. https://doi.org/10.1016/j.dsx.2020.04.020.
68. Jiang S, Hillyer C, Du L. Neutralizing antibodies against SARS-CoV-2 and other human Coronaviruses. Trends Immunol. 2020.
69. Song Z, Xu Y, Bao L, Zhang L, Yu P, Qu Y, Zhu H, Zhao W, Han Y, Qin C. From SARS to MERS, thrusting coronaviruses into the spotlight. Viruses. 2019;11(1):59.
70. https://coronavirus.jhu.edu/map.html. Accessed on 30 Mar 2021.
71. Mercer TR, Salit M. Testing at scale during the COVID-19 pandemic. Nat Rev Genet. 2021;22:415–26.
72. Surkova E, Nikolayevskyy V, Drobniewski F. False-positive COVID-19 results: hidden problems and costs. The Lancet. 2020;8(12):1167–8.
73. Medema G, Been F, Heijnen L, Petterson S. Implementation of environmental surveillance for SARS-CoV-2 virus to support public health decisions: Opportunities and challenges. Curr Opin Environ Sci Health. 2020;17:49–71.
74. Orive G, Lertxundi U, Barcelo D. Early SARS-CoV-2 outbreak detection by sewage-based epidemiology. Sci Total Environ. 2020;732:139298.
75. Bivins A, North D, Ahmad A, Ahmed W, Alm E, Been F, Bhattacharya P, Bijlsma L, Boehm AB, Brown J, Buttiglieri G, Calabro V, Carducci A, Castiglioni S, Gurol ZC, Chakraborty S, Costa F, Curcio S, Reyes FL, Vela JD, Farkas K, Fernandez-Casi X, Gerba C, Gerrity D, Girones R, Gonzalez R, Haramoto E, Harris A, Holden PA, Islam MdT, Jones DL, Kasprzyk-Hordern B, Kitajima M,

Kotlarz N, Kumar M, Kuroda K, Rosa G, Malpei F, Mautus M, McLellan SL, Medema G, Meschke JS, Mueller J, Newton RJ, Nilsson D, Noble RT, van Nuijs A, Peccia J, Perkins TA, Pickering AJ, Rose J, Sanchez G, Smith A, Stadler L, Stauber C, Thomas K, van der Voorn T, Wigginton K, Zhu K, Bibby K. Wastewater-based epidemiology: global collaborative to maximize contributions in the fight against COVID-19. 2020;54(13):7754–7757.

76. Lorenzo M, Pico Y. Wastewater-based epidemiology: current status and future prospects. Curr Opin Environ Sci Health. 2019;9:77–84.
77. Asghar OM, Diop G, Weldegebriel F, Malik S, Shetty L, El Bassioni, Akande AO, Al Maamoun E, Zaidi S, Adeniji AJ, Burns CC, Deshpande J, Oberste MS, Lowther SA. Environmental surveillance for polioviruses in the global polio eradication initiative. J Infect Dis. 210:294–303.
78. Liu P, Ibaraki M, VanTassell J, Geith K, Cavallo M, Kann R, Moe C. A novel COVID-19 early warning tool: Moore Swab method for wastewater surveillance at an institutional level. Prepr Server Health Sci. 2020.
79. Gao Z, Xu Y, Sun C, Wang X, Guo Y, Qui S, Ma K. A systematic review of asymptomatic infections with COVID-19. J Microbiol Immunol Infect. 2021;54(1):12–6.
80. Peirlinck M, Linka K, Costabal FS, Bhattacharya J, Bandavid E, Ioannidis JPA, Kuhl E. Visualizing the invisible: the effect of asymptomatic transmission on the outbreak dynamics of COVID-19. Comput Methods Appl Mech Eng. 2020;372(1):113410.
81. Daughton CG. Wastewater surveillance for population-wide Covid-19: the present and future. Sci Total Environ. 2020;736:139631.
82. Knoll RL, Klopp J, Bonewitz G, Gröndahl B, Hilbert K, Kohnen W, Weise K, Plachter B, Hitzler W, Kowalzik F, Runkel S, Zepp F, Winter J, Cacicedo ML, Gehring S. Containment of a large SARS-CoV-2 outbreak among healthcare workers in a pediatric intensive care unit. Pediatr Infect Dis J. 2020;39(11):336–339.
83. Wee LE, Sim XYJ, Conceicao EP, Aung MK, Goh JQ, Yeo DWT, Gan WH, Chua YY, Wijaya L, Tan TT, Tan BH, Ling ML, Venkatachalam I. Containment of COVID-19 cases among healthcare workers: the role of surveillance, early detection, and outbreak management. Infect Control Hosp Epidemiol. 2020;41(7):765–71.
84. Peccia J, Zulli A, Brackney DE, Grubaugh ND, Kaplan EH, Massana AC, Ko AI, Malik AA, Wang D, Wang M, Warren JL, Weinberger DM, Omer SB. SARS-CoV-2 RNA concentrations in primary municipal sewage sludge as a leading indicator of COVID-19 outbreak dynamics. 2020. https://doi.org/10.1101/2020.05.19.20105999.
85. Balboa S, Iglesias MM, Rodríguez S, Lamas LM, Vasallo FJ, Regueiro B, Lema JL. The fate of SARS-CoV-2 in wastewater treatment plants points out the sludge line as a suitable spot for incidence monitoring. 2020.
86. Tiwari SB, Gahlot P, Tyagi VK, Zhang L, Zhou Y, Kazmi AA, Kumar M. Surveillance of wastewater for early epidemic prediction (SWEEP): environmental and health security perspectives in the post COVID-19 anthropocene. Environ Res. 2021;195:110831.
87. Zhang D, Ling H, Huang X, Li J, Li W, Yi C, Zhang T, Jiang Y, He Y, Deng S, Zhang X, Wang X, Liu Y, Li G, Qu J. Potential spreading risks and disinfection challenges of medical wastewater by the presence of Severe Acute Respiratory Syndrome Coronavirus 2 (SARS-CoV-2) viral RNA in septic tanks of Fangcang Hospital. Sci Total Environ. 2020;741:140445.
88. Haramoto E, Kitajima M, Hata A, Torrey JR, Masago Y, Sano D, Katayama H. A review on recent progress in the detection methods and prevalence of human enteric viruses in water. Water Res. 2018;135:168–86.
89. Wu D, Koganti R, Lambe UP, Yadavalli T, Nandi SS, Shukla D. Vaccines and therapies in development for SARS-CoV-2 infections. J Clin Med. 2020;9(6):1885.
90. Randazzo W, Truchado P, Cuevas-Ferrando E, Simón P, Allende A, Sánchez G. SARS-CoV-2 RNA in wastewater anticipated COVID-19 occurrence in a low prevalence area. Water Res. 2020;181:115942.
91. Falman JC, Fagnant-Sperati CS, Kossik AL, Boyle DS, Meschke JS. Evaluation of secondary concentration methods for poliovirus detection in wastewater. Food Environ Virol. 2019;11(1):20–31. https://doi.org/10.1007/s12560-018-09364-y (PMID-30612304).
92. LaTurner ZW, Zong DM, Kalvapalle P, Gamas KR, Terwilliger A, Crosby T, Ali P et al. Evaluating recovery, cost, and throughput of different concentration methods for SARS-CoV-2 wastewater-based epidemiology. Water Res. 2021;197:117043.
93. Barril PA, Pianciola LA, Mazzeo M, Ousset MJ, Jaureguiberry MV, Alessandrello M, Sánchez G, Oteiza JM. Evaluation of viral concentration methods for SARS-CoV-2 recovery from wastewaters. Sci Total Environ. 2021;756:144105.
94. H10 Ahmed W, Bertsch P, Bivins A, Bibby K, Farkas K, Gathercole A, Haramoto E et al. Comparison of virus concentration methods for the RT-qPCR-based recovery of murine hepatitis virus, a surrogate for SARS-CoV-2 from untreated wastewater. Sci Total Environ. 2020;739:139960.
95. H11 Philo SE, Kemi EK, Swanstrom R, Ong A, Burnor EA, Kossik AL, Harrison JC et al. A comparison of SARS-CoV-2 wastewater concentration methods for environmental surveillance. Sci Total Environ. 2021;760:144215.
96. van Kasteren PB, van Der Veer B, van den Brink S, Wijsman L, de Jonge J, van den Brandt A, Molenkamp R, Reusken CBEM, Meijer A.

Comparison of seven commercial RT-PCR diagnostic kits for COVID-19. J Clin Virol. 2020;128:104412.
97. Jahn K, Dreifuss D, Topolsky I, Kull A, Ganesanandamoorthy P, Fernandez-Cassi X, Bänziger C et al. Detection of SARS-CoV-2 variants in Switzerland by genomic analysis of wastewater samples. medRxiv. 2021.
98. Landgraff C, Wang LYR, Buchanan C, Wells M, Schonfeld J, Bessonov K, Ali J, Robert E, Nadon C. Metagenomic sequencing of municipal wastewater provides a near-complete SARS-CoV-2 genome sequence identified as the B. 1.1. 7 variant of concern from a Canadian municipality concurrent with an outbreak. medRxiv. 2021.
99. Wurtzer S, Waldman P, Levert M, Mouchel JM, Gorgé O, Boni M, Maday Y, Marechal V, Moulin L. Monitoring the propagation of SARS CoV2 variants by tracking identified mutation in wastewater using specific RT-qPCR. medRxiv. 2021.
100. Lu D, Zhu DZ, Gan H, Yao Z, Fu Q, Zhang XJ. Prospects and challenges of using electrochemical immunosensors as an alternative detection method for SARS-CoV-2 wastewater-based epidemiology. Sci Total Environ. 2021:146239.
101. Eckerle LD, Becker MM, Halpin RA, Li K, Venter E, Lu X, Scherbakova S et al. Infidelity of SARS-CoV Nsp14-exonuclease mutant virus replication is revealed by complete genome sequencing. PLoS Pathog. 2010;6(5):e1000896.
102. Dilucca M, Forcelloni S, Georgakilas AG, Giansanti A, Pavlopoulou A. Codon usage and phenotypic divergences of SARS-CoV-2 genes. Viruses. 2020;12(5):498.
103. Hou YJ, Chiba S, Halfmann P, Ehre C, Kuroda M, Dinnon KH, Leist SR, Schäfer A, Nakajima N, Takahashi KJS. SARS-CoV-2 D614G variant exhibits efficient replication ex vivo and transmission in vivo. 2020;370(6523):1464–8.
104. Korber B, Fischer WM, Gnanakaran S, Yoon H, Theiler J, Abfalterer W, Hengartner N, Giorgi EE, Bhattacharya T, Foley BJC. Tracking changes in SARS-CoV-2 Spike: evidence that D614G increases infectivity of the COVID-19 virus. 2020;182(4):812–27. e819.
105. Gómez CE, Perdiguero B, Esteban MJV. Emerging SARS-CoV-2 variants and impact in global vaccination programs against SARS-CoV-2/COVID-19. 2021;9(3):243.
106. https://covid19.tarimorman.gov.tr.
107. Mao K, Zhang H, Zhugen Y. An integrated biosensor system with mobile health and wastewater-based epidemiology (iBMW) for COVID-19 pandemic. Biosens Bioelectron. 2020;169:112617.

Pharmaceuticals, Benzotriazoles and Polyfluoroalkyl Substances: Impacts and Potential Reduction Measures

15

Elke Fries, Manuela Helmecke, and Christoph Schulte

Abstract

Daily life products like medicines, dishwasher detergents or textiles contain various organic chemicals. According to use and disposal of such products pharmaceuticals, benzotriazoles, and polyfluoroalkyl substances (PFAS) appear among many other organic chemicals in municipal wastewaters. Carbamazepine, sulfamethoxazole, 4-methyl-1H-benzotriazole (4Me-BT), perfluoro carboxylic acids like perfluorooctanoic acid (PFOA) and perfluoro sulfonic acids like perfluorooctane sulfonic acid (PFOS) are not biodegradable and, therefore, classified as persistent in wastewater. Ibuprofen, paracetamol, 1H-benzotriazole (1H-BT) and 5-methyl-1H-benzotriazole (5Me-BT) are principally degradable, but they can be classified as pseudo-persistent when delivered continuously. Discharge of wastewater effluents is thus an important source of pharmaceuticals, benzotriazoles and PFAS, among other organic chemicals, in river water, where concentrations are usually at trace levels due to dilution. Thus, they belong to the so-called micro-contaminants. Bank filtration and irrigation of agricultural fields with reclaimed water or wastewater-impacted river water are potential sources of pharmaceuticals, benzotriazoles and PFAS in soil and groundwater resources. Paracetamol, sulfamethoxazole, 1H-BT and PFOA can be classified as mobile in the environment. Carbamazepine, ibuprofen, 4Me-BT, 5Me-BT and PFOS are less mobile and sorb principally better to solid matrices. In the environment, toxic or endocrine effects, development of antibiotic resistances, uptake by biota and plants, accumulation in the food chain, transfer into edible parts and fodder, impact on ecosystem functioning or drinking water contamination can be attributed to organic micro-contaminants posing risks to human and ecosystem health. Hence, the entry of pharmaceuticals, benzotriazoles and PFAS into the environment should be generally prevented. To avoid high expenses for water treatment and being in line with the precautionary principle, reduction at the sources is prioritized. Nevertheless, a cost-effective combination of measures is needed that also includes efforts targeting the use of substances and end-of-pipe solutions. Risk assessment should take into account ecological functions but also potential water usages. Finally yet importantly, stakeholder involvement is important to ensure data availability, the implementation and financing

E. Fries (✉)
Bundesanstalt für Geowissenschaften und Rohstoffe (GBR), Hannover, Germany
e-mail: elke.fries@bgr.de

M. Helmecke · C. Schulte
Umweltbundesamt, Dessau-Roßlau, Germany

M. Bahadir and A. Haarstrick (eds.), *Water and Wastewater Management*,
Water and Wastewater Management, https://doi.org/10.1007/978-3-030-95288-4_15

of measures as well as for the communication of potential risks.

Keywords

Antibiotics · Tolyltriazoles · Irrigation · PFOS · Wastewater

15.1 Introduction

Organic chemicals are manufactured and used in a variety of industries for daily life products around the globe. For example, medicines contain pharmaceutical active compounds, dishwasher detergents contain benzotriazoles as corrosion inhibitors and textiles contain poly- or perfluoroalkyl substances (PFAS) as water and grease repellent agents or coatings. According to product use and disposal, organic chemicals occur ubiquitously in municipal wastewater influents [1]. Input loads in wastewater are still increasing with population and standard of living. Municipal wastewater treatment plants can eliminate organic chemicals only to a limited extent, depending on treatment conditions [2], input loads, mobility, and biological degradability. Persistent organic chemicals are recalcitrant in conventional mechanical–biological water treatment processes, whereas pseudo-persistent ones are principally degradable. However, when delivery is continuous and residence times are lower than degradation rates, pseudo-persistent organic chemicals are also not removed completely from wastewater. In addition, transformation products might be even more stable. During wastewater treatment, organic chemicals, in particular those of high polarity, remain in the aqueous phase due to their low adsorption capacity to sewage sludge. Consequently, organic chemicals are ubiquitously present in municipal wastewater effluents. Surveys on organic chemicals in wastewater were published for the United States [3, 4], for Europe [5–10], and for Southern Mediterranean countries [11–14].

Figure 15.1 shows sources, sinks and effects of wastewater-borne organic chemicals in the environment. Discharge of wastewater effluents is an important source of organic chemicals in river water systems. In rivers, persistent compounds are poorly biodegradable, whereas pseudo-persistent chemicals are present due to continuous input from effluents. Persistent and mobile chemicals are in particular hazardous for the water cycle, because these substances might reach drinking water resources [7, 15]. In contrast to well-regulated persistent, bioaccumulative and toxic (PBT) chemicals, persistent and mobile chemicals are not removed from water by sorption processes due to an excellent water solubility (S) [15]. The partition coefficient between n-octanol and water (log P_{OW}) is an appropriate parameter to estimate the sorption tendency for uncharged compounds. In general, Log P_{OW} is inversely related to water solubility and directly proportional to molecular weight of neutral substances. Log P_{OW} provides a much better estimator for sediment–water partitioning than does the simple water solubility [16]. Chemicals with low log P_{OW} values show generally less tendency to sorb to solids like soils, suspended particles, and sediments. The mobility of ionic or dissociating compounds can be estimated, however, in a very simplified manner, by their acid dissociation constant (pKa) and pH value [15]. Compound mobility is expected to increases with decreasing pKa according to stronger dissociation.

Concentrations of organic chemicals in river water are usually lower than in effluents of wastewater treatment plants due to dilution with less contaminated upstream water. Because of typical concentration in the μg/L range and below, they are called micro-pollutants [17] or micro-contaminants [18]. In river water, mobile micro-contaminants remain preferentially in the aqueous phase, whereas less mobile ones sorb to sediments and to suspended particles. In particular, persistent polar ones are not fully retained by any of the natural or technical barriers, and

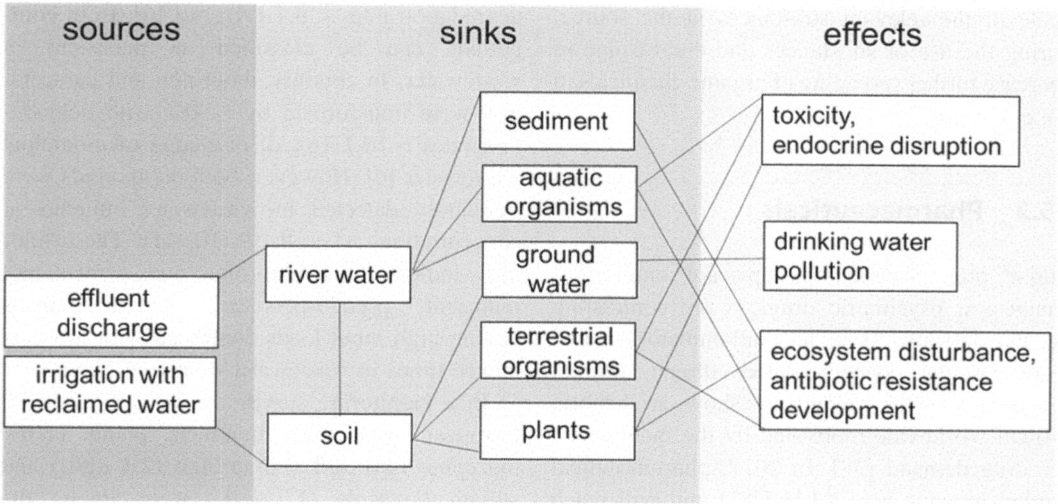

Fig. 15.1 Sources, sinks and effects of wastewater-borne organic chemicals in the environment

dilution may be the only mechanism, by which their concentrations decrease along the water path from wastewater to the raw water used for drinking water production [7]. However, one has to mention that in areas, where a high population and a high density of wastewater treatment plants hit low natural discharges due to low precipitation and high temperatures, rivers consist largely of wastewater effluents. During low flow conditions, wastewater contributions in European rivers are often around 50% or even higher [19, 20].

Since the last decade, significant advancement of knowledge on micro-contaminants in river water and groundwater has been achieved [21–26]. Bank filtration of river water might result in a transport of micro-contaminants into groundwater [27, 28]. Agricultural irrigation with reclaimed water impacts soil health and groundwater quality in general [29–32], but it is also a potential source of micro-contaminants [11–13, 33, 34]. Water reuse in agriculture poses several risks related to micro-contaminants [18]. In contrast to planned irrigation with reclaimed water, surface water irrigation is often uncontrolled and unregulated [19] resulting potentially in an input of micro-contaminants into groundwater and soil. In particular, mobile micro-contaminants can be quite problematic as they may travel along the water path to water used for drinking water production [7]. Usually, micro-contaminants are still not monitored routinely at irrigated sites.

Once discharged into the environment, micro-contaminants might have hazardous effects (Fig. 15.1). Some compounds are toxic, have endocrine effects or may foster the development of antibiotic resistances. Uptake of toxic or endocrine active chemicals by aquatic and terrestrial biota or plants—often followed by accumulation in the food chain—poses risks to human and ecosystem health. If groundwater is abstracted from wastewater-impacted wells for irrigation of other crops, micro-contaminants can be hazardous. Although uptake into plants cannot be excluded, impact on crop growth is only expected at concentrations much greater than would be currently expected in treated wastewater [35]. Ecosystem and soil functions might also be affected by wastewater reuse in irrigation [36]. To avoid all these impacts, the entry of organic chemicals into wastewater needs to be reduced to prevent the occurrence of micro-contaminants in the environment. This is particularly important in areas strongly affected by water scarcity, reported for example by the International Water Management Institute [37], where direct or indirect water reuse is highly important. Potential reduction measures can target different phases in the life cycle of the relevant substances at the source, during the use of substances and end-of-pipe to decrease further spreading of organic chemicals in the environment.

15.2 Pharmaceuticals

Major pharmaceutical groups are antibiotics, analgesics, psychiatric drugs, x-ray contrasting agents, anti-infectives, anti-inflammatories, and cardiovascular system-related drugs. Market revenues of pharmaceuticals show an historic growth worldwide motivated by the increase on the drug demand [38]. In 2017, pharmaceutical market revenue was 1,143 USD and will reach 1,462 billion USD in 2021 worldwide [39]. Pharmaceuticals and their degradation products can enter the environment mainly via excretion and disposal in wastewater [40]. In raw urban wastewater, the variability of analgesics/anti-inflammatories ranged between 0.0016 and 373 μg/L (ibuprofen) and of antibiotics between 0.001 and 32 μg/L (ofloxacin) [8]. Concentrations of paracetamol in wastewater influents were between 4.8 μg/L (United Kingdom) and 246 μg/L (Spain) [41].

Halling-Sorensen et al. [42] reviewed the occurrence, fate and effects of pharmaceuticals in the environment already at the end of the 20th century. Since then, research on pharmaceuticals in wastewaters, natural waters, soils, environmental fate, impacts and concerns, human health risks, and development of treatment technologies has been increased worldwide in particular in North America and Europe and lately in China [38].

Biological degradation of pharmaceuticals in municipal wastewater treatment plants was reported to be varying [6]. Information on the biological degradability in municipal wastewater treatment of paracetamol (analgesic), sulfamethoxazole (antibiotic), carbamazepine (antiepileptic drug) and ibuprofen (analgesics) is given in Table 15.1.

In sludge of municipal wastewater, carbamazepine and sulfamethoxazole were not removed to a significant extent (<20%) with degradation rates < 0.1 L/[g_{ss} d] [6]. Both compounds can be classified as persistent in wastewater. In contrast, ibuprofen and paracetamol were transformed by > 90% with degradation rates > 10 L/[g_{ss} d] in sludge of municipal wastewater [6]. However, both compounds were frequently detected in wastewater effluents at concentrations >1 μg/L [9, 10, 41]. These findings indicate that ibuprofen and paracetamol represent pseudo-persistent pharmaceuticals having high input loads combined with low residence times in wastewater treatment plants.

In a monitoring survey on effluents from 90 European wastewater treatment plants carbamazepine (4.6 μg/L), ibuprofen (2.1 μg/L) and sulfamethoxazole (1.7 μg/L) were among the emerging organic contaminants being detected with highest concentrations [9]. Concentrations of paracetamol in wastewater effluents ranged between 0.07 μg/L (Malaysia) and 6 μg/L (Germany) [41]. Nikolaou et al. [40] reviewed the occurrence of a variety of pharmaceuticals in sewage treatment plant effluents with concentrations between 0.01 μg/L and 2.97 μg/L (ibuprofen). In the review of Fatta-Kassinos et al. [45], concentrations of antibiotics in urban wastewater effluents were reported in the range of 0.02 μg/L and 9.46 μg/L (sulfamethoxazole). Concentrations of analgesics/anti-inflammatories and antibiotics in secondary effluents ranged between 0.001 and 57 μg/L (tramadol), and between 0.001 μg/L and 6.7 μg/L (trimethoprim), respectively [8]. In effluents from four selected municipal wastewater treatments plants in northeastern Tunisia, sulfamethoxazole, carbamazepine, ibuprofen, paracetamol, atenolol, naproxen, ketoprofene, diclophenac and furosemide were most abundant among 14 pharmaceuticals with concentrations >1 μg/L [13]. Transformation products of ibuprofen were detected frequently in wastewater [46, 47]. In municipal wastewater effluents in northeastern Tunisia, the two transformation products of ibuprofen 1-OH-ibuprofen and 2-OH-ibuprofen were detected at maximum concentrations of 0.421 μg/L and 11.06 μg/L, respectively. The concentration of 2-OH-ibuprofen exceeded even that of ibuprofen [13].

Table 15.1 Water solubility (S), n-octanol–water portioning coefficient (log P_{OW}), dissociation constant (pKa), and biological degradability of selected pharmaceuticals, benzotriazoles and PFAS selected compounds (ss: suspended solid, nm: not measurable). Physical–chemical properties are from https://pubchem.ncbi.nlm.nih.gov/, SRC PhysProp datasheets, ECHA website, and EPA Technical Fact Sheets

Compound	S (mg/L)	Log P_{OW}	pKa	Biological degradability
Paracetamol	14000	0.46	9.38	Degradation rate > 10 L/[g_{ss} d] [6]
Sulfamethoxazole	610	0.89	1.6 (pKa1) 5.7 (pKa2)	Degradation rate < 0.1 L/[g_{ss} d] [6]
1H-benzotriazole	19800	1.44	8.37	37% [43]
5-methyl-1H-benzotriazole	3070	1.71	8.85	11% [43]
4-methyl-1H-benzotriazole			9.15	No degradation [43]
Carbamazepine	18	2.45	13.9	Degradation rate < 0.1 L/[g_{ss} d] [6]
Ibuprofen	21	3.97	5.3	Degradation rate > 10 L/[g_{ss} d] [6]
Perfluorooctanoic acid	9500	nm	1.3	Persistent [44]
Perfluorooctane sulfonic acid	680	nm	<1	Persistent [44]

Discharge of wastewater effluents is an important source of pharmaceuticals in the aquatic environment. Pharmaceuticals and their transformation products occurred frequently in river water in different countries [14, 48–51]. In general, concentrations in surface waters are in the μg/L range [52]. Lower concentrations are attributed mainly to dilution. It is important to mention, that in certain areas, climate change results in decreasing river water levels, and hence in less dilution processes. A total of 57 pharmaceuticals were detected in the aquatic environment of ten Latin American countries [14]. Concentration of naproxen was highest at 75.8 μg/L in a Mexican river [53], followed by ibuprofen (36.8 μg/L) in Costa Rica [54] and acetaminophen (30.4 μg/L) in Brazil [55]. In the United States, a total of 93 pharmaceuticals have been reported to be present in surface water with concentrations between 0.0035 and 15 μg/L (sulfadimethoxine) [56]. In the Ebro river basin in Spain, concentrations of pharmaceuticals were between 0.0004 and 0.712 μg/L (acetaminophen) [57]. In Tunisia, 13 pharmaceutical compounds were detected in the Melian River, where concentrations of ibuprofen (1.024 μg/L) and paracetamol (2.073 μg/L) were highest [13]. These high surface water concentrations indicated a great proportion of wastewater in the river during periods of low precipitation. As reported for wastewater effluents, the concentration of 2-OH-ibuprofen (4.050 μg/L) exceeded that of ibuprofen in the river [13].

Pharmaceuticals have been also frequently detected in groundwater [21, 24, 58–62]. In ground water samples collected from 23 European Countries, carbamazepine (frequency of detection: 42%; maximum concentration: 0.390 μg/L) and sulfamethoxazole (frequency of detection: 24%; maximum concentration: 0.038 μg/L) were among the most relevant compounds [21]. Sulfamethoxazole is a critical trace pollutant in bank filtration due to its enhanced persistency compared with that of other antibiotics [42, 63]. Carbamazepine was also reported to be relatively persistent during subsurface flow [64–67]. In a study carried out at a riverbank filtration site in Poland, carbamazepine (max. 145 ng/L), sulfamethoxazole (max. 20 ng/L), diclofenac (max. 99 ng/L), naproxen (max. 21 ng/L) and iohexol (max. 146 ng/L) were detected in groundwater collected from a well located 82 m away from the river [28]. However, a significant attenuation of pharmaceuticals was observed at travel times of 40–50 days and distances of 60–80 m; in an observation well located 250 m away from the river only carbamazepine (max. 81 ng/L and iohexol (max. 184 ng/L) were detected [28]. Paracetamol was reported to be present in groundwater at concentrations between 0.01 μg/L (France) and 1.89 μg/L (United States) [41].

Agricultural irrigation using wastewater-impacted river water is a potential source of pharmaceuticals in soil and groundwater [68]. Some pharmaceuticals that are used also in veterinary medicines can be released directly to the agricultural environment due to manure application [69]. The fate and removal during soil passage depends strongly on hydrogeological conditions and irrigation practices as well as on compound persistency and mobility. In groundwater collected from a wastewater reuse site, carbamazepine concentrations showed high variability and ranged from 0.01 to 0.114 μg/L [70]. Two metabolites were also detected, but about tenfold and 20-fold lower for 10,11-dihydro-10,11-dihydroxycarbamazepine and carbamazepine-10,11-epoxide, respectively. In the same area, carbamazepine was still detected five years later with a maximum concentration of 0.155 μg/L [13]. Sulfamethoxazole was also detected in groundwater but with a much lower concentration of max. 0.046 μg/L whereas ibuprofen and paracetamol were absent. However, ibuprofen was present in the wastewater stored in a basin before irrigation at a maximum concentration of 3.041 μg/L [13]. This is an indication that ibuprofen was not removed during wastewater storage but during soil passage.

Concentration of pharmaceuticals found in soils in different countries ranged between 0.02 μg/kg (carbamazepine, China) and 60.1 μg/kg (trimethoprim, Mexico) [49]. Carbamazepine was the most frequently detected compound in soil among five studies. Carbamazepine was reported to accumulate in soils [71, 72]. Before irrigation, the carbamazepine concentration in soils was 0.28 μg/kg. An increase was observed with irrigation, and the concentration reached 0.94 μg/kg after 6 months. A slight decrease to 0.82 μg/kg was observed one month after the end of irrigation [70]. Carbamazepine and sulfamethoxazole were reported to be persistent in irrigated soils [58, 73], whereas ibuprofen was not [72, 73]. Barber et al. [61] suggested, using sulfamethoxazole as a tracer for subsurface contamination by wastewater because of natural attenuation is insignificant. In addition to sulfamethoxazole, carbamazepine was suggested as an indicator for pollution of groundwater resources at sites, where wastewater is reused for irrigation in agriculture [13].

The occurrence of pharmaceuticals was recognized as an environmental concern and human health risk [18, 38]. When contaminated groundwater is used again for irrigation of edible crops or for drinking water production, pharmaceuticals might pose risks to human health. However, public policy treaties, protocols and agreements still lack demands to decrease the presence of pharmaceutical in water, and there is not sufficient commitment to eliminate discharge or to generate environmentally friendly medicines. Key parameters for pollution of soil and groundwater resources are compound mobility and persistence. In addition to information on biodegradability, values of S, log P_{OW} and pKa of paracetamol, sulfamethoxazole, carbamazepine and ibuprofen are shown in Table 15.1 in order to give an overview on their mobility. Paracetamol and sulfamethoxazole have rather low values log P_{OW} of 0.46 and 0.89, respectively, indicating high mobility. Carbamazepine and ibuprofen have higher values of log P_{OW} of 3.45 and 3.97, respectively, indicating lower mobility. Regarding pKa, in particular for ionic forms of sulfamethoxazole, dissociation processes might additionally increase mobility. Carbamazepine had a low leaching potential in irrigated soils based on adsorption [74]. However, it is ubiquitously present in groundwater as mentioned above. According to its persistence, carbamazepine is a potential pollutant of soil and groundwater. Chen et al. [75] observed a high mobility of sulfamethoxazole in saturated porous media. Sulfamethoxazole has a low biodegradability and a high mobility (Table 15.1). As mentioned above, it occurs frequently in soil and groundwater. One can conclude that sulfamethoxazole is much likely to pollute soil and groundwater. During the soil passage, paracetamol did not efficiently sorb to soil particles and was still persistent [76]. Although, paracetamol is principally biodegradable (Table 15.1), it occurs frequently in groundwater as mentioned above. In conclusion, it has to be also considered as a potential groundwater pollutant. Only for ibuprofen, which is biodegradable and has low mobility (Table 15.1), pollution of soil and groundwater seems to be unlikely. Indeed, ibuprofen was mostly not detected in soil and groundwater resources.

Ecotoxicological effects of pharmaceuticals have to be also considered. Fent et al. [77] reviewed the ecotoxicology of pharmaceuticals and concluded that very little is known about their long-term effects to aquatic organisms. In order to assess risks posed by pharmaceuticals detected in surface water to living organisms, the risk quotient (RQ), a ratio of the highest concentrations of pharmaceuticals detected in surface water (PEC) and the predicted no-effect concentration (PNEC) has been established. Values of RQ were categorized into low risk ($RQ < 0.1$), medium risk ($0.1 \leq RQ < 1$), and high risk ($RQ \geq 1.0$) [56]. In the US, carbamazepine and sulfamethoxazole were among the compounds at medium risk with RQ values of 0.14 and 0.47, respectively. Ibuprofen was at low risk with an RQ of 0.06. In the study of Stuer-Lauridson [78], the PEC/PNEC ratio exceeded one for paracetamol and ibuprofen in an environmental risk assessment study in Denmark.

The occurrence of antibiotics in the environment additionally leads to the risk of forming antibiotic resistances. Wang et al. [79] found a positive correlation between antibiotics and antibiotic resistance genes commonly detected in wastewater worldwide. Antibiotics serve as a selective pressure to increase the abundance of resistance genes in soil communities as reported for sulfamethoxazole [80]. However, the occurrence of antibiotic resistance genes and antibiotic resistant bacteria in groundwater is not likely to pose human health risk, but their presence in groundwater, especially when used for drinking water provision, might contribute to the development of antibiotic resistance in humans [81].

15.3 Benzotriazoles

The corrosion inhibitors 1H-benzotriazole (1H-BT), 4-methyl-1H-benzotriazole (4Me-BT) and 5-methyl-1H-benzotriazole (5Me-BT), both isomers are often summarized to tolyltriazoles (TT), belong to the group of benzotriazoles. Benzotriazoles are mainly added to dishwasher detergents and aircraft anti-icing fluids to inhibit corrosion. They are also found in dry cleaning equipment and a number of plastics. Due to the widespread use of dishwasher detergents in households, benzotriazoles are disposed of in municipal wastewater treatment plants. When sewer systems of airports are connected to municipal sewer systems, loads of benzotriazoles in wastewater can be expected to be seasonally high according to high application volumes of aircraft anti-icing fluids. However, their use in dishwasher detergents is a main source of benzotriazoles in surface waters in Germany [82], Switzerland [83] and the United Kingdom [84]. The occurrence of 1H-BT and TT in urine samples from people resident in countries such as China, the USA, Korea and India provides a new insight on additional sources of benzotriazoles in wastewater [85].

The removal of benzotriazoles during wastewater treatment was reported to be incomplete [42, 86]. Information on the biological degradability in municipal wastewater treatment of 1H-BT, 4Me-BT and 5Me-BT is given in Table 15.1. 4Me-BT was of very low biodegradability resulting in an insignificant removal in conventional activated sludge municipal wastewater treatment [43]. 4Me-BT can be classified as a persistent organic chemical. In contrast, 1H-BT and 5Me-BT were degraded in conventional activated sludge municipal wastewater treatment to a certain extent of 37% and 11%, respectively [43]. Both compounds are pseudo-persistent organic chemicals when delivered continuously to wastewater treatment plants.

Benzotriazoles occurred widely in wastewater effluents in Europe with mean concentrations of 1H-BT and TT of 6.3 μg/ and 2.9 μg/L, respectively [9]. In Tunisia, concentrations of 1H-BT and TT in municipal wastewater effluents were varying with concentrations between 0.244 μg/L and 65.5 μg/L for 1H-BT and 0.139 μg/L and 10.4 μg/L for TT [13]. 1H-BT and TT were also detected frequently in river water [13, 82, 86, 87]. Concentrations in the Rivers Main, Hengstbach and Hegbach ranged from 0.038 to 4.521 μg/L for 1H-BT, from 0.024 to 1.766 μg/L for 5Me-BT, and from 0.025 to 4.345 μg/L for 4Me-BT [82, 87]. In winter, the concentrations of all three compounds were higher than in summer indicating an influence from the use of anti-icing fluids in the river catchment area [82]. In Tunisia, TT were measured in the estuary of the Meliane River with concentrations between 0.2 μg/L and 0.407 μg/L, whereas 1H-BT was not detected [13].

The entry of benzotriazoles in river water may have negative effects on aquatic organisms [88]. In vitro anti-estrogenic activity was reported for 1H-BT, but concentrations up to 1,000 μg/L used in the studies far exceeded those reported in surface waters [89]. Acute and chronic toxicity of benzotriazoles to aquatic organisms revealed lowest effect concentrations (EC_{10}) of 0.00097 μg/L for 1H-BT and of 0.0004 μg/L for 5MeBT indicating that these compounds pose no risk for aquatic ecosystems at regularly found concentrations [90]. However, due to the seasonality of input sources, field investigations are required to study exposure concentrations in rivers in particular under different temperature conditions.

1H-BT and TT can also affect groundwater quality. In a European groundwater monitoring study comprising of 23 European countries, 1H-BT (frequency of detection: 53%; maximum concentration: 1.032 μg/L) and TT (frequency of detection: 52%; maximum concentration: 0.516 μg/L) were among the most relevant chemicals found in groundwater in terms of frequency of detection and maximum concentrations [21]. 1H-BT was detected in groundwater collected from an area, where reclaimed water is used for irrigation, at concentrations up to 0.040 μg/L, whereas TT were absent. Hence, 1H-BT was suggested as an indicator for pollution of groundwater resources by benzotriazoles at sites, where wastewater is reused for agricultural irrigation [13]. Values of S, log P_{OW} and pKa of 1H-BT, 4Me-BT and 5Me-BT are shown in Table 15.1 in order to give an overview on their mobility. The log P_{OW} of 1H-BT (1.44) is lower than of 5Me-BT (1.71) indicating a higher mobility. The mobility of 4Me-BT is difficult to estimate, since information on log P_{OW} is lacking but might be similar to that of 5Me-BT according to its same molecular formula. 1H-BT was regarded as a polar compound, and according to Reemtsma et al. [7], belonging to the group, for which a later contamination of compartments of the water cycle is most likely. 1H-BT was estimated to have a high mobility in soil and groundwater [91]. Although 1H-BT is principally biodegradable, it occurs frequently in groundwater as mentioned above. In conclusion, it has to be also considered as a groundwater pollutant according to its mobility. It has to be focused on in particular, since it has been reported to inhibit nitrification [91].

15.4 Per- and Polyfluoroalkyl Substances (PFAS)

According to an OECD survey, approximately 4700 different per- and polyfluoroalkyl substances (PFAS) are manufactured worldwide [92]. Not all of them are of high importance for uses in consumer products like impregnating agents or non-stick cookware. Due to the high stability against temperature, chemical degradation and the surface active properties [93], many PFAS were used as such, in mixtures, and in various products like coatings to introduce the unique properties into these products. Certain PFAS, e.g. perfluorooctanoic acid (PFOA), ammonium-4,8-dioxa-3H-4,8-perfluornonanoat and hexafluoropropylene oxide dimer acid were also used to produce fluoropolymers like polytetrafluoroethylene (PTFE). PTFE and other fluoropolymers are materials for many technical applications in automotive, coatings, construction, medicine products, membranes in certain textiles or special lubricants, e.g., for bicycle chains. A recent study showed that approximately 1400 PFAS are on the market in a high variety of industries, uses and products [94], which could be categorized in more than 200 categories. Another study concluded a high commercial relevance of 256 PFAS [95]. The analysis of all of these substances in an aqueous environment is challenging [96]. Standard methods usually include certain perfluoroalkyl carboxylic acids (PFCAs) like PFOA and perfluoroalkyl sulfonic acids (PFSAs) as PFOS with a perfluorinated chain of 4–12 C-atoms, and some more individual substances. Hence, the risk of overlooking substances contributing to the complete picture of water contamination with PFAS is high. This is especially the case, because many polyfluorinated compounds (precursors) might degrade to stable products in the environment. When PFAS are released to the environment, abiotic and microbial degradation of polyfluoroalkyl moieties can result in PFCAs, such as PFOA [44]. PFAS in general, and also PFOA and PFOS are extremely persistent (Table 15.1) due to the high stability of the C–F bond [44, 97]. One method to include these precursors in analytical schemes is the Total Oxidizable Precursors (TOP) assay [98]. By a pretreatment of the samples, precursors were oxidized to the homological PFCAs with an identical chain length. Using this method, Göckener et al. [99] demonstrated increasing concentrations of PFCAs in samples of biota and soil indicating that precursors deliver a relevant contribution to the PFAS contamination of the samples. A similar effect of pre-treatment by the TOP assay was demonstrated for river water [100].

In wastewater treatment plants, especially PFCAs and PFSAs resist completely against degradation. Concentrations of PFCAs in the effluent were often higher than in the influent of the sewage treatment plants [96], because certain precursors were transformed during the treatment procedure to the respective PFCAs with an identical perfluoro chain length [101]. Depending on the chain length and active groups, PFAS are also distributed to the air (very short volatile PFASs) or to the sewage compartment of a treatment plant. This is especially the case for PFOS and other long chain PFSAs. Short chain PFAS like perfluorobutane sulfonic acid (PFBS) preferably persist in the aqueous phase and reach the river system with the effluents of the plant.

For PFOA and PFOS, log P_{OW} is not measurable because of multiple layer formation in an n-octanol–water mixture. Thus, they do not behave like traditional hydrophobic chemicals. Very low values of pKa indicate that they are highly mobile in the aquatic environment (Table 15.1). The undissociated acid and anionic forms of PFCAs and PFSAs are different with regard to physical and chemical properties. Hence, it is essential to distinguish between the undissociated acid and the anionic form to select the appropriate physical and chemical parameters for fate and transport modelling [102].

Standard monitoring programs of rivers and other surface waters in Europe usually include PFOS as a priority hazardous substance of the Water Framework Directive 2000/60/EC and Directive 2008/105/EC [103], setting environmental quality standards in the field of water policy. The study of Loos et al. [9] with 90 wastewater treatment plants all over Europe analyzed 156 organic chemicals including seven environmentally relevant PFAS. In 99% of the effluents, PFOA was detected, and also other PFCAs included in the study (perfluoro hexanoic acid—PFHxA, perfluoro heptanoic acid—PFHpA, perfluorononanoic acid—PFNA, perfluorododecanoic acid -PFDA) were found in more than 70% of the samples exceeding the detection limits of 0.001 µg/L (PFHxA, PFOA, PFDA, PFHxS), respectively, 0.0005 µg/L (PFHxA, PFNA, PFOS) with maximum concentrations of 23.9 µg/L for PFHxA and 15.9 µg/L for PFOA. The PFSAs were dominated by PFOS in 93% of the samples with a maximum concentration of 2.1 µg/L [9].

In a study analyzing effluents of 49 municipal wastewater treatment plants of different magnitudes representing the plants in Germany, 16

PFAS were analyzed using solid phase extraction and HPLC–MS–MS [104]. Seven substances were detected at concentrations exceeding the quantification limit of 0.01 μg/L.

According to Buck et al. [105], the PFCAs like PFBA, and PFHxA, and PFBS as s PFAS were defined as short chain PFAS. These three substances were determined in 89–98% of the effluents at concentrations up to 0.092 μg/L (PFBA), 0.067 μg/L (PFHxA) and 0.17 (PFBS), respectively. Also PFOA as a long chain PFCA was found in 90% of the samples up to 0.11 μg/L. The long-chain PFOS was determined in 84% of the effluents at concentrations up to 0.17 μg/L. Due to the sorptive properties, this substance was also found in the sludge compartment in 33 of 57 samples up to 50 μg/kg The detection frequency of the longer-chain PFCAs and PFPeS, PFHxS, PFHpS as short chain PFSAs in effluents was low, probably because these substances are only used in few processes and not primarily in the production of consumer products. These studies indicate that most of the PFAS included in the analysis were able to pass wastewater treatment plant in the water phase without degradation or were transformed to stable PFCAs. For certain PFCAs like PFHxA, the concentrations in the effluents (0.72 μg/L) were up to one magnitude higher than in the influents (0.067 μg/L), indicating for transforming processes of precursors in the treatment process. Only PFOS was distributed to the sludge compartment in relevant amounts due to adsorption, and hence eliminated from the water phase to a certain degree.

With regard to use treated water for irrigation purposes, the further fate and distribution of PFAS need to be considered in order to assess the risk for the environment and the human health. While the more adsorptive substances of the group are expected to accumulate in soil, the mobile short chain PFAS will leach into groundwater or could be up-taken by plants and distributed also in edible parts of the plants [106–108]. In general, due to the extreme stability of PFAS and their behavior during wastewater treatment in combination with the toxic properties of certain PFAS, emissions of PFAS into the environment need to be restricted.

15.5 Potential Reduction Measures

A comprehensive approach to protect water resources from organic chemicals needs a combination of multiple measures along the relevant pathways to minimize the entry into the environment [109]. An overview on potential measures to limit emissions of micro-contaminants into surface and groundwater is given in Fig. 15.2. This approach includes measures at the source of emissions (e.g., manufacturing of substances, preparation of formulation and mixtures), during the intended use, technologies for reducing downstream emissions as well as regulatory and economic measures.

There has been a growing understanding that reduction efforts need to be implemented at the source, and that wide ranges of stakeholders (e.g., producers, authorities, users, water providers, etc.) need to be engaged and committed. Acknowledging this, in Germany, the Federal Ministry for the Environment has established a stakeholder dialog, which brings relevant stakeholders together in order to identify and to implement voluntary measures to reduce micro-contaminants [110]. Similarly, The Netherlands has established a "Chain Approach" to reduce pharmaceutical residues in water involving regional authorities and stakeholders, representing the healthcare, pharmaceutical and water sectors [111]. The activities can complement regulatory measures, but restrictions, threshold values or economic instruments are nevertheless important tools.

In line with the precautionary principle, due to limitations of wastewater treatment and also taking cost-effectiveness into account, measures at the source are key to reduce organic contaminants in the aquatic environment [109, 112, 113]. This includes legal restrictions of manufacture, import, distribution and uses of certain substances or substance groups, or incentives for

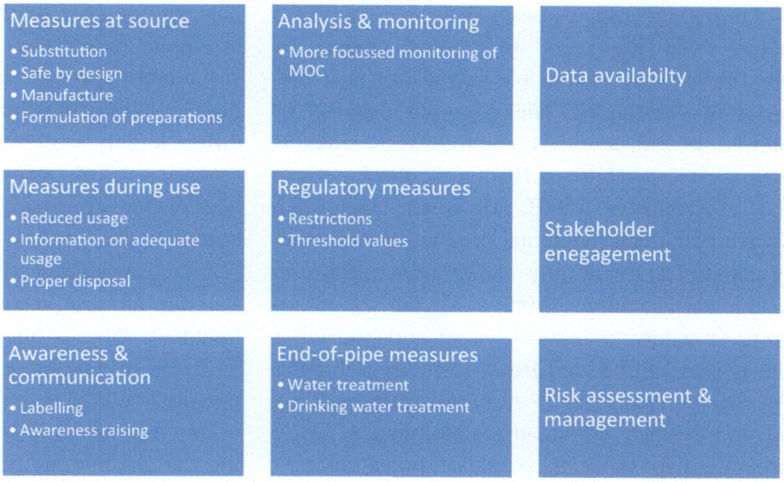

Fig. 15.2 Overview on potential measures to limit emissions of micro-contaminants into surface and groundwater (MOC: Mobile organic compounds)

their substitution and reduction, aiming especially at uses of chemicals of concern identified as non-essential. This can be fostered by awareness raising campaigns, e.g., in the health care sector, by product fees or by environmental information systems as on pharmaceuticals [114–116]. When users are aware about the potential environmental impacts of chemicals in consumer products, they can favor more environmentally friendly options, e.g., dishwashing detergents free from benzotriazoles or water-repellent textiles without PFAS. Approaches for substitution are supported by new developments within green and sustainable chemistry, which aim to design organic chemicals and products in a way that allows for biodegradability during treatment processes or in the environment. Consequently, the release of hazardous chemicals can be reduced. A better reuse and recycling can also be enabled. These concepts include "benign by design", "sustainable chemistry", "safe and sustainable-by-design chemicals" or "green pharmacy" [113, 116–119].

These approaches might be effective for chemicals in many consumer products. However, limitations exist for chemicals of concern in professional uses or complex use chains including uses in preparations or mixtures with other substances. Also, for uses of chemicals identified as "essential" due to missing effective alternatives [120], voluntary measures and information of consumers are not helpful enough. Examples are specific uses of certain PFAS in semiconductor production, metal plating, medicinal products, workers protection or lightweight construction in the high-tech sector. To foster the development of sustainable alternatives, substantial funding of research and incentives are needed [109, 116]. Current developments within the EU as, e.g., outlined in the "Chemicals Strategy for Sustainability - Towards a Toxic-Free Environment" can contribute to that [121, 122].

Further potential measures at the source relate to the regulatory context, e.g., the establishment of environmental quality standards or the prescription of effective risk reduction measures during authorization or restriction processes [109]. Reemtsma et al. [15] identified a regulatory gap for persistent mobile organic chemicals in the EU; closing this is another step towards reduction. Often, limited monitoring, analytical challenges and lack of data on substance properties hamper further reduction measures, if relevant sources, usages and entry pathways to the aquatic environment are not well understood. The German Environment Agency (Umweltbundesamt) stated that improved access to environmental data from substance authorization processes, especially for pharmaceuticals could help in this regard [109].

During the usage phase, there is a need for better awareness and information about the appropriate and sustainable application and disposal of substances. The improper disposal of unused pharmaceuticals via toilets or sinks is estimated to attribute to 10% of the discharges to the environment [109]. However, the main source of human medicinal products is excretion of patients. Overall, only 15–20% are released in health care facilities [109, 113], but for radiocontrast agents, certain antibiotics and cytostatics the portion introduced from hospitals is higher. As radiocontrast agents are hardly eliminated by advanced water treatment technologies [123], a separate disposal offers reduction potentials.

Municipal wastewater treatments plants are a main entry pathway for micro-contaminants into water bodies. Thus, advanced wastewater treatment plays a significant role to reduce inputs. Common advanced treatment methods include ozonation and the use of activated carbon. However, depending on their specific properties, there are limits to eliminate organic chemicals. Rizzo et al. [123] compiled data for removal in advanced treatment processes for selected contaminants of emerging concern. While for carbamazepine, abatement is >80% with ozone and powdered activated carbon (PAC), abatement for the benzotriazoles 1H-BT and TT was reported 50–80% for both treatment technologies. For PFOA and PFOS, both PAC treatment and ozonation were found to be inefficient. Considering the variety of organic chemicals in municipal wastewaters, usually a combination of several treatment methods is needed. The decision to upgrade wastewater treatment plants should take into account the size of the treatment plant and the receiving water bodies. Especially in large water treatment plants and those discharging to waterbodies used for drinking water supply, an advanced treatment can significantly reduce the emissions of organic chemicals. However, these treatment steps are often energy and cost-intensive. With regard to additional costs that occur for advanced treatment, the concept of extended producer responsibility gains more attention, e.g., charges in relation to substance properties can secure financial contributions for additional treatment processes or create incentives for substitution or more sustainable applications [124]. One has also to bear in mind that the level of water treatment differs largely worldwide. While in high-income countries, 70% of municipal and industrial wastewater are being treated, in low-income countries this applies to only 7% of wastewater [125]. Considering the infrastructure, political will and financial resources needed, an advanced treatment stage is often far away from reality.

Even in favorable conditions, none of the potential reduction measures will directly result in an elimination of all organic chemicals entering the aquatic environment; necessary changes often require a longer time span to be implemented. Water safety planning and systematic risk management play a crucial role to prevent adverse impacts of micro-contaminants in drinking water, but also in other uses as irrigation. The concept of the World Health Organization (WHO) on Water Safety Planning [126] allows analyzing potential hazards in the system, assessing risks, and deriving measures and monitoring criteria. This is especially relevant for cases of so-called "unplanned water reuse", where rivers with high contributions of treated effluent are sources of irrigation and of drinking water. Regulations or threshold values considering the wastewater impact are often missing [19]. For the assessment of risks in bank filtration, Karakurt et al. [20] highlighted the need to understand better site-specific hydrogeological factors. Using oxypurinol, valsartanic acid and carbamazepine as indicator chemicals, which are highly persistent and commonly present in wastewater influenced rivers, it was shown that even low contributions of bank filtrate to raw water can be critical, when wastewater effluent contributions in the respective river are high [20].

15.6 Conclusions

Organic chemicals like pharmaceuticals, benzotriazoles and PFAS are present in municipal wastewaters worldwide. According to high input variations as well as varying performances and technical states of wastewater treatment plants in different countries, estimations of effluent loads are difficult. Irrigation with wastewater effluents and river water are common responses to combat water scarcity strengthened by climate change. In addition to other human activities, e.g., application of sewage sludge and manure or pesticides usage, irrigation of reclaimed water can be an important source of micro-contaminants in natural resources. At irrigated sites, persistent but also pseudo-persistent micro-contaminants might occur in soil and groundwater. In general, soil passage can decrease or even remove micro-contaminants. However, this is only true for healthy soils. Consequently, irrigation should be accompanied and monitored carefully in order to sustain healthy soils and thus groundwater quality. Risk assessment needs to consider fate and transport characteristics of both, precursors and transformation products. Representative substances of different persistence and mobility characteristics should be used as indicators for the potential of contamination of soil and groundwater resources, in particular when irrigation of wastewater effluents is practiced or river water or groundwater is used without quality control. Carbamazepine, sulfamethoxazole, 1H-benzotriazole and selected PFAS are suggested as indicator substances combining (i) persistence with a certain adsorption capacity, (ii) persistence with mobility, and (iii) pseudo-persistence with mobility. Future transport and fate studies should also address transformation products of pharmaceuticals as well as resistance genes. Uptake of micro-contaminants by plants and their accumulation in the food chain should be also studied more in detail in the future.

Treatment procedures based on adsorption are not effective to eliminate especially short chain PFAS, and they will be translocated with (irrigation) water through the soil compartment to groundwater or taken up by plants, where they might enrich in edible parts. Hence, appropriate reduction strategies for this group of hazardous PFAS need to be developed further. One way forward is the intended restriction of manufacture, import, distribution, sale and use of the complete group under the REACH regulations (Registration, Evaluation, Authorisation and Restriction of Chemicals) allowing only a minority of uses identified as essential.

Measures at the source generally play a crucial role to reduce the entry of micro-contaminants into the environment. They should be complemented by measures along the whole life cycle. Depending on the specific substance properties, entry pathways as well as site-specific conditions, potential hazards and risks, a cost-effective combination of measures could include efforts to substance substitution, awareness rising campaigns, voluntary commitments by producers or decentralized treatment or disposal. While advanced wastewater treatment can significantly reduce pressures especially for sensitive waters or those receiving high loads of effluents, there are also limitations (and costs) that call for the engagement of several stakeholders including producers, authorities, users. This is further encouraged by the precautionary and polluter pays principle. In general, a transformation to a more sustainable chemistry is needed in future.

References

1. Deblonde T, Cossu-Leguille C, Hartemann P. Emerging pollutants in wastewater: a review of the literature. Int J Hyg Environ Health. 2011;214:442–8.
2. Margot J, Rossi L, Holliger C. A review of the fate of micropollutants in wastewater treatment plants. Wiley Interdiscliplinary Rev Water. 2015;2:457–87.
3. Boyd GR, Reemtsma H, Grimm DA, Mitra S. Pharmaceuticals and personal care products (PPCPs) in surface and treated waters of Louisiana, USA and Ontario, Canada. Sci Total Environ. 2006;311:135–49.
4. Kostich MS, Batt AL, Lazorchak JM. Concentrations of prioritized pharmaceuticals in effluents from 50 large wastewater treatment plants in the US and implications for risk estimation. Environ Pollut. 2014;184:354–9.
5. Heberer T. Tracking persistent pharmaceutical residues from municipal sewage to drinking water. J Hydrol. 2002;266:175–89.
6. Joss A, Zabaczynski S, Göbel A, Hoffmann B, Löffler D, McArdell CS, Ternes TA, Thomsen A, Siegrist H. Biological degradation of pharmaceuticals in municipal wastewater treatment: proposing a classification scheme. Water Res. 2006;40:1686–96.

7. Reemtsma T, Weiss S, Mueller J, Petrovic M, Gonzalez S, Barcelo D, Ventura F, Knepper T. Polar pollutants entry into the water cycle by municipal wastewater: a European perspective. Environ Sci Technol. 2006;40:5451–8.
8. Verlicchi P, Al Aukidy M, Zambellom E. Occurrence of pharmaceutical compounds in urban wastewater: removal, mass load and environmental risk after a secondary treatment—a review. Sci Total Environ. 2012;429:123–55.
9. Loos R, Carvalho R, António DC, Comero S, Locoro G, Tavazzi S, Paracchini B, Ghiani M, Lettieri T, Blaha L, Jarosova B, Voorspoels S, Servaes K, Haglund P, Fick J, Lindberg RH, Schwesig D, Gawlik BM. EU-wide monitoring survey on emerging polar organic contaminants in wastewater treatment plant effluents. Water Res. 2013;47:6475–87.
10. Luo Y, Guo W, Ngo HH, Nghiem LD, Ibney Hai F, Zhang J, Liang S, Wang XC. Review on the occurrence of micropollutants in the aquatic environment and their fate and removal during wastewater treatment. Sci Total Environ. 2014;473–474:619–41.
11. Mahjoub O, Leclercq M, Bachelot M, Casellas C, Escande A, Balaguerc B, Bahri A, Gomez E, Fenet H. Estrogen, aryl hydrocarbon and pregnane X receptors activities in reclaimed water and irrigated soils in Oued Souhil area (Nabeul, Tunisia). Desalination. 2009;246:425–34.
12. Fenet H, Mathieu O, Mahjoub O, Li Z, Hillaire-Buys D, Casellas C, Gomez E. Carbamazepine, carbamazepine epoxide and dihydroxycarbamazepine sorption to soil and occurrence in a wastewater reuse site in Tunisia. Chemosphere. 2012;88:49–54.
13. Fries E, Mahjoub O, Mahjoub B, Berrehouc A, Lions J, Bahadir M. Occurrence of contaminants of emerging concern (CEC) in conventional and non-conventional water resources in Tunisia. Fresenius Environ Bull. 2016;25:3317–39.
14. Valdez-Carrillo M, Abrell L, Ramírez-Hernández J, Reyes-López JA, Carreón-Diazconti C. Pharmaceuticals as emerging contaminants in the aquatic environment of Latin America: a review. Environ Sci Pollut Res. 2020;27(36):44863–91. https://doi.org/10.1007/s11356-020-10842-9.
15. Reemtsma T, Berger U, Arp HPH, Gallard H, Knepper TP, Neumann M, Quintana JB, de Voogt P. (2016) Mind the gap: persistent and mobile organic compounds—water contaminants that slip through. Environ Sci Technol. 2016;50:10308–15.
16. Karickhoff SW, Brown DS, Scott TA. Sorption of hydrophobic pollutants on natural sediments. Water Res. 1979;13:241–8.
17. Kümmerer K. Commentary: emerging contaminants versus micro-pollutants. Clean: Soil, Air, Water. 2011;39:889–90.
18. Helmecke M, Fries E, Schulte C. Regulating water reuse for agricultural irrigation: risks related to organic micro-contaminants. Environ Sci Europe. 2020;32. https://enveurope.springeropen.com/articles/10.1186/s12302-019-0283-0.
19. Drewes J, Hübner U, Zhiteneva V, Karakurt S. Characterization of unplanned water reuse in the EU. European Commission; 2017. https://ec.europa.eu/environment/water/pdf/Report-UnplannedReuse_TUM_FINAL_Oct-2017.pdf.
20. Karakurt S, Schmid L, Hübner U, Drewes JE. Dynamics of wastewater effluent contributions in streams and impacts on drinking water supply via riverbank filtration in Germany—a national reconnaissance. Environ Sci Technol. 2019;53(11):6154–61. https://doi.org/10.1021/acs.est.8b07216.
21. Loos R, Locoro G, Comero S, Contini S, Schwesig D, Werres F, Balsaa P, Gans O, Weiss S, Blaha L, Bolchi M, Gawlik BM. Pan-European survey on the occurrence of selected polarorganic persistent pollutants in ground water. Water Res. 2010;44:4115–26.
22. Boxall ABA, et al. Pharmaceuticals and personal care products in the environment: what are the big questions? Environ Health Perspect. 2012;120 (9):1221–9.
23. Lapworth DJ, Baran N, Stuart ME, Ward RS. Emerging organic contaminants in groundwater: a review of sources, fate and occurrence. Environ Pollut. 2012;163:287–303.
24. Lopez B, Ollivier P, Togola A, Baran N, Ghestem J-P. Screening of French groundwater for regulated and emerging contaminants. Sci Total Environ. 2015;518–519:562–73.
25. Draženka S, Zrinka D, Siniša R, Rebok K, Jordanova M. Broad spectrum screening of 463 organic contaminants in rivers in Macedonia. Ecotoxicol Environ Saf. 2017;135:48–59.
26. Ginebreda A, Sabater-Liesa L, Rico A, Focks A, Barceló D. Reconciling monitoring and modeling: an appraisal of river monitoring networks based on a spatial autocorrelation approach—emerging pollutants in the Danube River as a case study. Sci Total Environ. 2018;618:323–35.
27. Fries E, Püttmann W. Monitoring of the organophosphate esters TBP, TCEP and TBEP in river water and ground water (Oder, Germany). J Environ Monit. 2003;5:346–52.
28. Kruć R, Krzysztof Dragon K, Górski J. Migration of pharmaceuticals from the Warta Riverto the aquifer at a riverbank filtration site in Krajkowo (Poland). Water. 2019;11:2238.
29. WHO. Guidelines for the safe use of wastewater, excreta and greywater. World Health Organization; 2006.
30. El Ayni F, Cherif S, Jrad A, Trabelsi-Ayadi M. Impact of treated wastewater reuse on agriculture and aquifer recharge in a coastal area: Korba Case Study. Water Resour Manage. 2011;2:2251–65.

31. Jemai I, Ben Aissa N, Gallali T, Chenini F. Effects of municipal reclaimed wastewater irrigation on organic and inorganic composition of soil and groundwater in Souhil Wadi Area (Nabeul, Tunisia). Hydrol Curr Res. 2013;4:1–17.
32. Lazarova V, Asano T, Bahri A, Anderson J. Milestones in water reuse. IWA publishing; 2013.
33. Calderón-Preciado D, Jiménez-Cartagena C, Matamoros V, Bayona JM. Screening of 47 organic microcontaminants in agricultural irrigation waters and their soil loading. Water Res. 2011;45:221–31.
34. Malakar A, Snow DD, Ray C. Irrigation water quality—a contemporary perspective. Water. 2019;11:1482.
35. Poustie A, Yang Y, Verburg P, Pagilla K, HaniganD. Reclaimed wastewater as a viable water source for agricultural irrigation: a review of food crop growth inhibition and promotion in the context of environmental change. Sci Total Environ. 2020;739:139756.
36. Becerra-Castro C, Lopes AR, Vaz-Moreira I, Silva F, Manaia CM, Nunes OC. Wastewater reuse in irrigation: a microbiological perspective on implications in soil fertility and human and environmental health. Environ Int. 2015;75:117–135. ISSN 0160-4120.
37. IWMI (International Water Management Institute). World water supply and demand. Colombo, Sri Lanka; 2000. ISBN 92-9090-400-3.
38. González Peña OI, López Zavala MA, Cabral Ruelas H. Pharmaceuticals market, consumption trends and disease incidence are not driving the pharmaceutical research on water and wastewater. Int J Environ Res Public Health. 2021;18(5).
39. Global Pharmaceutical Industry. Statista: Dossier. Hamburg, Germany; 2018.
40. Nikolaou A, Meric S, Fatta D. Occurrence patterns of pharmaceuticals in water and wastewater environments. Anal Bioanal Chem. 2007;387:1225–34.
41. Al-Kaf AG, Naji KM, Abdullah QYM, Edrees WHA. Occurrence of paracetamol in aquatic environments and transformation by microorganisms: a review. Chron Pharm Sci. 2017;1:341–55.
42. Halling-Sorensen B, Nielsen SN, Lanzky PF, Ingerslev F, Lutzhoft HCH, Jorgensen SE. Occurrence, fate and effects of pharmaceutical substances in the environment—a review. Chemosphere. 1998;36:357–94.
43. Weiss S, Jakobs J, Reemtsma T. Discharge of three benzotriazoles corrosion inhibitors with municipal wastewater and improvements by membrane bioreactor treatment and ozonation. Environ Sci Technol. 2006;40:7193–9.
44. Liu J, Mejia Avendaño S. Microbial degradation of polyfluoroalkyl chemicals in the environment: a review. Environ Int. 2013;61:98–114. https://doi.org/10.1016/j.envint.2013.08.022.
45. Fatta-Kassinos D, Meric S, Nikolaou A. Pharmaceutical residues in environmental waters and wastewater: current state of knowledge and future research. Anal Bioanal Chem. 2011;399:251–75.
46. Borges KB, odrigo Moraes de Oliveira A, Barth T, Aparecida Polizel Jabor V, Tallarico Pupo M, Sueli Bonato P. LC–MS–MS determination of ibuprofen, 2-hydroxyibuprofen enantiomers, and carboxyibuprofen stereoisomers for application in biotransformation studies employing endophytic fungi. Anal Bioanal Chem. 2011;399:915–925.
47. Ferrando-Climent L, Collado N, Buttiglieri G, Gros M, Rodriguez-Roda I, Rodriguez-Mozaz S, Barceló D. Comprehensive study of ibuprofen and its metabolites in activated sludge batch experiments and aquatic environment. Sci Total Environ. 2012;438:404–13.
48. Kümmerer K. The presence of pharmaceuticals in the environment due to human use—present knowledge and future challenges. J Environ Manage. 2009;90:2354–66.
49. Li WC. Occurrence, sources, and fate of pharmaceuticals in aquatic environment and soil. Environ Pollut. 2014;187:193–201.
50. UBA. Pharmaceuticals in the environment: global occurrence and potential cooperative action under the Strategic Approach to International Chemicals Management (SAICM). UBA Texte. 2016;67:95.
51. UBA. The database “Pharmaceuticals in the Environment”—update and new analysis. Final report UBA Texte. 2019;67.
52. Xu WH, Zhang G, Zou SC, Li XD, Liu YC. Determination of selected antibiotics in the Victoria Harbour and the Pearl River, South China using high performance liquid chromatography-electrospray ionization tandem mass spectrometry. Environ Pollut. 2007;145:672–9.
53. Cruz ES. Fármacos y disruptores endócrinos en cuerpos de agua superficial impactadas por descargas de aguas residuales de Tapachula Chiapas, México. Dissertation, El Colegio de la Frontera Sur (Spanish); 2013.
54. Spongberg AL, Witter JD, Acuña J, Vargas J, Murillo M, Umaña G, Gómez E, Perez G. Reconnaissance of selected PPCP compounds in Costa Rican surface waters. Water Res. 2011;45:6709–17.
55. Campanha MB, Awan AT, de Sousa DN, Grosseli GM, Mozeto AA, Fadini PS. A 3-year study on occurrence of emerging contaminants in an urban stream of São Paulo State of Southeast Brazil. Environ Sci Pollut Res. 2015.
56. Deo RP. Pharmaceuticals in the surface water of the USA: a review. Curr Environ Health Rep. 2014;1:113–22.
57. López-Serna R, Petrović M, Barceló D. Occurrence and distribution of multi-class pharmaceuticals and their active metabolites and transformation products in the Ebro River basin (NE Spain). Sci Total Environ. 2012;44:280–9.
58. Clara M, Strenn B, Kreuzinger N. Carbamazepine as a possible anthropogenic marker in the aquatic

environment: investigations on the behaviour of carbamazepine in wastewater treatment and during groundwater infiltration. Water Res. 2004;38:947–54.
59. Benotti MJ, Fisher SC, Terracciano SA. Occurrence of pharmaceuticals in shallow ground water of Suffolk County, New York, 2002–2005. Open-File Report 2006–1297 U.S. Department of the Interior. U.S. Geological Survey; 2006.
60. Avisa D, Lester Y, Ronen D. Sulfamethoxazole contamination of a deep phreatic aquifer. Sci Total Environ. 2009;407:4278–82.
61. Barber LB, Keefe SH, LeBlan DR, Bradlay PM, Chapelles FH, Meyer MT, Loftin KA, Rubio F. Fate of sulfamethoxazole, 4-nonylphenol, and 17 betaestradiol in groundwater contaminated by wastewater treatment plant effluent. Environ Sci Technol. 2009;43:4843–50.
62. Qian S, Xuqi C, Shuguang L, Wentao Z, Zhaofu Q, Gang Y. Occurrence, sources and fate of pharmaceuticals and personal care products in the groundwater: a review. Emerg Contam. 2015;1:14–24.
63. Jjemba PK. The potential impact of veterinary and human therapeutic agents in manure and biosolids on plants grown on arable land: a review. Agr Ecosyst Environ. 2002;93(1–3):267–78.
64. Nagy-Kovács Z, László B, Fleit E, Czihat-Mártonné K, Till G, Börnick H, Adomat Y, Grischek T. Behavior of organic micropollutants during river bank filtration in Budapest, Hungary. Water. 2018;10:1861.
65. Maeng SK, Ameda E, Sharma SK, Grutzmacher G, Amy GL. Organic micropollutant removal from wastewater effluent-impacted drinking water sources during bank filtration and artificial recharge. Water Res. 2010;44:4003–14.
66. Burke V, Schneider L, Greskowiak J, Zerball-van Baar P, Sperlich A, Dünnbier U, Massmann G. Trace organic removal during river bank filtration for two types of sediment. Water. 2018;10:1736.
67. Driezum IH, Derx J, Oudega TJ, Zessner M, Naus FL, Saracevic E, Kirschner AKT, Sommer R, Farnleitner AH, Blaschke AP. Spatiotemporal resolved sampling for the interpretation of micropollutantremoval during riverbank filtration. Sci Total Environ. 2019;649:212–23.
68. Vodyanitskii YN, Yakovlev AS. Contamination of soils and groundwater with new organic micropollutants: a review. Eurasian Soil Sci. 2016;49:560–9.
69. Boxall ABA. New and emerging water pollutants arising from agriculture. OECD. COM/TAD/CA/ENV/EPOC(2010)17/FINAL. 2012; 2010.
70. Fenet H, Mathieu O, Mahjoub O, Li Z, Hillaire-Buys D, Casellasm C, Gomez E. Carbamazepine, carbamazepine epoxide and dihydroxycarbamazepine sorption to soil and occurrence in a wastewater reuse site in Tunisia. Chemosphere. 2011;88:49–54.
71. Kinney CA, Furlong ET, Werner SL, Cahill JD. Presence and distribution of wastewater-derived pharmaceuticals in soil irrigated with reclaimed water. Environ Toxicol Chem. 2006;25:317–26.
72. Williams CF, McLain JET. Soil persistence and fate of carbamazepine, lincomycin, caffeine, and ibuprofen from wastewater reuse. J Environ Qual. 2012;41:1473–80.
73. Grossberger A, Hadar Y, Borch T, Chefetz B. Biodegradability of pharmaceutical compounds in agricultural soils irrigated with treated wastewater. Environ Pollut. 2014;185:168–77.
74. Williams CF, Adamsen FJ. Sorption-desorption of carbamazepine from irrigated soils. J Environ Qual. 2006;35:1779–83.
75. Chen H, Gao B, Li H, Ma LQ. Effects of pH and ionic strength on sulfamethoxazole and ciprofloxacin transport in saturated porous media. J Contam Hydrol. 2011;126:29–36.
76. Heberer T. Occurrence, fate and removal of pharmaceutical residues in the aquatic environment: a review of recent research data. Toxicol Lett. 2002;131(1–2):5–17.
77. Fent K, Weston AA, Caminada D. Review: ecotoxicology of human pharmaceuticals. Aquat Toxicol. 2006;76:122–59.
78. Stuer-Lauridsen F, Birkved M, Hansen LP, Lützhøft HC, Halling-Sørensen B. Environmental risk assessment of human pharmaceuticals in Denmark after normal therapeutic use. Chemosphere. 2000;40 (7):783–93.
79. Wang J, Chu L, Wojnárovits L, Takács E. Occurrence and fate of antibiotics, antibiotic resistant genes (ARGs) and antibiotic resistant bacteria (ARB) in municipal wastewater treatment plant: an overview. Sci Total Environ. 2020;744. https://doi.org/10.1016/j.scitotenv.2020.140997.
80. Carey DE, McNamara PJ. The impact of triclosan on the spread of antibiotic resistance in the environment. Front Microbiol. 2014;5(780):1–11.
81. Zainab SM, Junaid M, Xu N, Malik RN. Antibiotics and antibiotic resistant genes (ARGs) in groundwater: a global review on dissemination, sources, interactions, environmental and human health risks. Water Res. 2020;187:116455.
82. Kiss A, Fries E. Seasonal source influence on river mass flows of benzotriazoles. J Environ Monit. 2012;14:697–703.
83. Giger W, Schaffner C, Kohler H-PE. Benzotriazole and tolyltriazole as aquatic contaminants. 1. Input and occurrence in rivers and lakes. Environ Sci Technol. 2006;40:7186–92.
84. Janna H, Scrimshaw MD, Williams RJ, Churchley J, Stumpter JP. From dishwasher to tap? Xenobiotic substances benzotriazole and tolyltriazole in the environment. Environ Sci Technol. 2011;45: 3858–64.
85. Asimakopoulos AG, Wang L, Thomaidis NS, Kanna K. Benzotriazoles and benzothiazoles in human urine from several countries: a perspective on occurrence, biotransformation, and human exposure. Environ Int. 2013;59:274–81.

86. Voutsa D, Hartmann P, Schaffner C, Giger W. Benzotriazoles, alkylphenols and bisphenol A in municipal wastewaters and in the Glatt River, Switzerland. Environ Sci Pollut Res. 2006;13:333–41.
87. Kiss A, Fries E. Occurrence of benzotriazoles in the rivers main, Hengstbach and Hegbach (Germany). In: Environmental science and pollution research. Special Series: Chemical and biological environmental monitoring, vol 16; 2009. pp 702–710.
88. Corsi SR, Geis SW, Loyo-Rosales JE, Rice CP. Aquatic toxicity of nine aircraft deicer and anti-icer formulations and relative toxicity of additive package ingredients alkylphenol ethoxylates and 4,5-Methyl-1H-benzotriazoles. Environ Sci Technol. 2006;40:7409–15.
89. Harris CA, Routledge EJ, Schaffner C, Brian JV, Giger W. Benzotrizole is antiestrogenic in vitro but not in vivo. Environ Toxicol Chem. 2007;26:2367–72.
90. Seeland A, Oetken M, Kiss A, Fries E, Oehlmann J. Acute and chronic toxicity of benzotriazoles to aquatic organisms. Environ Sci Pollut Res. 2011;19:1781–90.
91. Breedveld GD, Roseth R, Sparrevik M, Hartnik T, Hem LJ. Persistence of the de-icing additive benzotriazole at an abandoned airport. Water Air Soil Pollut Focus. 2003;3:91–101.
92. OECD. Toward a new comprehensive global database of per- and polyfluoroalkyl substances (PFASs): Summary report on updating the OECD 2007 list; 2018. https://www.oecd.org/chemicalsafety/portal-perfluorinated-chemicals/.
93. Kissa E, Schick M, Hubbard A. Fluorinated surfactants and repellents. 2nd ed. Weinheim, Germany: Wiley; 1997.
94. Glüge J, Scheringer M, Cousins IT, DeWitt JC, Goldenman G, Herzke D, Lohmann R, Ng CA, Trier X, Wang Z. An overview of the uses of per- and polyfluoroalkyl substances (PFAS). Environ Sci Process Impacts. 2020;22:2345–73. https://doi.org/10.1039/D0EM00291G.
95. Buck RC, Korzeniowski SH, Laganis E, Adamsky F. Identification and classification of commercially relevant per- and poly-fluoroalkyl substances (PFAS). Integr Environ Assess Manag. 2021;1–11. https://doi.org/10.1002/ieam.4450.
96. Gremmel C, Frömel T, Knepper TP. HPLC–MS/MS methods for the determination of 52 perfluoroalkyl and polyfluoroalkyl substances in aqueous samples. Anal Bioanal Chem. 2017;409:1643–55. https://doi.org/10.1007/s00216-016-0110-z.
97. Coggan TL, Moodie D, Kolobaric A, Szabo D, Shimeta J, Crosbie ND, Lee E, Fernandes M, Clarke BO. An investigation into per- and polyfluoroalkyl substances (PFAS) in nineteen Australian wastewater treatment plants (WWTPs). Heliyon. 2019;5:e02316.
98. Houtz EF, Sedlak DL. Oxidative conversion as a means of detecting precursors to perfluoroalkyl acids in urban runoff. Environ Sci Technol. 2012;46:9342–9. https://doi.org/10.1021/es302274g.
99. Göckener B, Fliedner A, Rüdel H, Fettig I, Koschorreck J. Exploring unknown per- and polyfluoroalkyl substances in the German environment—the total oxidizable precursor assay as helpful tool in research and regulation. Sci Total Environ. 2021;782. https://doi.org/10.1016/j.scitotenv.2021.146825.
100. Joerss H, Schramm T-R, Sun L, Guo C, Tang J, Ebinghaus R. Per- and polyfluoroalkyl substances in Chinese and German river water—point source- and country-specific fingerprints including unknown precursors. Environ Pollut. 2020;267. https://doi.org/10.1016/j.envpol.2020.115567.
101. UBA. Investigations on the presence and behavior of precursors to perfluoroalkyl substances in the environment as a preparation of regulatory measures. UBA-Texte 08/2016. https://www.umweltbundesamt.de/sites/default/files/medien/378/publikationen/texte_08_2016_investigations_on_the_presence_and_behavior.pdf. ISSN 1862-4804; 2016.
102. ITRC (Interstate Technology & Regulatory Council). PFAS technical and regulatory guidance document and fact sheets PFAS-1. Washington; 2020. https://pfas-1.itrcwe.org/.
103. Comber SDW, Gardner MJ, Ellor B. Perfluorinated alkyl substances: sewage treatment and implications for receiving waters. Sci Total Environ. 2021;791. https://doi.org/10.1016/j.scitotenv.2021.148391.
104. UBA. Prioritäre Stoffe in kommunalen Kläranlagen Ein deutschlandweit harmonisiertes Monitoring. UBA-Texte 173/2020. https://www.umweltbundesamt.de/sites/default/files/medien/5750/publikationen/2020_09_25_texte_173-2020_prioritaere_stoffe_in_kommunalen_klaeranlagen.pdf. ISSN 1862-4804; 2020.
105. Buck RC, Franklin J, Berger U, Conder JM, Cousins IT, de Voogt P, Jensen AA, Kannan K, Mabury SA, van Leeuwen SPJ. Perfluoroalkyl and polyfluoroalkyl substances in the environment: Terminology, classification, and origins. Integr Environ Assess Manag. 2011;7:513–41. https://doi.org/10.1002/ieam.258.
106. Blaine AC, Rich CD, Hundal LS, Lau C, Mills MA, Harris KM, Higgins CP. Uptake of perfluoroalkyl acids into edible crops via land applied biosolids: field and greenhouse studies. Environ Sci Technol. 2013;47:14062–9.
107. Felizeter S, McLachlan MS, De Voogt P. Uptake of perfluorinated alkyl acids by hydroponically grown lettuce (Lactuca sativa). Environ Sci Technol. 2012;46:11735–43.

108. Felizeter S, McLachlan MS, De Voogt P. Root uptake and translocation of perfluorinated alkyl acids by three hydroponically grown crops. J Agric Food Chem. 2014;62:3334–42.
109. UBA. Recommendations for reducing micropollutants in waters. UBA Hintergrundpapier/2018; 2018. https://www.umweltbundesamt.de/sites/default/files/medien/1410/publikationen/180709_uba_pos_mikroverunreinigung_en_bf.pdf.
110. BMU/UBA (ed). Recommendations from the multi-stakeholder dialogue on the trace substance strategy of the German Federal Government; to reduce trace substance inputs to the aquatic environment. Policy Paper; 2017. http://www.bmu.de/fileadmin/Daten_BMU/Download_PDF/Binnengewaesser/spurenstoffstrategie_policy_paper_en_bf.pdf12.
111. NL Government – Government of the Netherlands. Reducing pharmaceutical residues in water: a chain approach. Implementation programme 2018–2022; 2019. https://www.government.nl/binaries/government/documents/policy-notes/2019/02/12/reducing-pharmaceutical-residues-in-water-a-chain-approach/EN+Implementation+programme+pharmaceutical+residues.pdf.
112. Kümmerer K, Dionysiou D, Olsson O, Fatta-Kassinos D. Reducing aquatic micropollutants—increasing the focus on input prevention and integrated emission management. Sci Total Environ. 2019;652:836–50. https://doi.org/10.1016/j.scitotenv.2018.10.219.
113. OECD. Pharmaceutical residues in freshwater: hazards and policy responses. In: OECD studies on water. OECD Publishing, Paris; 2019.https://doi.org/10.1787/c936f42d-en.
114. fass.se. Environmental classification of pharmaceuticals at www.fass.se. Guidence for pharmaceutical companies; 2012. http://www.fass.se/pdf/Environmental_classification_of_pharmaceuticals-120816.pdf.
115. janusinfo.se. Electronic means of communication of the drug therapeutic committee and the health and medical care administration of the Stockholm County Council, Sweden; 2021.
116. UBA. Maßnahmen zur Verminderung des Eintrages von Mikroschadstoffen in die Gewässer. UBA-Texte 85/2014; 2014. http://www.umweltbundesamt.de/publikationen/massnahmen-zur-verminderung-des-eintrages-von. ISSN 1862-4804.
117. Leder C, Rastogi T, Kümmerer K. Putting benign by design into practice-novel concepts for green and sustainable pharmacy: designing green drug derivatives by non-targeted synthesis and screening for biodegradability. Sustain Chem Pharm; 2015.
118. Kümmerer K. Sustainable from the very beginning: rational design of molecules by life cycle engineering as an important approach for green pharmacy and green chemistry. Green Chem. 2007;9(8):899. https://doi.org/10.1039/b618298b.
119. Kümmerer K. Sustainable chemistry: a future guiding principle. Angew Chem Int Ed. 2017;56:16420–1.
120. Cousins IT, Goldenman G, Herzke D, Lohmann R, Miller M, Ng CA, Patton S, Scheringer M, Trier X, Vierke L, Wang Z, DeWitt JC. The concept of essential use for determining when uses of PFASs can be phased out. Environ Sci Process Impacts. 2019;21:1803–15. https://doi.org/10.1039/C9EM00163H.
121. EU COM—European Commission. Chemicals strategy for sustainability towards a toxic-free environment. COM/2020/667 final; 2020. https://eur-lex.europa.eu/legal-content/EN/TXT/?uri=COM%3A2020%3A667%3AFIN.
122. EEA, European Environment Agency. Designing safe and sustainable products requires a new approach for chemicals; 2021. https://www.eea.europa.eu/themes/human/chemicals/delivering-products-that-are-safe.
123. Rizzo L, Malato S, Antakyali D, Beretsou VG, Đolić MB, Gernjak W, Heath E, Ivancev-Tumbas I, Karaolia P, Lado Ribeiro AR, Mascolo G, McArdell CS, Schaar H, Silva AMT, Fatta-Kassinos D. Consolidated vs new advanced treatment methods for the removal of contaminants of emerging concern from urban wastewater. Sci Total Environ. 2019;655:986–1008. ISSN 0048-9697. https://doi.org/10.1016/j.scitotenv.2018.11.265.
124. Deloitte. Study on the feasibility of applying extended producer responsibility to micropollutants and microplastics emitted in the aquatic environment from products during their life cycle; 2020. https://library.wur.nl/WebQuery/edepot/512256.
125. UN. UN World Water Development Report, 22 March, 2017, wastewater: an untapped resource; 2017. https://www.unwater.org/publications/world-water-development-report-2017/.
126. WHO. Water safety plan manual: step-by-step risk management for drinking-water suppliers. Genevax: World Health Organization; 2009.

Wastewater-Based Epidemiology: Overview of Covid-19 Tracking in Brazil

16

Juliana Calabria de Araújo, Andreas Haarstrick, Sávia Gavazza, Lourdinha Florêncio, and Elvis Carissimi

Abstract

It is undoubtedly that Covid-19 pandemic disrupted massively our earthly lives. Besides the washing of hands, use of masks, social distancing and constant application of hand sanitizers, this pandemic accentuated the sanitation problems in developing countries, where water and sewage collection and/or treatment lack in many areas. Sewage can reveal true scale of the population contamination outbreak. Thus, in places, where clinical testing is deficient, SARS-CoV-2 sewage monitoring can be of paramount importance to proceed with accurate public health policies to prevent the spread of the contamination. In this paper, the main goal was to overview the Covid-19 tracking in Brazil among the distinct research networks formed in this continental country. An overview of the beginning and the status regarding this disease tracking in Brazil via wastewater and environmental monitoring was discussed. Results showed that at least seven research groups' leaders in four of the five regions of Brazil are conducting decentralized monitoring of covid-19 in sewage trough Wastewater Based Epidemiology (WBE) and monitoring the affected areas. All official information were centralized in the federal government agency, National Agency for Water and Sanitation. A consolidated and centralized center with this information is important to pave the way for the development of guidelines for a future and permanent National Wastewater Surveillance Plan. This information is especially useful to precede future needs for hospitalizations and establish a rigorous control in terms of biosafety protocol and lockdown. These efforts and research along with continuous monitoring showed also to be of paramount importance to

J. C. de Araújo (✉)
Water and Wastewater Microbiology Laboratory, Department of Sanitary and Environmental Engineering, Universidade Federal de Minas Gerais (UFMG), Belo Horizonte, MG, Brazil
e-mail: juliana@desa.ufmg.br

A. Haarstrick
Leichtweiss-Institut für Wasserbau, Exceed, Technische Universität Braunschweig, Braunschweig, Germany

S. Gavazza · L. Florêncio
Laboratory of Environmental Sanitation, Department of Civil and Environmental Engineering, Universidade Federal de Pernambuco (UFPE), Recife, PE, Brazil

E. Carissimi
Environmental and Engineering Laboratory, Department of Sanitary and Environmental Engineering, Universidade Federal de Santa Maria (UFSM), Santa Maria, RS, Brazil

M. Bahadir and A. Haarstrick (eds.), *Water and Wastewater Management*,
Water and Wastewater Management, https://doi.org/10.1007/978-3-030-95288-4_16

verify vaccine effectiveness and to detect outbreaks in specific areas.

Keywords

Covid-19 · Sanitation · Epidemiology · Water · Wastewater

16.1 Introduction

Wastewater contains a wide variety of microorganisms, some of which are pathogens that cause the so-called waterborne diseases, such as hepatitis A, polio, typhoid and paratyphoid fever, cholera, salmonellosis, giardiasis, amoebiasis, gastroenteritis, cryptosporidiosis, and worms. These diseases can debilitate or even lead to death, especially in the most vulnerable groups (elderly and children). Therefore, it has become worldwide a good practice in water distribution companies to monitor the presence of enteric microorganisms. However, wastewater can carry many other constituents, besides those traditionally expected to be excreted in feces and urine. For this reason, wastewater-based epidemiology has been used in several countries as an important tool to evaluate the circulation of illicit drugs and their metabolites, pathogenic enteric microorganisms, and other hazardous constituents.

In the recent COVID-19 pandemic caused by the SARS-CoV-2 coronavirus, several studies have revealed that the main route of transmission is carried by droplets and aerosols from the respiratory secretions of infected people [1]. The presence of SARS-CoV-2 viral RNA was detected in stool samples from both symptomatic and asymptomatic patients [2–4]. However, to date, no infectious SARS-CoV-2 has been found in feces and sewage samples. The earliest studies published in 2020 in several countries, such as The Netherlands [5], Spain [6], and France [7] reported that viral particles of SARS-CoV-2 were detected in wastewater even before the detection of clinical cases of COVID-19. The virus was excreted in the feces of the infected people (symptomatic and asymptomatic) since the beginning of the infection and then reached the sewage system.

The dynamics of contamination of the population by SARS-CoV-2 in Brazil showed that the limitation of data quality has been evident, as their collection has been more in an emergency context (incidence, hospitalizations, deaths, lethality, among others), thus, not necessarily reflecting the reality of the pandemic. It would be desirable to apply mass clinical testing in combination with contact tracking and isolation of those infected people (symptomatic and asymptomatic). But there are practical and economic limitations that preclude this approach, which is focused on individual testing.

Monitoring the presence of viral particles in wastewater can provide valuable information on the prevalence of COVID-19 in the population of a particular sewershed, including those asymptomatic and underreported by the health system; focus is, therefore, on the collectivity. For this reason, monitoring of SARS-CoV-2 in sewage as an epidemiological tool (Wastewater Based Epidemiology—WBE) has been used in many parts of the world such as in Germany [8], Australia [9], Spain [6, 10], United States [11–15], France [7, 16], The Netherlands [5], and Italy [17, 18], among others, and in several cities in Brazil (Belo Horizonte, Brasília, Foz do Iguaçu, Niterói, Porto Alegre, Recife, Rio de Janeiro and São Paulo) that will be discussed in this work. Some research groups have included urban drainage waters, as these are contaminated by sanitary sewage, since the level of service provided by public sewage and treatment systems in Brazil is still low.

Sewage monitoring works as an initial warning system for peaks in cases and complements the information from the tests carried out by the health system on those infected. With this approach, it is possible to follow the epidemic evolution or the resurgence of cases in the different regions of the cities and in places of large circulation of the population (bus and train stations, airports, shopping malls, and universities, among others). The Brazilian experience will be detailed in the Sect. 16.3 of this chapter.

16.2 Sanitation in Brazil

Brazil has a huge availability of surface water, which is estimated to be at least 12% of the total world reserves. SNIS [19] data shows that Brazil has a total of 83.7% of water supply coverage, however, 16.3% of the population still does not have access to clean drinking water, which means an estimated population of 34 million inhabitants. The situation of sewage collection and treatment is shown in Table 16.1. According to the data in this Table, 78.5% of the sewage is collected, but only 49.1% is treated, which means that more than 50% of the sewage produced is basically discharged into water resources. In the Southeast and Midwest regions, sewage treatment rates are the highest and reach above 55% of the urban population. By contrast, in the Amazon region the values are below 22%. However, this inequality in sanitation infrastructure is also observed in rural and peri-urban areas of all Brazilian cities. This precarious situation, due to the lack of investments in basic sanitation infrastructure, is more evident with the COVID-19 pandemic, and simple measures such as recurrent hand washing can be a great challenge for those who do not have access to clean water [19, 20].

The absence of sanitation infrastructure increases the transmission of waterborne diseases, such as infectious gastrointestinal diseases, yellow fever, dengue, leptospirosis, malaria and schistosomiasis, whose prevalence is higher in the poorest population. In 2019, the number of cases due to waterborne diseases reached 258,826 hospitalizations, resulting in 2,340 deaths, shown in Table 16.2 [19]. These data reveal that Brazil has serious deficiencies in both drinking water supply and sewage treatment systems.

To overcome the deficit in sanitation services, the Brazilian government created an action plan (National Basic Sanitation Plan—Plansab) to promote by 2033 the supply of drinking water and sewage treatment to 99% and 90% of the population, respectively. For this, new legislation was established (Brazilian Regulatory Framework for Sanitation—Law 14,026/2020 [21]), which provides and encourages competition and privatization of state sanitation companies. One of the fundamental principles adopted was the regionalization of basic sanitation services, considered as an important step towards achieving universal service. With this, it will be possible to achieve gains in scale and technical and economic-financial feasibility in order to serve several municipalities at the same time. Figure 16.1 summarizes the goals of the new Sanitation Regulatory Framework (SRF).

16.3 Brazilian Research Group Experiences: Case Studies

In low-income countries, such as Brazil, due to its continental size, economic diversity, poor coverage by sanitation and health services, and where the clinical testing is deficient, WBE can be used as a complementary and possibly early tool to detect pathogens in a community,

Table 16.1 Water and sewage service levels of municipalities, according to geographic macro-region and Brazil (*Source* National Sanitation Information System—SNIS [19])

Macroregion	Water Supply (%)		Urban sewage rate (%)	
	Total area (rural and urban)	Urban area only	Treatment	Collection
North	57.5	70.4	22.0	82.8
Northeast	73.9	88.2	33.7	82.7
Southeast	91.1	95.9	55.5	73.4
South	90.5	98.7	47.0	94.6
Midwest	89.7	97.6	56.8	93.2
Brazil	83.7	92.9	49.1	78.5

Table 16.2 Number of hospitalizations and deaths due to waterborne diseases in Brazil (*Source* National Sanitation Information System—SNIS [19])

	Hospitalizations	Deaths
Brazil	258,826	2340
Northern (N)	44,984	198
Northeast (NE)	124,609	928
Midwest (CO)	41,904	752
Southern (S)	28,474	325

Fig. 16.1 Goals of the new Sanitation Regulatory Framework (Law 14,026 [21]) to improve sanitation services in Brazil until 2033

considering that specific and representative sampling points in areas with and without sewerage systems are chosen [22]. Specifically in urban agglomerations, such as slums and confined communities, where the sanitation coverage is very low, WBE has enormous potential to be used as an epidemiological tool, and sometimes the main tool to understand SARS-CoV-2 circulation in a population [22]. In addition, due to scarce resources, sewage monitoring in developing countries should consider methodologies that are easy to use in any laboratory. In this sense, not all laboratories are capable of carrying out viral concentration using ultracentrifugation (a method commonly used in virology laboratories, which requires expensive equipment); and therefore, simpler methods such as RNA extraction method using electronegative membrane can be a suitable option for SARS-CoV-2 concentration requiring standard laboratory equipment found in sanitary and environmental engineering labs, being therefore feasible for low-income countries [22, 23].

Table 16.3 shows that there are at least seven different research groups in Brazil conducting SARS-CoV-2 sewage monitoring. Some of the results and initiatives are described below in chronological order. In April 2020, three groups started collecting sewage in different locations (Belo Horizonte-MG, Niteroi-RJ and São Paulo-SP) to monitor viral loads over time and to assess virus circulation in different regions of the respective cities [23, 24]. The strategy adopted in these monitoring studies was that of decentralized monitoring (considering points in the sewer network as well as sewage pumping stations), in addition to centralized monitoring the Sewage Treatment Plants (STPs). This strategy is especially important in developing countries, such as Brazil, because in these, it is quite common that a significant part of the sewage generated by the community does not reach the STPs, making it difficult to extrapolate the results of centralized monitoring to the entire population. In this way, decentralized sewage monitoring approach allows to assess the circulation and prevalence of the virus in different areas and neighborhoods, and indirectly, the number of contaminated people in these different regions—a kind of "indirect testing" that includes both symptomatic and asymptomatic Covid-19 carriers. This information is of vital importance in the Brazilian context, given the excessively low number of clinical testing, extreme social differences, and restricted access to water supply and sanitation services [25, 26].

In the case of Belo Horizonte (Minas Gerais), the SARS-CoV-2 RNA sewage monitoring program was started as a Pilot Project called: Detection and quantification of the new coronavirus in sewage samples in the cities of Belo Horizonte and Contagem, a joint initiative of the National Agency for Water and Sanitation (ANA) and the National Institute of Science and Technology in Sustainable Sewage Treatment Plants (INCT Sustainable STPs—UFMG), in partnership with Sanitation Company for Minas Gerais (Copasa), Minas Gerais Institute for Water Management (Igam) and Minas Gerais State Health Authority (SES). This monitoring program started in April 2020, with sewage collection from the inlet of two STPs and from a total of 15 different points in the sewers [25, 26], representing the sewage generated by 1.5 million inhabitants chosen to collect sewage from regions with different socioeconomic and health vulnerability indicators. The objective was to assess temporal and spatial information on viral load in different neighborhoods to help local health authorities in decision-making through weekly bulletins (available online at www.ana.gov.br) [26]. The results of one year of sewage monitoring (from May 2020 to March 2021) revealed that the viral concentration in sewage followed the same trend as the COVID-19 cases (reported cases), shown in Fig. 16.2. This is shown in the first peak of the epidemic curve (epidemiological weeks 24 to 33, from May to July 2020), and in the reduction of cases (weeks 33 to 42, from August to September 2020), when viral levels also dropped. However, since October 2020 (epidemiological week 42), the viral signal in sewage has increased several days (7 to 14 days) before the resurgence of COVID -19 cases (see the distance between the peaks) and the collapse of the local health system, which led authorities to adopt stricter circulation measures to prevent the spread of the virus (June and July 2020 and mid-January 2021). Therefore, the Covid Sewage Project in Belo Horizonte showed that sewage monitoring can be used as an important epidemiological tool [26], which can give an early warning for spikes of cases, thus helping health authorities to implement control measures of the pandemic before the uncontrolled increase in cases and potentially preventing the collapse of the health system. All project results can be accessed on the dashboard (Painel Dinâmico Monitoramento Covid Esgotos) and in the published weekly reports (see also Report No. 34, available in Portuguese at: https://www.gov.br/ana/pt-br/assuntos/acontece-na-ana/monitoramento-covid-esgotos).

In the case of Sars-CoV-2 sewage monitoring in the municipality of Niterói (RJ), twelve sewage samples were weekly collected from two STPs and from 17 sewer pipes (SP) from surrounding neighborhoods and slums throughout 20 weeks (April 15th to August 25th, 2020) [26]. SARS-CoV-2 RNA was detected in 84.3% of samples with a positive rate ranging from 42% in the first week of monitoring to 100% during the peak of epidemic. Positive rates were higher in STPs, when compared with SP, being a useful tool for monitoring trends in the evolution of the COVID-19 curve, while SP data were more effective, when health public interventions were needed. Heat maps based on SARS-CoV-2 data from sewage samples were built (and were weekly updated) and available online to the general population as an indicator of the ongoing epidemic situation in the Niterói city [23, 26].

Table 16.3 Groups and municipalities in Brazil where SARSCoV2 sewage monitoring are being conducted to date

	Municipality/Institution leader	Sampling locations	SARS-CoV-2 concentration (copies/L)	Remarks
Northeast	Fortaleza-CE/UFC	4 STPs, 5 pumping stations, 1 sewage interceptor	10^4 to 10^6	(1)
	Recife-PE/FPE	4 STPs, 4 rainwater drainage system	10^4 to 10^6	(2)
Midwest	Brasilia-DF/UNB		10^5 to 10^6	(3)
Southeast	Belo Horizonte-MG/UFMG	2 STPs, 15 sewer pipes points, 2 rivers (points before STP), sewage from airport, bus station, university and 2 shopping malls	10^3 to 10^5	(1)
	Rio de Janeiro-RJ (metropolitan region)/UFRJ	8 STPs, 2 pumping stations	10^4 to 10^5	(1, 3)
	Niterói-RJ/ FIOCRUZ	2 STPs, 17 sewer pipes points from neighborhoods and slums	10^4 to 10^8	(4)
	São Paulo-SP (metropolitan region)/CETESB	5 STPs, 5 rivers, 5 drainage channels located in 5 communities (slums)	10^2 to 10^6 (sewage); 10^3 to 10^6 copies/day (river water)	(5)
	Santo André-SP; São Bernardo-SP/UFABC	STPs and SPs	10^4 to 10^6	(6)
Southern	Curitiba-PR/UFPR	5 STPs (including sewage from the airport)	10^4 to 10^5	(1)
	Metropolitan region of Porto Alegre-RS (Porto Alegre, Navegantes, Novo Hamburgo, São Leopoldo, etc.)/FEEVALE	10 STPs, 3 pumping stations, 4 hospitals, 6 rivers impacted by sewage discharge	10^5 to 10^6	(7)

In case of monitoring carried out by CETESB (Environmental Company of the State of São Paulo), this also started in April 2020, and continues until present. Samples of raw sewage and treated effluent from the five main STPs in the metropolitan region of São Paulo (Barueri, Parque Novo Mundo, São Miguel, ABC and Suzano) are being monitored, in addition to the ETE of Paraguaçu Paulista in the interior of the state and EPC Rebouças do Santos Submarine Outfall, which receives sewage from the cities of Santos and São Vicente in Baixada Santista. Surface waters of the Tietê, Tamanduateí and Juqueti rivers (in some points) are also being monitored, which correspond to the main sanitary sewage basins in the metropolitan region of São Paulo, to consider the portion of the population that does not have access to the sewage collection network.

Drainage sub-basins (sewage) of the communities (slums) of Heliópolis, Paraisópolis, Cidade Tiradentes, Brasilândia and São Remo are also being monitored, which even presented high rates of COVID-19. All results are available on the CETESB website (https://cetesb.sp.gov.br/sars-cov-2/).

In the following months, other initiatives emerged in Brazil. In the Northeast Region, in Recife (PE), since May 2020, the Environmental Sanitation Laboratory of the Federal University of Pernambuco (LSA-UFPE) with support from the Keizo Asami Immunopathology Laboratory

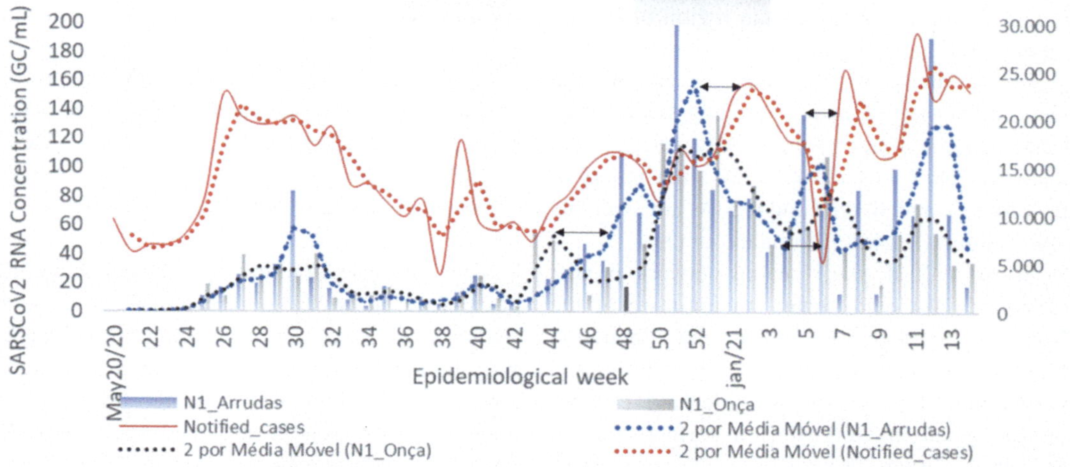

Fig. 16.2 Temporal evolution of SARS-CoV-2 RNA concentration (N1 genomic copies/mL) in sewage of Arrudas STP and Onça STP and new notified cases of COVID-19 throughout 1 year monitoring (from May 2020 till March 2021) in Belo Horizonte-MG (Brazil)

(LIKA-UFPE) for 16 weeks began to monitor the genome of SARS-CoV-2 in various types of water. For this purpose, partnerships were established with Sanitation Company for Pernambuco (Compesa), BRK Ambiental, the State Environment Agency (CPRH) and the Pernambuco Institute for Risk and Disaster Reduction (IRRD). The monitoring points were as follows: three wastewater treatment plants, a hospital and eight points in the urban drainage rainwater network, since less than 50% of the population is connected to public sewage systems. The results revealed that the virus genome was present in all types of water evaluated with an average concentration of 3.2×10^6 copies/L ($\rho < 0.05$). In the extreme southern of the country, in Rio Grande do Sul (RS) also in May 2020, another collaborative environmental monitoring network project of SARS-CoV-2 was started with the participation of federal, state, and municipal public agencies, in addition to public and community universities. The study monitors the presence of SARS-CoV-2 in raw water and sewage in the municipalities Alvorada, Cachoeirinha, Canoas, Gravataí, Novo Hamburgo, Porto Alegre, and São Leopoldo. Results for the period from May 11th to August 19th, 2020, showed that of the 116 samples collected in 22 different places 48.9% were positive, 45.7% were negative (for the presence of the RNA of SARS-CoV-2) and 5.3% were still awaiting the results (according to Monitoring Bulletin No. 3 of the State Center for Health Surveillance (CEVS) of the Health Department of RS [27, 28].

In June 2020, two other groups started the same sewage monitoring in the municipalities of Santo André and São Bernardo Campo-SP [28, 29] and in Brasília (DF-federal capital) [24]. In the municipality of Santo André, 80% of the samples collected at the ABC-STP tested were positive for the presence of the SARS-CoV-2 RNA during the sampling period (June 6 till July 11, 2020 [28]). The coronavirus monitoring Project in the sewage from the ABC region is conducted by UFABC and supported by the Virus Network, financed by MCTI and Brazilian Health Ministry (MS). Data are available at:

https://www.ufabc.edu.br/noticias/monitoramento-do-coronavirus-nos-esgotos-antecipa-aumento-de-casos-na-regiao-do-abc.

In September 2020, another research group led by Federal University of Rio de Janeiro (UFRJ) and CEDAE, in collaboration with FIOCRUZ, ABES-RJ and SES-RJ, started the spatial and temporal SARS-CoV-2 sewage monitoring in the metropolitan region of Rio de Janeiro -RJ, by sampling eight sewage treatment plants located in different areas of the metropolitan region and 2 pumping stations (located in Leblon and Pavuna). First results were made available in October 2020 (first report of the project), and all data and reports are available online (https://monitoracoronarj.com.br).

In April 2021, the research network project called REDE-Covid began, an initiative of the National Agency for Water and Sanitation (ANA), the INCT-Sustainable STPs-UFMG and the National Council for Scientific and Technological Development (CNPq), in partnership with Sanitation Service Providers, Health Secretariats and Federal Universities located in the following capitals: Belo Horizonte, Brasília, Curitiba, Fortaleza, Recife and Rio de Janeiro. The research network project aims to monitor the SARS-CoV-2 in sewage in different locations by replicating the experiences of Pilot Project Covid Sewage Monitoring to other states and cities in Brazil in order to consider different regional realities and to assess virus circulation in different cities and regions. The first report of Rede-Covid project was launched on May 26, 2021 and is available on the ANA website (https://www.gov.br/ana/pt-br/assuntos/acontece-na-ana/monitoramento-covid-esgotos). Table 16.3 summarizes the different groups and municipalities in Brazil that are conducting SARS-CoV-2 sewage monitoring to date.

UFMG (Federal University of Minas Gerais); UFPE (Federal University of Pernambuco); UFC (Federal University of Ceará); UNB (Nacional University of Brasília); UFRJ (Federal University of Rio de Janeiro); FIOCRUZ (Oswaldo Cruz Foundation); UFPR (Federal University of Paraná); FEEVALE (FEEVALE University);1-Report N.01/2021 of the REDE Covid research network available at:

- https://www.gov.br/ana/pt-br/assuntos/acontece-na-ana/monitoramento-covid-esgotos;
- https://sites.ufpe.br/lsa/covid-19/
- https://monitoracoronarj.com.br
- Prado et al., 2021;
- https://cetesb.sp.gov.br/sars-cov-2/;
- https://www.ufabc.edu.br/noticias/monitoramento-do-coronavirus-nos-esgotos-antecipa-aumento-de-casos-na-regiao-do-abc;
- Reports N. 8/2021 and N. 09/2021 of SARS-CoV-2 environmental monitoring at https://coronavirus.rs.gov.br/boletim-ambiental

Brazil has an extraordinary technical-scientific competence to work in research networks and tradition in virus environmental surveillance including polio virus, enteric viruses and more recently SARS—CoV-2 using the WBE approach. However, most consolidated research groups are at universities and depend directly on support and funding from federal and state research agencies. Up to now, monitoring results have helped decision makers and health authorities in fighting the COVID-19 pandemic, which can later be used to verify the effectiveness of vaccination of the Brazilian population in areas, where sewage is monitored. Thus, the combination of academic interests with those of health surveillance will contribute to develop a National Plan for wastewater-based epidemiology, not only for monitoring SARS-CoV-2, but also for other bacterial and viral pathogens of importance to public health.

16.4 Final Remarks

This chapter presented an overview of Brazilian research groups that track SARS-COV-2 viral particles in sewage, stormwater, and surface waters to support local health authorities in making the best decisions with limited resources to prevent Covid-19 spread in monitored areas. In almost all Brazilian regions, there are research institutions or health research centers conducting investigations into the WBE for Covid-19 surveillance, except for the North region (Amazon region), which does not yet have any research group published in WBE. The results show that the poorest areas, where there is a lack of drinking water supply and sewage collection and treatment system, are those with a high number of infected people. All official reports and communications generated by the research groups are being centralized by National Agency for Water and Sanitation to serve as guidelines for the development of a future National Wastewater Surveillance Program and are available for use by public authorities and ordinary citizens.

Acknowledgements Authors would like to thank for the support and funding received by the following Agencies: National Agency for Water and Sanitation (ANA), National Council for Scientific and Technological Development (CNPq), Minas Gerais Research Foundation (FAPEMIG), Minas Gerais Sanitation Company (Copasa), Minas Gerais Institute of Water Management (Igam), Minas Gerais State Health Authority (SES), Pernambuco Sanitation Company (Compesa), BRK Ambiental, Pernambuco Environmental Agency (CPRH), National Institute of Science and Technology in Sustainable WTPs (INCT) and Pernambuco Research Foundation (FACEPE). Authors also would like to express their gratitude to DAAD to support the Exceed-Swindon Project that enabled to form a strong network among researchers, to improve the quality of science information and to promote a better quality of life for the population in developing countries.

References

1. Stadnytskyi V, Bax CE, Bax A, Anfinrud P. The airborne lifetime of small speech droplets and their potential importance in SARS-CoV-2 transmission. Proc Natl Acad Sci. 2020;117(22):11875–11877. https://doi.org/10.1073/pnas.2006874117.
2. Chen Y, Chen L, Deng Q, Zhang G, Wu K, Ni L, Yang Y, Liu B, Wang W, Wei C, Yang J, Ye G, Cheng Z. The presence of SARS-CoV-2 RNA in the feces of COVID-19 patients. J Med Virol. 2020;92 (7):833–40. https://doi.org/10.1002/jmv.25825.
3. Wang W, Xu Y, Gao R, Lu R, Han K, Wu G, Tan W. Detection of SARS-CoV-2 in different types of clinical specimens. J Am Med Assoc. 2020;323:1843–4. https://doi.org/10.1001/jama.2020.3786.
4. Wu Y, Guo C, Tang L, Hong Z, Zhou J, Dong X, Yin H, Xiao Q, Tang Y, Qu X, Kuang L, Fang X, Mishra N, Lu J, Shan H, Jiang G, Huang X. Prolonged presence of SARS-CoV-2 viral RNA in faecal samples. Lancet Gastroenterol Hepatol. 2020;5 (5):434–5. https://doi.org/10.1016/S2468-1253(20)30083-2.
5. Medema G, Heijnen L, Elsinga G, Italiaander R, Brouwer A. Presence of SARS- Coronavirus-2 RNA in sewage and correlation with reported COVID-19 prevalence in the early stage of the epidemic in the Netherlands. Environ Sci Technol Lett. 2020.
6. Randazzo W, Cuevas-Fernando E, Sanjuan R, Domingo Calap P, Sanchez G. Metropolitan wastewater analysis for COVID-19 epidemiological surveillance. Int J Hyg Environ Health. 2020;230. https://doi.org/10.1016/j.ijheh.2020.113621.
7. Wurtzer S, Marechal V, Mouchel JM, Maday Y, Teyssou R, Richard E, Almayrac JL, Moulin L. Evaluation of lockdown impact on SARS-CoV-2 dynamics through viral genome quantification in Paris wastewaters. medRxiv. 2000. 10.11.101/202.04.12.20062679.
8. Schiwy S, Linnemann V, Brinkmann M, Wider M, Greve C, Janke A, Hollert H, Wintgens T, Ciesek S. Detection of SARS-CoV-2 in raw and treated wastewater in Germany-Suitability for COVID-19 surveillance and potential transmission risks. Sci Total Environ. 2021;751. https://doi.org/10.1016/j.scitotenv.2020.141750.
9. Ahmed W, Angel N, Edson J, Bibby K, Bivins A, O'Brien JW, Choi PM, Kitajima M, Simpson SL, Li J, Tscharke B, Verhagen R, Smith WJM, Zaugg J, Dierens L, Hugenholtz P, Thomas KV, Mueller JF. First confirmed detection of SARS-CoV-2 in untreated wastewater in Australia: a proof of concept for the wastewater surveillance of COVID-19 in the community. Sci Total Environ. 2020;728(138764):1–8. https://doi.org/10.1016/j.scitotenv.2020.138764.
10. Randazzo W, Truchado P, Cuevas-Fernando E, Simon P, Allende A, Sanchez G. SARS-CoV-2 RNA titers in wastewater anticipated COVID-19 occurrence in a low prevalence area. Water Res. 2020;181. https://doi.org/10.1016/j.watres.2020.115942.
11. Nemudryi A, Nemudraia A, Surya K, Wiegand T, Buyukyoruk M, Wilkinson R, Wiedenhelft B. Temporal detection and phylogenetic assessment of SARSCoV-2 in municipal wastewater. Cell Rep Med. 2020;1(6). https://doi.org/10.1016/j.xcrm.2020.100098.
12. Peccia J, Zulli A, Brackney DE, Grubaugh ND, Kaplan E, Casanovas-Massana A, Ko AI, Malik AA, Wang D, Wang M, Weinberger DM, Omer SB. Measurement of SARS-CoV-2 RNA in wastewater tracks community infection dynamics. Nat Biotechnol. 2020;38:1164–7.
13. Sherchan SP, Shahin S, Ward L, Tandukar S, Aw TG, Schmitz B, Ahmed W, Kitajima M. First detection of SARS-CoV-2 RNA in wastewater in North America: a study in Louisiana, USA. Sci Total Environ. 2020;743. https://doi.org/10.1016/j.scitotenv.2020.140621.
14. Gonzales R, Curtis K, Bivins A, Bibby K, Weir M, Yetka K, Thompson H, Keeling D, Mitchell J, Gonzalez D. COVID-19 surveillance in Southeastern Virginia using wastewater-based epidemiology. Water Res. 2020;186:116296.

15. Weidhass J, Aanderub ZT, Roper DK, VanDerslice J, Brown Gaddis E, Ostermiller J, Hoffman K, Jamal R, Heck P, Zhang Y, Torgersen K, Vander Laan J, LaCross N. Correlation of SARS-CoV-2 RNA in wastewater with COVID-19 disease burden in sewersheds. Sci Total Environ. 2021;775:145790.
16. Wurtzer S, Marechal V, Mouchel JM, Moulin L. Time course quantitative detection of SARS-CoV-2 in Parisian wastewaters correlates with COVID-19 confirmed cases. medRxiv. 2020. https://doi.org/10.1101/2020.04.12.20062679.
17. La Rosa G, Iaconelli M, Mancini P, Ferraro GB, Veneri C, Bonadonna L, Lucentini L, Suffredini E. First detection of SARS-CoV-2 in untreated wastewaters in Italy. Sci Total Environ. 2020;736. https://doi.org/10.1016/j.scitotenv.2020.139652.
18. Rimoldi AG, Stefani F, Gigantiello A, Polesello S, Comandatore F, Mileto D, Maresca M, Longobardi C, Mancon A, Romeri F, Pagani C, Moja L, Gismondo M, Salerno F. Presence and infectivity of SARS-CoV-2 virus in wastewaters and rivers. Sci Total Environ. 2020;744. https://doi.org/10.1016/j.scitotenv.2020.140911.
19. SNIS. Diagnóstico dos Serviços de Água e Esgoto. Ministério do Desenvolvimento Regional. In Portuguese; 2019. http://www.snis.gov.br/downloads/diagnosticos/ae/2019/Diagn%C3%B3stico_SNIS_AE_2019_Republicacao_31032021.pdf.
20. WHO. Inheriting a sustainable world: Atlas on children's health and the environment. World Health Organization; 2017. https://www.who.int/ceh/publications/inheriting-a-sustainable-world/en/.
21. Law 14026/2020—Brazilian regulatory framework for sanitation. In Portuguese. http://www.planalto.gov.br/ccivil_03/_ato2019-2022/2020/lei/l14026.htm.
22. Araujo JC, Gavazza S, Leão T, Florencio L. SARS-CoV-2 sewage surveillance in low-income countries: potential and challenges. J Water Health. 2021;19:1. https://doi.org/10.2166/wh.2020.168.
23. Prado T, Fumian T, Mannarino CF, Maranhão AG, Siqueira MM, Miagostovich MP. Preliminary results of SARS-CoV-2 detection in sewerage system in Niterói municipality, Rio de Janeiro, Brazil. Memórias Instituto Oswaldo Cruz, Rio de Janeiro, vol 115. 2020. pp e200196. 1|3. https://doi.org/10.1590/0074-02760200196.
24. Sodre FF, Brandão CCS, Vizzotto CS, Maldaner AO. Wastewater-based epidemiology as a strategy for community monitoring, mapping of hotspots and early warning systems for COVID-19 (in Portuguese). Quimica Nova. 2020;43(4):515–519. https://doi.org/10.21577/0100-4042.20170545.
25. Chernicharo CAL, Araújo JC, Mota Filho CR, Bressani-Ribeiro T, Chamhum-Silva LA, Leal CD, Leroy D, Machado E, Cordero MFE, Azevedo LS, Fernandes L, Leão T, Laguardia F, Reis MTP, Melo MC, Ayrimoraes SR. Monitoramento do esgoto como ferramenta de vigilância epidemiológica para controle da COVID-19: estudo de caso na cidade de Belo Horizonte. Engenharia Sanitária e Ambiental (In Portuguese). 2020.
26. Mota CR, Bressani-Ribeiro T, Araújo JC, Leal CD, Leroy-Freitas D, Machado EC, Espinosa MF, Fwernandes L, Leão TL, Chamum-Silva L, Azevedo L, Morandi T, Freitas GTO, Costa MS, Carvalho BO, Reis MT, Melo MC, Ayrimoraes S, Chernicharo CAL. Assessing spatial distribution of COVID-19 prevalence in Brazil using decentralized sewage monitoring. Water Res. 2021;202:117388.
27. Prado T, Fumian TM, Mannarino CF, Resende PC, Motta FC, Eppinghaus AF, Vale VHC, Braz RMS, Andrade JSR, Maranhão AG, Miagostovich MP. Wastewater-based epidemiology as a useful tool to track SARS-CoV-2 and support health policies at municipal level in Brazil. Water Res. 2021;191:116810.
28. Centre of Sanitation Vigilance in Rio Grande do Sul (Brazil). Secretaria da Saúde do Rio Grande do Sul (SES-RS), Centro Estadual de Vigilância em Saúde (CEVS-RS), et al. Monitoramento Ambiental de SARS-CoV-2 no Rio Grande do Sul: Boletim de acompanhamento nº 3. Monitoramento Ambiental de SARS-CoV-2 [Internet]. Aug 30; vol 1. 2020. pp 3–18. https://www.cevs.rs.gov.br/upload/arquivos/202008/31173720-boletim-informativo-n-3-versao-final.pdf.
29. Cabral A, Mantovani Claro I, Ribeiro Augusto M, Friolani VN, Bezerra CE, Graciosa MCP, Fonseca FJA, Speranca MA, Freitas Bueno R. Standardization of the method of concentration and extraction of nucleic acids in wastewater samples: A low-cost tool for use in a surveillance SARS-CoV-2. Engenharia Sanitária e Ambiental. 2021.

Investigation of Microplastics in Water and Wastewater: A Review

17

Souad El Hajjaji, Abdelmalek Dahchour, Jamal Mabrouki, Youssef Assou, Nasser Alanssari, Latifa Elfarissi, and Hanane Benqlilou

Abstract

Microplastics (MPs) are emerging pollutants, which are continuously released due to human activities into the environment through several pathways. Their cycle is still a mysterious one, but their quantity is enormous, their presence has an impact on the environment, organisms, and human health. Studies show that wastewater treatment plants (WWTPs) could constitute an important source of MPs in the environment. Indeed, they come for the most part from the fragmentation of plastics in the sea, constituting a very varied universe of forms, colours, and composition, and reflecting the various uses of man. Then MPs adsorb persistent organic pollutants and heavy metals or bind with pathogenic microorganisms from wastewater and sludge and become more dangerous. There are studies on the detection and understanding of the occurrence and fate of MPs in the treatment plant, but remediation strategies and their elimination mechanisms are limited. Therefore, it is important to update the status of detection, occurrence, and removal of MPs in WWTPs. This chapter thus summarizes the sources and types of MPs, discusses the impacts of MPs on the environment and human health as well as shows the current MPs detection practices and treatment solutions. More specifically, the different techniques used to collect MPs from wastewater and sewage sludge as well as their pre-treatment and characterization methods are reviewed and analyzed. Key aspects regarding the presence of MPs in WWTPs, such as concentrations, total releases, materials, shapes, and sizes are summarized and compared. The elimination of MPs at different stages of the treatment and their retention in the sewage sludge are analyzed. Potential treatment technologies targeting MPs are also presented.

Keywords

Effluents · Microplastic pollution · Wastewater treatment plant

S. El Hajjaji (✉) · A. Dahchour · J. Mabrouki
CERN2D, Faculty of Science, Mohammed V University in Rabat, Rabat, Morocco
e-mail: souad.elhajjaji@fsr.um5.ac.ma

Y. Assou · N. Alanssari · L. Elfarissi
Technical Center of Plastics and Rubber, Complex of Industrial Technical Centers, Casablanca, Morocco

H. Benqlilou
International Institute for Water and Sanitation, IEA Under National Office of Electricity and Drinking Water ONEE, Rabat, Morocco

M. Bahadir and A. Haarstrick (eds.), *Water and Wastewater Management*,
Water and Wastewater Management, https://doi.org/10.1007/978-3-030-95288-4_17

17.1 Introduction

Over the past decade, the introduction of microplastics (MPs) in aquatic systems has become an emerging issue [1]. There is an increasing interest to understand their environmental impact due to their involvement in many biochemical processes (by the release of chemicals through their degradation) and their participation in the food web chain [2]. MPs are found in a variety of products ranging from cosmetics over synthetic garments to plastic bags and bottles that easily enter the environment because of pollution in form of solid waste, and remain both in sea water, freshwater, and wastewater. Research over the last years has identified the presence of MPs mainly in large water systems (e.g., oceans and open seas), as well as in variable surface water receptors (e.g., rivers) [3]. Primary contamination of MPs consists of manufactured raw plastic material that enter the aquatic environments. Secondary contamination of MPs occurs, when primary MPs undergo mechanical, photo (oxidative) and/or biological degradation to smaller fragments or even other substances. Thus, MPs are a sink and a source of chemical contaminants. Hence, the leaching of chemicals used in plastic production (e.g., plasticizers, heavy metals) and the effects of MP co-contaminants against aquatic organisms in water ecosystems is an issue of paramount importance.

A recent study indicated that wastewater treatment plants (WWTPs) can be an important source of releasing MPs to the environment [4]. Depending on the type of treatment, the WWTP may remove some of the MPs. However, it has been shown that MPs could bypass the WWTP, entering the aquatic water bodies and finally accumulated in the environment. A large proportion of studies are conducted on the development of methods for sampling, identifying and treatment of MPs in WWTPs.

This chapter attempts to describe (1) the presence and abundance of MPs in wastewater and sludge, (2) the impacts of MPs on the environment and human heath, and (3) sampling and characterization of MPs in wastewater and sludge. The remediation strategies to reduce the quantity of MPs in WWTPs will be also discussed.

17.2 Effects of Microplastics (MPs)

Recently, MPs have become global contaminants' concern for human and ecological health. They are not biodegradable and are found in a variety of environments, where they accumulate and persist, and remain bioavailable because of their small size for thousands of species from almost every trophic level, because they are often mistaken for food. Globally, several studies have been conducted on the potential toxic effects of MPs.

17.2.1 Socio-Economic Impacts

Despite the difficulties concerning the evaluation of the socio-economic impacts, the first evidence concerns the consequences linked to the pollution of coastal areas, beaches, and foreshores by plastic. The heritage value of the sites is largely affected, and the economic stakes linked to tourism can be strongly affected (recent closures of very touristy beaches for example) [5]. These impacts are often of an aesthetic nature and are translated and quantified by the costs incurred by the cleaning. Along the coastline, aquaculture activities can be the cause of significant inputs of plastics to the marine environment, particularly in shellfish production areas (oysters, mussels, and other shellfish) due to the loss of materials, whether unintentional or not. Socio-economic impacts also concern underwater interventions on the bottom of ports or along the coastline as well as environmental awareness and education programs [6]. At sea, plastic constitutes an economic pressure on navigation, in particular pleasure boating, due to frequent accidents (encountering obstacles, entanglement of nets or plastic sheets in boat propellers or in the cooling systems). These impacts are also significant for fishing vessels, with, for the latter, additional costs for

Table 17.1 Main socio-economic consequences related to plastic waste at sea [9]

Sector economic	Types of impacts	Significant costs
Managers (municipalities, local authorities)	Injuries on the beach, aesthetic impact, negative publicity, labelling	Cleaning and treatment
Tourism	Aesthetic impact, negative publicity, loss of revenue income, loss of loss of enjoyment	Decrease in revenue
Associations	Volunteering, operational costs, management	Volunteer time
Industry	Maintenance, waste disposal, damage to equipment	Maintenance, repair
Navigation	Damage to ships, rescue operation, repairs, legal obligations	Damage to ship/repair
Fishing	Damage to gear repair/replacement replacements, fishing time, cleaning, stock alteration	Cleaning, repair of machines
Aquaculture	Cleaning of nets, maintenance	Cleaning and maintenance
Ecosystem services	Biodiversity, costs of degradation	Poorly estimated costs

cleaning and repairing nets or lines [7]. This issue of ghost nets is particularly critical in certain regions of Europe (South Brittany, North Adriatic, Gulf of Lion), where stock losses can reach 2–3% of the entire population of certain species [8]. More globally, the socio-economic impacts are extremely diverse (Table 17.1), and the associated costs are still poorly known, with costs estimated at nearly 260 million euros for marine waste in European waters alone [9]. For the whole of the world's oceans, the financial damage is estimated at about 12 billion euros per year.

17.2.2 Environmental Impacts

The impacts of plastic waste at sea can be presented according to two main types: global impacts on the scale of ecosystems, mainly linked to the transport of species, and impacts on the scale of organisms and populations. At the ecosystem level, plastic waste constitutes a new habitat for many species, benthic macro-organisms such as arthropods, molluscs, hydraires, bryozoans, and many microorganisms, bacteria, viruses, fungi, microalgae of the dinoflagellate genus and diatoms [10]. These species will rapidly colonize the plastic waste at sea, by settling and even developing, and will constitute the Plastisphere [11]. Since plastics are persistent and highly mobile materials, they will have the capacity to transport these species over large scales of space and time, thus creating a "raft" effect. For example, following the 2011 tsunami, nearly 300 species were transported from the Japanese coast to the West American coast mostly on plastic debris [12]. These species can then settle in or even become invasive to the detriment of endemic species, leading to a disruption of marine communities and, therefore, of the ecosystem. In addition, some species identified on the surface of plastic waste at sea are harmful, toxic, or potentially pathogenic, as suggested by the detection of large bacterial families, of which some strains (Campylobacteraceae, Flavobacteriaceae, Aeromonadaceae) are responsible for diseases in humans, certain fish and shellfish species [13]. The question is then to know, if these species disseminated on the surface of plastic wastes can transmit diseases, which is a current research topic of the scientific community. Finally, the prolonged contact times between species, especially bacterial species, on this new support could encourage exchanges of genetic material and contribute to the propagation of multiple antibiotic resistances across bacterial genera [14].

At the level of individual organisms, the impacts generated by plastic waste at sea are particularly visible on large marine animals, including seabirds, mammals and turtles trapped in large plastic waste such as ghost nets. But this is only the tip of the iceberg. Indeed, compared with macroplastics, MPs are much more numerous and affect more widely the whole marine food chain. Because of their small size, they are easily ingested by a very large number of species. Once ingested, these MPs can either obstruct the digestive system [15]. However, the smallest particles, such as nanoplastics, can also pass through the digestive membranes and migrate into the circulatory system and even into other organs, as has been observed in the laboratory in the brain of fish, with the effect of modifying their swimming behavior. Circulatory system and even in other organs, as has been observed in the laboratory in the brain of fish, resulting in changes in their swimming behavior. In any case, even a simple transit of MPs in the digestive tract induces major changes in the biology of the animal that ingested them: changes in digestion that disrupt energy intake via the diet [16], direct source of cellular stress, with disturbances on the major physiological functions such as growth, immune defenses [17], and reproduction [18]. Moreover, the additives contained in plastics can also be released under the conditions of the digestive tract during transit and cause a chemical disturbance [19], for example with an associated endocrine disruption. The entire life cycle of an organism can thus be affected with transgenerational repercussions. For example, by exposing oysters to MPs or nanoplastics in the laboratory, a reduction of half the number of gametes produced and a 20% delay in the growth of the offspring have been observed as well as a decrease in fertilization and development of embryos and larvae in connection with the appearance of numerous malformations [20].

The feeding of bivalves is not spared since the life cycle of diatoms (phytoplanktonic microalgae that play a major role in marine primary production) has been shown to be disrupted by exposure to polystyrene nanoballs in laboratory [21], demonstrating the importance of understanding the effects of plastics at the complex scale of ecosystems. Very few studies have initiated this questioning. Among them, Green [22] showed that a natural reef shaped by the flat oyster, an ecosystem engineer species that builds a habitat rich in biodiversity, was altered when exposed to MPs. Changes in oyster filtration affected the diversity of associated microorganism communities, from microfauna to cyanobacteria, potentially leading to changes in biogeochemical cycles in the sediment. This type of chronic and realistic impact study on the scale of diverse and interconnected communities requires approaching the diversity of the ecosystem but also the diversity of plastic waste (strong variability of size, shape, roughness, type of polymers and associated additives [23]), because this has a strong influence on their fate and behavior in the sea and, therefore, on their toxicity. This is a research priority that will be an essential support for decision-making, but it requires bringing together the diversity of scientific communities working on plastic waste in the aquatic environment by promoting multi- and transdisciplinary approaches.

17.2.3 Impacts on Health

It is now proven that MPs are present in all compartments of the environment and that they have penetrated human food chain. MPs presence has been shown in everyday seafood products such as salts with variable quantities depending on the geographical location [24]. It is important to underline the presence of significant quantities of MPs in table salt from the sea [25]. MPs have also been found in other shellfish such as oysters and clams as well as in crustaceans, scampi, shrimp, crabs, and spider crabs. They are also present in many species of fish, mainly in their digestive system, and very exceptionally in the muscle. In addition, many other foods have revealed the presence of MPs: beer, sugar, honey, water, and for the first time, plastics have been identified in human feces. Moreover, human exposure to MPs is not limited to the food chain; one can also be exposed through inhalation of

MPs and airborne fibers. This route of entry may even be more important than through the food, but it is highly variable depending on the environment, and often associated with certain work environments, potentially inducing respiratory inflammatory reactions. Concerning the impact on the health of consumers of products containing MPs, there is still little knowledge, and several questions currently arise:

- On the composition of MPs, polymers, and additives. Indeed, some monomers are dangerous on their own, such as those used in polyurethane or PVC [26]. It is important to also consider the additives, which can represent a significant quantity of the composition of plastics, because they can be released by leaching. They could also cause cocktail effects by association [27]. The main substances found in the environment are phthalates, bisphenol A, brominated flame retardants and nonylphenols, and these substances are known to be toxic.
- On the carrying chemical and biological contaminants thanks to the adsorption properties of MPs. Indeed, according to the displacements and the buoyancy of MPs, persistent organic pollutants can be fixed on their surface due to their hydrophobic properties. One can find PCBs (polychlorinated biphenyls) or PAHs (polycyclic aromatic hydrocarbons) [28].

Studies have also shown the presence of heavy metals (Hg, Cd, Pb, etc.) [29] and pesticides on the surface of MPs. However, even if the accumulation of persistent organic pollutants has been demonstrated in some organisms (mussels, fish), MPs are apparently not the main vectors compared with other particles suspended in the seas and oceans, but they do not influence bioaccumulation in marine organisms, which does not seem to be preponderant in human food [30]. There remain the biological contaminants, the MPs being able to be colonized by bacteria and in particular bacteria pathogenic for the human like that of the kind Vibrio [31].

- On the possibilities of translocation according to the size of the particles found, especially concerning nanoplastics that could enter all organs. As far as MPs are concerned, it seems that there is no absorption possible in any compartment for MPs of sizes greater than about 150 μm [32]. Beyond this size, translocation could be dependent on the hydrophobicity and the charges of the particles [33].
- On the interactions of plastics in the digestive system. Indeed, the particles during transit could mechanically generate by abrasion localized inflammatory reactions, but they have also impacted on the microbiota and go as far as creating a dysbiosis [34]. At the level of the digestive system, different mechanisms can intervene thanks to Payer's patches and the possibilities of persorption of the intestinal epithelium. According to the current knowledge, one can underline that the particles arriving the organism can also trigger immunological responses as noted for the PET (polyethylene terephthalate) or the PE (polyethylene). More globally, to know the long-term impacts of MPs on human health, many research still must be conducted, considering the diversity of compositions, sizes, shapes, and asperity, and considering MPs by the specific study of congeners including their morphological characteristics that are important for a particulate contaminant, and not as a as a homogeneous family of contaminants.

17.3 Occurrences of MPs in Wasterwater Treatments Plants (WWTPs)

Wastewater treatment plants contribute to eliminate various pollutants including MPs. However, the efficacy of removal seems to be low as MPs are detected with higher count in the field near or downstream of WWTPs. These amounts could vary in the range of 50 between WWTPs [35, 36].

Higher removal efficiency was observed in the WWTP employing primary clarification, suggesting that retrofitting secondary plants with primary clarifiers could improve MPs removal. Upgrading plants to include primary clarification is dependent on different factors [37].

In a critical review on the efficiency of removal of MPs in WWTPs, the authors indicated that MPs are removed at 88% and 94% in secondary and tertiary WWTPs, respectively; 72% on average were removed during preliminary and primary treatment. Primary sedimentation removes spherical particles of commonly used polymers > 27–149 μm in diameter [38].

Sorption processes of three model compounds perfluorooctane sulfonic acid (PFOS), benzo[a] pyrene (BaP) and oxybenzone (BP3) on different sizes and different industrial polymers showed differences in responses (kinetics and sorption efficiency), depending on the particle sizes as well as the chemical properties of the compounds.

Study of the vectoring of these 3 model compounds adsorbed on MPs and their toxicities using the early stages, juveniles, and adults of zebrafish, through direct exposure or via trophic exposure were evaluated according to the OECD 236 guideline as well as via chronic trophic exposure, starting at the larval stage up to 5 months. The main results were consistent with low acute toxicity on early developmental stages (embryos and larvae) exposed to particulate matter, organic extracts but also leachates, from artificial or environmental MPs. Nevertheless, ingestion of MPs during trophic exposure led to long-term effects, with different intensities related to the chemical compounds adsorbed to MPs [39, 40]. Study of MPs concentrations in two Oregon bivalve species, spatial, temporal, and species variability showed that toxic effects included alterations in growth (fish size and weight), reproductive effects, behavioral changes in exposed adults, and changes in larval swimming activity [41].

There are several hundred different types of polymers and polymer blends in commercial production, but the market is dominated by polyethylene (both high density, HDPE and low density, LDPE), polypropylene (PP), polyvinyl chloride (PVC), polyurethane (PUR), polystyrene (PS) and polyethylene terephthalate (PET). These six polymers account for about 80% of plastics production and are likely to form a large portion of most marine litter [42, 43]. The most common man-made and petroleum-derived polymers found in MPs are listed in Table 17.2 [44].

Different physical and chemical properties of MPs could be considered in a prioritization framework. These are particle size, particle shape, surface area, polymer type and crystallinity, chemical composition and additive compounds as well as changes in surface properties, e.g., adsorbed microorganisms, xenobiotics, and dirt and mud [45, 46].

WWTPs are a principal barrier to the direct discharge of waterborne MPs pollution into the aquatic environment. However, only a limited number of studies have examined MPs removal through the various treatment processes [47].

Table 17.2 Main polymers found in MPs [44]

Polymers	Abbreviation	Molecular formula
Polyethylene	PE	$(C_2H_4)n$
Polypropylene	PP	$(C_3H_6)n$
Polystyrene expanded	EPS	$(C_8H_8)n$
Polyethylene Terephthalate	PET	$(C_{10}H_8O_4)n$
Polymethylmetacrylate	PMMA	$[(CH_2C(CH_3)(CO_2CH_3)]n$
Polytetrafluoroethylene	PTFE	$(CF2CF_2)n$
Polyamide (nylon)	PA	$C_{23}H_{26}N_2O_4$
Polyurethane	PU	$C_3H_8N_2O$

A review of 18 studies on the occurrence of MPs in wastewater found that typical WWTP effluent has a lower median concentration of MPs particles compared with the influent, although the range in effluent concentrations varied significantly [48]. This may be an indication that some WWTPs have ineffective treatment practices or are not designed for optimal removal of MPs [49]. In these cases, discharges from WWTPs can represent routes for MPs to enter fresh waters and then possibly into drinking-water [50].

In many countries, significant efforts have been made to increase the quality of WWTP effluents to meet higher quality targets for surface water. Where such receiving waters are used as a drinking-water source, the MPs load originating from the WWTP is expected to be significantly reduced. However, in low- and middle-income countries, only 33% of the population have sewer connections. Wastewater for the remaining 67% of the population is collected and treated in onsite systems or discharged directly to soil and water bodies. In addition, approximately 20% of household wastewater collected in sewers does not undergo at least secondary treatment. In these cases, the contribution of MPs into the receiving water body is expected to be higher.

17.4 Analytical Methods for MPs in Wastewater and Sludge Samples

A recent systematic review of the literature identified 50 studies detecting MPs in fresh water, drinking-water, or wastewater [48]. The lack of standard methods for sampling and analyses of MPs in the environment means that comparisons across various studies are difficult.

17.4.1 Chemical Sources

Polymerization reactions during plastic production do not generally proceed to full completion, resulting in small proportions of monomers such as 1,3-butadiene, ethylene oxide and vinyl chloride, that can migrate into the environment. Residual monomers may also arise because of biodegradation and weathering of plastics. However, the extent to which this occurs is uncertain. It is likely that unbound monomers resulting from these scenarios would be released into the environment.

Additives such as phthalate plasticizers and polybrominated diphenyl ether flame retardants are, for the most part, not covalently bound to the polymer backbone and can often easily migrate into the environment. Migration can also be impacted by the molecular weight of the additives. Small and low molecular weight additive molecules generally migrate at faster rates than larger ones. Aging and weathering likely strongly influence the migration process, the overall impact of which is not well understood. However, relative to other emission routes of additives to the environment, it is anticipated that leaching from MPs will be relatively small.

The hydrophobic nature of MPs implies that they have the potential to accumulate hydrophobic persistent organic pollutants (POPs), such as PCB, PAH, and organochlorine pesticides (OCP). POPs indiscriminately sorb on organic carbon in the environment and, therefore, the fraction of POPs sorbed on MPs will be small relative to other environmental media containing organic carbon such as sediment, algae, and the lipid fraction of aquatic organisms.

17.4.2 Sampling

MPs samples can be acquired using trawl nets (typically 300 μm) drawn across the surface of the water, or through collection of water samples, from which the particles are extracted later. Initial sample purification usually involves filtration, followed by some sort of extraction process such as density separation, in which samples are mixed with a liquid of defined density, allowing MPs particles to float and heavier particles to sink (see Table 2.1 for a list of plastics and their densities). Further purification may require chemical or enzymatic methods to remove organic or inorganic contaminants (biofouling). The extent of the preparation is dependent on the nature of the samples: dirtier samples will require more preparation steps than cleaner ones.

Sampling of MPs in water surface remains the most important step in the analytical process. It is usually performed with neuston, or plankton nets supported by a flow meter in order to determine the exact volume. Mesh sizes range from 50 to 3,000 μm, while 300 μm is the most used mesh leading to particle sizes <300 μm [51].

In sediments, sampling can differ according to sampling tool used. Results are expressed on basis of volume or mass of sample collected, or on the sis of the surface grilled [52]. The separation of MPs from water seems easier than from sediment. In water samples, the techniques are based on the low density of MPs that allow floating the MPs. In sediments, addition of some salts of sodium and/or zinc are used [53, 54]. Samples generally contain some debris that could prevent separation. They are destroyed using chemical or enzymatic hydrogen peroxide, mixtures of hydrogen peroxide and sulfuric acid [55].

17.4.3 Analysis

Large particles of MPs are identified with the naked eye, while small MPs using binocular microscopes or scanning electron microscopy (SEM) [55, 56]. Spectroscopic identification methods of MPs include Fourier Transform Infrared Spectroscopy (FTIR) and Raman Spectroscopy to detect characteristic functional groups of the polymer particles. For larger particles (> 500 μm), FTIR can be carried out using an Attenuated Transverse Reflection Unit (ATR) [26, 57]. Coupling FTIR to microscopes as reflectance or transmission micro-FTIR allows the detection of smaller MPs [52, 58]. Further identifications are made with thermal degradation techniques that specifies each polymer [59].

MPs are recovered from the supernatant and filtered or sieved. The concentrate may be visually sorted before quantification by microscopic counting with or without tagging using dyes, but neither of these methods can unambiguously confirm the particles are plastics.

Three different approaches are available to determine the chemical composition and/or size of plastic particles: spectroscopic, thermoanalytical, and chemical analyses. These methods are described briefly below. For further information about these methods, including capabilities and limitations related to detection levels and the ability to detect particle dimensions, refer to [60].

Spectroscopic methods are used to identify the specific chemical structure of polymers by comparing their absorption or emission spectra with reference spectra. FTIR is a well-established, relatively fast, and reliable spectroscopic method that, when coupled with microscopy, can identify particles to about 10–20 μm. However, biofilms, if not removed, can interfere with the detection of MPs. Hyphenated FTIR/microscopy technique also requires expensive instrumentation that is not available in many laboratories. Microscopy coupled with Raman spectroscopy can identify particles in the range of 1–20 μm. However, it can be subject to interferences, may be slow, and requires expensive instrumentation.

With thermoanalytical methods [60], the sample is pyrolyzed under inert conditions, so that specific decomposition products of the individual polymers can be analyzed. These methods tend to require larger particle masses compared with spectroscopic methods. Pyrolysis–gas chromatography/mass spectrometry (Py-GC/MS) can provide information on additives as well as on the polymer, and if the sample amount is large enough, it can identify the polymer composition of nanoplastic particles.

Other spectrometric methods such as Inductively Coupled Plasma Mass Spectrometry (ICP-MS) can be used to decompose the samples and detect specific fragments of polymers or elements. Again, these tend to require larger particle masses [61].

Software packages are often used in both tagging and spectroscopic studies to recognize and to count particles, and to characterize particles by comparing them with library spectra.

17.5 Treatment Technologies for Removing MPs from Wastewater

Wastewater treatment systems, where they exist, are considered highly effective in removing particles with characteristics like those of MPs [61–66]. Properties relevant to MPs removal in the water treatment include size, density, and surface charge. According to available data, wastewater treatment can effectively remove more than 90% of MPs from wastewater, with the highest removals from tertiary treatment such as filtration. Conventional treatment, when optimized to produce treated water of low turbidity, can remove particles smaller than a micrometer through processes of coagulation, flocculation, sedimentation/flotation, and filtration. Advanced treatment can remove smaller particles. For example, nanofiltration can remove particles >0.001 μm, while ultrafiltration can remove particles >0.01 μm. These facts combined with well-understood removal mechanisms point to the rational conclusion that water treatment processes can effectively remove MPs.

An important consideration is that wastewater and drinking-water treatment is not available nor optimized in many countries. Approximately 67% of the population in low- and middle-income countries lack access to sewage connections, and about 20% of household wastewater collected in sewers do not undergo at least a secondary treatment [66]. At these places, MPs may exist in greater concentrations in freshwater sources of drinking-water. However, the health risks associated with exposure to pathogens present in untreated or inadequately treated water will be by far much greater. By addressing the bigger problem of exposure to untreated water, communities can simultaneously address the smaller concern related to MPs in surface water and other drinking-water supplies.

Another factor to consider is how treatment waste is handled. Plastics are not usually destroyed, but rather transferred from one phase to another. For this reason, water treatment waste needs to be considered as a potential source of MPs contamination in the environment. There are currently limited data available on how treatment wastes are handled and the impact they may have on the environment.

17.6 Recommendations to Improve Sampling and Analytical Methods for PMs

There is a general need to improve sampling and analysis of MPs in water samples. The following improvements are particularly important:

- Studies should provide complete information about the method of sampling so that it can be reproduced.
- The sample volumes will depend on the nature of the water being sampled and size of the particles being analyzed, which in turn is determined by the filter or mesh size being used. Sample volumes should be sufficiently large to reliably detect low MPs concentrations.
- Wherever possible, plastic material should be avoided for sampling and analyses. If plastic material must be used, it should be characterized and reported.
- Materials should be rinsed with filtered water to avoid contamination.
- Sampling and sample processing should be carried out by trained professionals, or the quality of samples collected or processed by volunteers should be (quantitatively) validated against results obtained by professionals.
- If preservatives are used, their ability to affect polymer mass or particle shapes should be tested, either in the context of the study or via literature support.
- Laboratory surfaces should be thoroughly cleaned with filtered water to avoid contamination.
- All samples should be handled in a laminar-flow hood or in a clean-air laboratory.

- Blanks should be run per day or per series at least in triplicate to verify and to correct for contamination, and results should be corrected against blanks.
- Positive controls should be used to verify the recovery of particles during digestion, density separation and filtration steps.
- Digestion should be applied when necessary. Usually, digestion is not necessary for drinking-water from a treated source. However, for surface water and wastewater samples, where high organic matter concentrations hamper the selection and (visual) identification of particles, a digestion step is inevitably required.
- Polymer identification is required for a representative subsample of the entire sample.
- Data should be reported as number of particles/L and mass/L together with their detection limits. Minimum and maximum particle sizes and when possible, morphologies should be specified. All these characteristics may inform the risk assessment.
- Standard methods of sampling and analysis should be developed, but they may vary for the different media being sampled. For example, sediment methods may be different from seawater methods, which itself will also vary from drinking-water methods. As far as possible the same principles are needed to be followed.

In general, the use of plastics in everyday life cannot be overlooked or eliminated due to their recyclability, functionality, and light weight, so the unique characteristics of plastics allow them to play a major role in achieving a more sustainable and resource efficient future. On the one hand, they can help us to save essential resources like energy and water in key sectors like packaging, automotive and renewable energy, on the other hand, the daily use of plastics in a wide range of applications cause an increase in the generation of plastic waste that ultimately results in MPs.

However, if one wants to maintain the environment, and at the same time, to improve the circularity of plastics [11], it is essential to recover more and more plastic waste to avoid it ending up in landfills or in the environment. Especially in recent years, several recycling methods have been developed, such as primary recycling, where used plastics are recovered by extrusion, generating materials being like the original ones. Secondary mechanical recycling involves the collection, sorting and washing of the waste. Then the plastics are directly melted and molded into a new shape or turned into pellets. In addition to primary and secondary recycling, tertiary recycling is chemical recycling. In this type of recycling, plastics are converted into smaller molecules, usually liquids or gases, such as pyrolysis oil or syngas, which are commonly used as raw materials to obtain chemical products (e.g., methanol, olefins, alcohols, insecticides, and fungicides). Finally, the recovery of plastic waste is the set of operations, whose aim is to give this waste a new use value.

17.7 Conclusions

Microplastics are ubiquitous in the environment, including in the water cycle. They have been detected in marine water, wastewater, fresh water, and both tap and bottled water. However, the quality and quantity of data vary across different water types.

Important sources of MPs into fresh water are surface run-off and wastewater effluents, but there are insufficient data to quantify these inputs and to determine more specifically the primary sources. Further, some contamination may also occur during treatment, distribution or bottling processes of drinking water.

Study results should be interpreted in the context of the methods used. For example, smaller mesh sizes are generally used in drinking-water studies compared with freshwater studies, leading to higher particle counts. In general, there is a need to improve, to standardize and to harmonize MPs sampling and analyses in water. Most studies conducted to date are not considered fully reliable.

References

1. World watch Institute: Global plastic production rises. Recycling lags. 2015. http://www.worldwatch.org/global-plasticproduction-rises-recycling-lags. Accessed Mar 2018.
2. Rist S, Carney AB, Hartmann NB, Karlsson TM. A critical perspective on early communications concerning human health aspects of microplastics. Sci Total Environ. 2018;626:720–6.
3. Browne MA, Crump P, Niven SJ, Teuten E, Tonkin A, Galloway T, Thompson R. Accumulation of microplastic on shorelines worldwide: sources and sinks. Environ Sci Technol. 2011;45:9175–9.
4. Frias JP, do Sul JAI, Panti C, Lima AR. Microplastics in the marine environment: sources, distribution, biological effects and socio-economic impacts; 2021.
5. Van der Meulen MD, Devriese L, Lee J, Maes T, Van Dalfsen JA, Huvet A, Vethaak AD. Socio-economic impact of microplastics in the 2 Seas, Channel and France Manche Region: an initial risk assessment. MICRO Interreg project IVa. Mededeling ILVO; 2015. p 177.
6. Mouat J, Lopez RL, Bateson H. Economic impacts of marine litter: assessment and priorities for response. Report of the OSPAR Commission. ISBN 978-1-906840-26-6, 2010. p 117.
7. Wu Q, Liu S, Chen P, Liu M, Cheng SY, Ke H, ... Cai M. Microplastics in seawater and two sides of the Taiwan Strait: reflection of the social-economic development. Marine Pollut Bull. 2021;169:112588.
8. UNEP MAP. Marine litter assessment in the mediterranean. UNEP MAP publication. ISBN 978-92-807-3564-2; 2015. p 88.
9. Reisser J, Shaw J, Hallegraeff G, Proietti M, Barnes DKA, Thums M, Wilcox C, Denise Hardesty B, Pattiaratchi C. Millimeter-sized marine plastics: a new pelagic habitat for microorganisms and invertebrates. PLoS One. 2014;9:e100289.
10. Gall M, Wiener M, de Oliveira CC, Lang RW, Hansen EG. Building a circular plastics economy with informal waste pickers: recyclate quality, business model, and societal impacts. Resour Conserv Recycl. 2020;156:104685.
11. James T, Carlton JW, Geller JB, Miller JA, Carlton DA, McCuller MI, Treneman NC, Steves BP, Ruiz GM. Tsunami-driven rafting: transoceanic species dispersal and implications for marine biogeography. Science. 2017;357:1402–6.
12. Dussud C, Hudec C, George M, Fabre P, Higgs P, Bruzaud S, Delort A-M, Eyheraguibel B, Meistertzheim A-L, Jacquin J, Cheng J, Callac N, Odobel C. Colonization of non-biodegradable and biodegradable plastics by marine microorganisms. Front Microbiol. 2018;9:1571.
13. Laganà P, Caruso G, Corsi I, Bergami E, Venuti V, Majolino D, La Ferla R, Azzaro M. Simone Cappello-Do plastics serve as a possible vector for the spread of antibiotic resistance? Intern. J Hyg Environ Health. 2019;222(1):89–100.
14. David M, Bruno E, Patrick Q, Armelle S, Christine H, Lauriane M, Olivier M, Philippe S, Johan R, Arnaud H, Jose-Luis Z. Evaluation of the impact of polyethylene microbeads ingestion in European sea bas (Dicentrarchus labrax) larvae. Mar Environ Res. 2015;112:78–85.
15. Sussarellu R, Suquet M, Thomas Y, Lambert C, Fabioux C, Eve Julie Pernet M, Goïc NL, Quillien V, Mingant C, Epelboin Y, Corporeau C, Guyomarch J, Robbens J, Paul-Pont I, Soudant P, Huvet A. Oyster reproduction is affected by exposure to polystyrene microplastics. Proc Natl Acad Sci USA. 2016;113:2430–5.
16. Kim D, Chae Y, An YJ. Mixture toxicity of nickel and microplastics with different functional groups on Daphnia magna. Environ Sci Technol. 2017;51(21):12852–8.
17. Lenz R, Enders K, Nielsen TG. Microplastic exposure studies should be environmentally realistic. Proc Natl Acad Sci. 2016;113(29):E4121–2.
18. Koelmans AA. Aquatic ecology and water quality management group, Department of Environmental Sciences, Wageningen University, P.O. Box 47, 6700 AA Wageningen, The Netherlands, Microplastic as a vector for chemicals in the aquatic Environment: critical review and model-supported reinterpretation of empirical studies. Environ Sci Technol. 2016;50:3315–26.
19. Tallec K, Huvet A, Poi CD, González-Fernández C, Lambert C, Petton B, Le Goïc N, Berchel M, Soudant P, Paul-Pont I. Nanoplastics impaired oyster free living stages, gametes and embryos. Environ Pollut. 2018;242:1226–35.
20. González-Fernández C, Toullec J, Lambert C, Le Goïc N, Seoane M, Moriceau B, Huvet A, Berchel M, Vincent D, Courcot L, Soudant P, Paul-Pont I. Do transparent exopolymeric particles (TEP) affect the toxicity of nanoplastics on Chaetoceros neogracile? Environ Pollut. 2019;250:873–82.
21. Green D. Effects of microplastics on European flat oysters, Ostrea edulis and their associated benthic communities. Environ Pollut. 2016;2016(216):95–103.
22. GESAMP. Sources, fate and effects of microplastics in the marine environment (Kershaw PJ, Rochman CM, eds). Rep. Stud. GESAMP 2016, no 93, 220; 2016.
23. el Hermabessiere D. Microplastic contamination and pollutant levels in mussels and cockles collected along the Channel coasts. Environ Pollut. 2019;250:807–19.
24. Peixoto D, Pinheiro C, Amorim J, Oliv-Teles L, Guilhermino L, Natividа deVieira M. Microplastic pollution in commercial salt for human consumption: a review. Estuar Coast Shelf Sci. 2019;219:161–8.
25. Lithner D, Larsson A, Dave G. Environmental and health hazard ranking, and assessment of plastic

polymers based on chemical composition. Sci Total Environ. 2011;409:3309–24.
26. Sridharan S, Kumar M, Bolan NS, Singh L, Kumar S, Kumar R, YouS. Are microplastics destabilizing the global network of terrestrial and aquatic ecosystem services? Environ Res. 2021:111243.
27. Gauquie J, Devriese L, Robbens J, De Witte B. A qualitative screening and quantitative measurement of organic contaminants on different types of marine plastic debris. Chemosphere. 2015;138:348–56.
28. Turner A. Heavy metals, metalloids and other hazardous elements in marine plastic litter. Mar Pollut Bull. 2016;111:136–42.
29. Barboza LGA. Marine microplastic debris: An emerging issue for food security, food safety and human health. Mar Pollut Bull. 2018;133:336–48.
30. Keswani A, Oliver DM, Tony G, Quilliam R. Microbial hitchhikers on marine plastic debris: Human exposure risks at bathing waters and beach environments. Mar Environ Res. 2016;118:10–9.
31. Soares J, Miguel I, Venâncio C, Lopes I, Oliveira M. Perspectives on micro (nano) plastics in the marine environment: biological and societal considerations. Water. 2020;12(11):3208.
32. Fackelman G, Sommer S. Microplastics and the gut microbiome: how chronically exposed species may suffer from gut dysbiosis. Mar Pollut Bull. 2019;143:193–203.
33. Wright S, Kelly F. Plastic and human health: a micro issue? Environ Sci Technol. 2017;51:6634–47.
34. Prata JC. Microplastics in wastewater: state of the knowledge on sources, fate, and solutions. Mar Pollut Bull. 2016;129:262–5.
35. Estahbanati S, Fahrenfeld NL. Influence of wastewater treatment plant discharges on microplastic concentrations in surface water. Chemosphere. 2016;162:277–84.
36. Conley K, Clum A, Deepe J, Lane H, Beckingham B. Wastewater treatment plants as a source of microplastics to an urban estuary: removal efficiencies and loading per capita over one year. Water Res. 2019; X 3:100030.
37. Iyare PU, Oukia SK, Bond T. Microplastics removal in wastewater treatment plants: a critical review. Environ Sci Water Res Technol. 2020:10.
38. Stolte A, Forster S, Gerdts G, Schubert H. Microplastic concentrations in beach sediments along the German Baltic coast. Mar Pollut Bull. 2015;99(1–2):216–29.
39. Baechler BR, Granek EF, Hunter MV, Conn KE. Microplastic concentrations in two Oregon bivalve species: Spatial, temporal, and species variability. Limnol Oceanogr Lett. 2020;5:54–65.
40. Claessens M, De Meester S, Van Landuyt L, De Clerck K, Janssen CR. Occurrence and distribution of microplastics in marine sediments along the Belgian coast. Mar Pollut Bull. 2011;62(10):2199–204.
41. Kershaw P, Turra A, Galgani F. Guidelines for the monitoring and assessment of plastic litter in the ocean-GESAMP reports and studies no 99. GESAMP Reports and Studies; 2019.
42. Yonkos LT, Friedel EA, Perez-Reyes AC, Ghosal S, Arthur CD. Microplastics in four estuarine rivers in the Chesapeake Bay, USA. Environ Technol. 2014;48(24):14195–202.
43. Dris R, Lahens L, Rocher V, Gasperi J, Tassin B. Premières investigations sur les microplastiques en Seine; 2016. https://doi.org/10.1051/tsm/201512025.
44. Huang JC, Shetty AS, Wang MS. Biodegradable plastics: a review. Adv Polym Technol. 1990;10 (1):23–30.
45. St. Louis D, Apply A, Michel D, Emmanuel E. Microplastiques et santé environnementale : identification des dangers environnementaux en Haïti; 2021. https://doi.org/10.13140/RG.2.2.30822.96321.
46. Sun J, Dai X, Wang Q, van Loosdrecht MCM, Ni B-J. Microplastics in wastewater treatment plants: detection, occurrence and removal. Water Res. 2019;152:21–37.
47. Koelmans AA, Nor NHM, Hermsen E, Kooi M, Mintenig SM, De France J. Microplastics in freshwaters and drinking water: critical review and assessment of data quality. Water Res. 2019;155:410–22.
48. Magnusson K, Norén F. Screening of microplastic particles in and down-stream a wastewater treatment plant. Report C55, Swedish Environmental Research Institute, Stockholm; 2014.
49. Paul K, Robert H, Isobel M, Bajic L, McKenna N. Wastewater treatment plants as a source of microplastics in river catchments. Environ Sci Pollut Res. 2018;25:20264–7.
50. Sascha K, Dimzon IK, Eubeler J, Knepper TP. Analysis, occurrence, and degradation of microplastics in the aqueous environment. In: Lambert S, Wagner M (eds) Freshwater microplastics emerging environmental contaminants? Springer; 2017.
51. Thompson RC, Olsen Y, Mitchell RP, Davis A, Rowland SJ, John AWG, McGonigle D, Russell AE. Lost at sea: where is all the plastic? Science. 2014;304(5672):838.
52. Nuelle MT, Dekiff JH, Remy D, Fries E. A new analytical approach for monitoring microplastics in marine sediments. Environ Pollut. 2014;184:161–9.
53. Karapanagioti HK, Klontza I. Investigating the properties of plastic resin pellets found in the coastal areas of lesvos island. Global Nest J. 2007;9(1):71–6.
54. Gilfillan LR, Ohman MD, Doyle MJ, Watson W. Occurrence of plastic micro-debris in the Southern California current system. Cal Coop Ocean Fish. 2009;50:123–33.
55. Rasta M, Sattaria M, Taleshi MS, Namin JI. Identification and distribution of microplastics in the sediments and surface waters of Anzali Wetland in the Southwest Caspian Sea, Northern Iran. Mar Pollut Bull. 2020;160:111541.
56. Li J, Liu H, Paul Chen J. Microplastics in freshwater systems: a review on occurrence, environmental

effects, and methods for microplastics detection. Water Res. 2018;137:362e374.

57. Doyle MJ, Watson W, Bowlin NM, Sheavly SB. Plastic particles in coastal pelagicecosystems of the Northeast Pacific ocean. Mar Environ Res. 2011;71 (1):41–52.
58. Eubeler JP, Bernhard M, Knepper TP. Environmental biodegradation of syntheticpolymers II. Biodegradation of different polymer groups. Trac Trends Anal Chem. 2010;29(1):84–100.
59. Mallow O, Spacek S, Schwarzböck T, Fellner J. A new thermoanalytical method for the quantification of microplastics in industrial wastewater. Environ Pollut. 2020;259:113862.
60. Huppertsberg S, Knepper TP. Instrumental analysis of microplastics—benefits and challenges. Anal Bioanal Chem. 2018;410:6343–52.
61. Kay P, Hiscoe R, Moberley I, Bajic L, McKenna N. Wastewater treatment plants as a source of microplastics in river catchments. Environ Sci Pollut Res. 2018;25:20264–7.
62. Conley K, Clum A, Deepe J. Wastewater treatment plants as a source of microplastics to an urban estuary: removal efficiencies and loading per capita over one year. Water Res X. 2019;3(1):1–9.
63. Joana CP. Microplastics in wastewater: state of the knowledge on sources, fate and solutions. Mar Pollut Bull. 2018;129:262–5.
64. Bayo J, Olmos S, Lopez-Castellanos J. Microplastics in an urban wastewater treatment plant: the influence of physicochemical parameters and environmental factors. Chemosphere. 2020;238:124593.
65. Carr SA, Liu J, Tesoro AG. Transport: fate of microplastic particles in wastewater treatment plants. Water Res. 2016;91:174–82.
66. Microplastics in drinking-water, World Health Organization 2019.

Consequences of Heavy Metals in Water and Wastewater for the Environment and Human Health

18

Fatma Beduk, Senar Aydin, Mehmet Emin Aydin, and Müfit Bahadir

Abstract

The widespread distribution of heavy metals in water sources and wastewaster is of great concern because of their highly toxic properties. But, some metals are essential for normal growth of plants and animals. Considerable effort is being made to identify and to reduce sources of heavy metals in aquatic environments. The intensity of industrial activities using heavy metals results in continued heavy metal load in water media. The toxic effects of heavy metals depend on their intrinsic properties, water solubilities and bioavailabilities. In this chapter, the authors aimed to give an overview about speciation of heavy metals and the state-of-the-art of resource, behavior and fate of heavy metals in the environment.

Keywords

Heavy metals · Non-essential elements · Toxicity · Transportation · Occurrence

18.1 Introduction

Heavy metals are among the most hazardous contaminants in the environment because of their high toxicity. The therm "heavy metal" refers to metals and metalloids with an atomic density higher than 5 g/cm^3. Heavy metals derive from both natural and anthropogenic sources, however, latter have more significant effect on environmental pollution. Natural sources include magmatic and metamorphic rocks, and soil formation, that pass through water sources by hydrological cycle [1]. Heavy metal pollution caused by anthropogenic sources may consist of mainly mining, electroplating, disposal of high metal containing wastes, leachate, leaded gasoline and paints, land application of fertilizers, sewage sludge, pesticides, wastewater irrigation, coal combustion residues, spillage of petrochemicals, and atmospheric deposition [2, 3]. While high toxicity of some heavy metals is of great concern, some others are essential for normal growth and development of plants and animal, however, essential elements can be toxic if present in excess [4, 5]. In alphabetical order, aluminum (Al), arsenic (As), beryllium (Be),

F. Beduk (✉) · S. Aydin
Engineering Faculty, Environmental Engineering Department, Necmettin Erbakan University, Konya, Turkey
e-mail: fabeduk@erbakan.edu.tr

M. E. Aydin
Engineering Faculty Civil Engineering Department, Necmettin Erbakan University, Konya, Turkey

M. Bahadir
Institut für Ökologische und Nachhaltige Chemie, Technische Universität Braunschweig, Braunschweig, Germany

M. Bahadir and A. Haarstrick (eds.), *Water and Wastewater Management*,
Water and Wastewater Management, https://doi.org/10.1007/978-3-030-95288-4_18

boron (B), cadmium (Cd), chromium (Cr), cobalt (Co), copper (Cu), iron (Fe), lead (Pb), lithium (Li), manganese (Mn), mercury (Hg), molybdenum (Mo), nickel (Ni), selenium (Se), vanadium (V), and zinc (Zn) are among the most toxic, and therefore, regulated metals and metaloids. The heavy metals like (B), Co, Cu, Fe, Mn, Mo, Ni, and Zn are biologically essential micro nutrients required by plant growth, but harmful to the plant beyond the permissible limit values [6]. For example, exceeding 30 mg/kg recommended value of Cu fosters oxidative stress and promotes leaf chlorosis in plants [7].

Essential elements for human health include Cr, Co, Cu, Fe, Zn, Mo, and Se, which have adverse health effects in case of deficiencies. Beneficial health effects of these elements in case of degenerative diseases encourage people to take them as supplements. So as to determine the most ideal dose of essential elements among toxicity and deficiency, acceptable range of oral intake (AROI) concept has been used.

People are exposed to high amounts of heavy metals by consuming water and food contaminated with heavy metals, or by breathing. As, Cd, Cr(VI), Pb, and Hg are non-essential toxic heavy metals, which exert toxicity at very low concentrations [8]. These elements are reported as major chemicals of public concern by World Health Organization [9]. Heavy metals are not biodegradable, can bioaccumulate in tissues, and being concentrated in food chain. Since these metals are present in every compartment of the environment, human exposure is not readily preventable. As, Cd, Cr(VI), Be, and Ni are human carcinogens [10]. As has been found to be human carcinogen as well, even et very low exposure levels. Due to the consumption of As polluted groundwater in many parts of the world, people have been exposed to the risk of As poisoning. It is estimated that over 320,000 people will die from As induced cancer over the next 50 years [11]. There are sufficient data for As exposure to evaluate its carcinogenic effect, however, single metal exposure of human is not generally definable for other heavy metals. Animal experiments have been used to identify metal induced tumors [10].

In this study, natural and antropogenic sources of the most important and, therefore, regulated heavy metals are given. It is aimed to review recent literature in order to give an overall picture about heavy metal pollution, transport and sinks in the environment. Besides, information about toxic effects on human and aqueous organisms is also given.

18.2 Non-Essential Toxic Heavy Metals

18.2.1 Arsenic (As)

As exists in both organic and inorganic forms, and in four oxidation states: elemental arsenic (0), arsenite (+3), arsenate (+5), and arsine (−3). Oxidation state determines the level of toxicity and mobility. Arsenite (As^{3+}) is more toxic and mobile than arsenate (As^{5+}), and inorganic forms of As are more toxic than the organic ones (majorly seafood) [12]. As mostly occurs in the As^{3+} oxidation state in groundwater that makes it a serious water security problem. While As^{5+} is present as negatively charged anion under pH conditions of groundwater ($H_2AsO_4^-$ or $HAsO_4^{2-}$), arsenite occurs uncharged (H_3AsO^3). Due to the difference between their charges, adsorption and desorption reactions of these two types differ from each other. Redox reactions also affect the mobility of As [13].

Natural and anthropogenic sources can cause As contamination in water bodies. According to deep rock types, minerals, and ore structures, As pollution occurs naturally in well waters. Anthropogenic pollution sources include mining activities, the use of pesticides, burning of As containing coals, etc.

Due to the consumption of As polluted groundwater in many parts of the world, people have been exposed to the risk of As poisoning. Taking the toxic effects of As into account, a limit value of 10 μg/L for drinking water was determined by US EPA [14], and WHO [9]. The European Union, USA, Taiwan, Vietnam, and India have also regulated As at 10 μg/L value, while it is still 50 μg/L in China, Bangladesh, Argentina, Nepal, and Mexico, suffering from highly As contaminated water sources from aquifer sediments [15]. The low permitted levels for As in drinking water made efficient removal methods necessary. In Bangladesh, approximately 30–40 million people are exposed to As contamination, with an expected 2.5 million people developing some forms of arsenicosis symptoms in their lifetime [16].

18.2.2 Cadmium (Cd)

Cd is a highly toxic heavy metal, which occurs naturally as cadmium sulfide, and is usually in association with Zn. Cd has both occupational and environmental exposure routes. The highest amount of Cd is used for battery production. It is also used for electroplating, coating, pigment industries, etc. Dissolution of Cd minerals in water and deposition of atmospheric Cd are natural sources of water contamination, but Cd pollution is mainly caused by industrial discharges and agricultural runoff. Phosphorus fertilizers, wastewater irrigation, and sewage sludge applications are agricultural sources of Cd contamination [17]. Cd can be transported through the roots, stems, and leaves of the plant, which is controlled by plant species. Boavailability of Cd in soil depends on soil pH and the content of organic matter [18, 19]. Cd contamination of rice is well documented. A 20–40 μg daily intake of Cd from rice was reported for Asia [20].

18.2.3 Chromium (Cr)

The most common and highly stable oxidation states of Cr are Cr(III) and Cr(VI). Cr(VI) is an important environmental concern due to its high toxicity and mobility. As a result of its high solubility, it is not fisible to remove Cr(VI) from wastewater by precipitation techniques. Unlike Cr(VI), Cr(III) has low solubility and almost no toxicity [21]. Cr(VI) is used in large scale in industry, including stainless steel industries, dyes and leather tanneries, mining and smelting, and electroplating processes. It is also used as anti-corrosive agent. The highest exposure route is in occupational settings. Cr(VI) exerts high toxicity on human and animal, causing kidney dysfunction, diarrhea, ulcers, and lung carcinom [22, 23].

18.2.4 Lead (Pb)

Pb is released to the environment from the atmospheric sources, such as automobiles' exhausts using leaded gasoline, lead containing paint, emissions of smelters, etc. This element is found in air, water and food. The atmospheric release of lead has decreased in recent years as a result of removing lead from gasolines. Lead solder has been a water contamination source in water distribution networks. Old drinking water pipes in Europe used to be often made of lead. Once a human exposed to Pb, it mainly stores in the bone and affects kidney, central nervous system and liver. It is transferred through the placenta and has deleterious biological effects on fetus [24]. Wasewater irrigation is a widespread application in Marrakech, Morocco. Wastewater irrigation pose Pb contamination risk in soil and crop in the region. Chaoua et al. (2019) [25] reported about 1.417 ± 0.32 mg/L Pb for wastewater samples in discharge point, 57.3 ± 10.8 mg/kg Pb for wastewater irrigated soil samples, and 52.9 ± 22.2 mg/kg Pb for *Avena sativa* leaves cultivated in the area. All of the agricultural samples taken in the field exceeded 5 mg/kg limit value set for plants by WHO/FAO [26], which pose a risk for people consuming these food staff.

18.2.5 Mercury (Hg)

Hg enters aquatic media in three forms: Firstly, in "elementel mercury form" in zero-oxidation state (Hg^0), such as mercury vapor, as the only metal in liquid form at room temperature; Secondly, in "inorganic mercury form" such as mercuric chloride; Thirdly, in "organic mercury form" such as methyl-mercury, phenyl-mercury, etc. [27]. In the past, Hg compounds were widely used in agriculture, industry and medicine. Today, people expose to Hg in the occupational settings of chloralkali industry, production of fluorescent lights and thermometers, and from dental amalgams [1]. Another important exposure route is the consumption of fish and seafood containing methyl mercury [27, 28]. Inorganic mercury is being biomethylated to methyl mercury by microorganisms present in aquatic media, and enters aquatic food chain from zooplankton to fish. Consumption of fish is the main route of methyl mercury exposure for human. Methyl mercury is bioaccumulated in human body, and has relatively long half life, and higher toxicity when compared with inorganic mercury [29]. Jung et al. [30] analyzed 400 blood samples taken from South Korean population, consuming fish in their diet. 6.35 μg/L and 4.44 μg/L geometric means of total mercury and methyl mercury were determined, respectively. 71.9% of total mercury was present as methyl mercury. Besides, higher concentrations of both mercury forms were determined for blood samples taken from coastal region residents when compaired with terrestrial region.

18.3 Occurrence and Transportation of Heavy Metals in the Environment

Industrial discharges are major environmental concern due to the variety and high concentration of heavy metals. Industrial discharges to sewerage system without enough or no treatment leads to high pollution load in municipal treatment plants and cause accumulation in sewage sludge that adversely affect its agricultural use. The water, soil and air quality have seriously affected by industrial discharges and emissions. Aydin et al. [31] reported unauthorized industrial discharges of Cr, Zn and Ni to sewerage system from valve producers and galvanization workshops located in the city centre of Konya, Turkey. High load of heavy metals on urban Wastewater Treatment Plant (WWTP) was detected by passive sampling of slime in the sewerage system. Similar research results from China showed that, even though strict discharge limits for heavy metals were set, industrial discharges increased heavy metal loads to WWTPs in the country. Zhou et al. [32] analysed samples taken from 800 WWTPs in nine provinces from China in the means of heavy metal loads. While Hg and As were mostly detected heavy metals, total Cr, Cr(VI), Pb, and Ni were determined to exceed limit values set by national standards. Removal rates of heavy metals in WWTPs were minimum 25,6% for Be and maximum 69,8% for Cr, which were comparable to other countries.

Depending on raw wastewater inflow to WWTP, sewage sludge contains a variaty of heavy metals, which restrict its agricultural use. Heavy metals form complexes with different mobility fractions in sewage sludge. These are (I) carbonate fraction; (II) iron and manganese oxide fraction; (III) organic matter and sulphides fraction, and (IV) silicate fraction [33, 34]. While fractions (I) and (II) are the most mobile forms, fraction (IV) is the least mobile one. Mobility of fraction (III) depends on certain conditions, such as the organic matter content. Sewage sludge pretreatment methods, such as dewatering, stabilization, composting, hygienization, etc. do not reduce its heavy metal content. Sewage sludge samples taken from an industrialized zone in Poland were analyzed for seven heavy metals, namely Cd, Cr, Cu, Hg, Ni, Pb, and Zn. It was found that sludge samples were environmentally hazardous in the means of Zn, Ni, and Cd, while posed a potential ecological risk in the means of Hg, Cd, and Cu [33]. Since heavy metal accumulation in sewage sludge is an inevitable end for conventioanl WWTPs taking industrial discharges, it is important to control heavy metal sources in order to take end-of-pipe actions. Besides, conventional wastewater treatment methods are insufficient for complete removal of heavy metals, while investment costs obstacle large-scale applications of new and more effective treatment technologies.

In arid and semi-arid countries, treated or untreated domestic wastewater is widely used for agricultural irrigation. Many studies reveal both soil and agricultural product contamination by heavy metals in every part of the world. The use of wastewater for irrigation of agricultural areas in Turkey dates long back. Aydin et al. [35] reported strong Cd pollution in soil samples (8.23–11.6 mg/kg) compared with non-irrigated controls, taken from Konya, Turkey, after

40 years of irrigation with untreated municipal wastewater. Heavy metal contamination of wheat grain, cultivated in the same area was also investigated. 45% of the areas irrigated with wastewater in China were also contaminated with high levels of heavy metals. Even 14 years after industrial wastewater irrigation, it was determined that Cd, Zn and Pb concentrations exceeded the limit values given for soil use in China. The results of the study once again revealed the continuity of metal pollution in soils polluted with heavy metals, even after many years [36].

Sediments may be reservoirs for heavy metals in wastewater receiving water bodies. Heavy metals are adsorbed to geogenic components of sediments [37]. Accumulated heavy metals in sediments may not be permanent. They can be transfered from sediments to aqueous phase, when the environmental conditions are changed; so, heavy metals in sediments have long-term harmful effects on water media [38]. Sediments also give temporal patterns of heavy metal pollution in aquatic media. Maximum pollution concentration was determined for Pb (1 mg/g dry weight) for sediment samples taken from Lake Lucerne in Central Switzerland. Pollution attributed to industrial coal combustion was going back to the fist part of the twentieth century [39].

Beacuse of certain solubility of heavy metals in aquatic environment, they can be taken up by living organisms. It was documented that green algea species *Spirogyra* and *Cladophora* have biosorption capacity of Pb and Cu in aqueous media [40]. High levels of Cd and Cr were found in the tissues of *Genus barbus* caught from Tigris River, which is the most popular food fish in Iraq [41]. Milenkovic et al. [42] analyzed heavy metals in sea fish samples taken from Serbian markets. Cd was determined in the range of 0.01–0.81 mg/kg, while Hg was determined in the range of 0.01–1.47 mg/kg, which was the highest in sharks. 6.56 m/kg of Pb was determined in *Atlantic mackerel* samples. Researches reveal the relationship of fish size with heavy metal accumulation, such as for Hg [43].

18.4 Ecotoxicological Effects

The aquatic toxicity of metals depend on bioavailability of the metal, which means that the organism does not uptake the total amount of the metal in its habitat. Physicochemical characteristics of the environment, chemical characteristics of water, chemical species of the metal, and tolerance of the organism influence the ecotoxicity. Many ecotoxicological studies focus on solubility of the metal species [44].

Agency for Toxic Substances and Disease Registry (ATSDR) ranked toxic heavy metals as As at 1st place; Pb at 2nd place, Hg at 3rd place, Cd at 7th place, and Cr (VI) at 17th place, in the hazardous substances list for toxicology [45]. The list was based on occurence frequency, toxicity and human exposure potential of heavy metals, but not specific for the aqueos media. In the marine environment, As(V) is the dominant form of As. The more toxic form of As(III) accounts for roughly 20% of total As in seawater [46]. As shows affinity towards proteins, lipids, and other cellular components. Aquatic invertebrates, crustaceans and mollusks uptake As by their mussels. Formation of arsenoribosides in marine invertebrates and algae was identified [47]. However, As is not bioaccumulative for fish. Therefore, it does not biomagnify in the food chain. Water solubility is the key factor about As toxicity. While As (0) has low solubility and lowest toxicity, As(V) easily enter body fluids as a result of its virtual solubility, that result in higher toxicity [1].

Pajany et al. [48] conducted a research study on toxic effects of As contaminated sediment samples, taken from French Mediterranean ports on oyster larvae (*Crassostrea gigas*) and *Vibrio fischeri*. It was determined that toxic effects of sediments on the test organisms had some correlation with the As content. While sediment samples showed acute toxic effects on oyster larvae, they presented sub-chronic toxic effects on *Vibrio fischeri*.

Main route of Pb exposure is occupational exposure in the industry. Food, water, soil, and household dust are other pathways. Pb toxicity is defined as chronic, which causes neurodevelopmental deficits for children [49]. Pb also causes dyesfunction of kidney, brain, and reproductive system [50]. Pb causes oxidative stress in living organisms and is a potential carcinogen [51]. Algae, which are in the bottom of the aquatic food chain, are the main organisms that uptake Pb in water media [40]. Kim et al. (2020) [52] made a research study about toxic effect of Pb on various sizes of Zebra fish. Results showed that Pd induced toxicity on all experimented sizes of fish samples and caused behavioral changes in the organizm.

There is an increasing concern about Hg toxicity on marine and surface water ecosystems. Hg has neurotoxic, genotoxic, and immunotoxic effects, depending on its metallic, inorganic and organometallic structure. Since Hg is tranferred and accumulated through food chain, it biomagnifies in the aquatic ecosystem. It generally enters the food chain via bacteria and phytoplankton, which are the primary producers. Methy mercury and other organic mercury compounds are highly liphofilic, therefore, they accumulate in the tissues of the fish [53] that led to the Minamata disease in Japan mid of 1950s [54].

Bioavailability of Cd is related to free Cd ion. Cd forms chlorocomplexes in marine environment that makes it less available for aquatic organisms. Cd accumulates mainly in the kidneys, creates oxidative stress in tissues and causes pulmonary emphysema, and renal tubular damage. Toxicity of Cd is also correlated with some types of cancer [55]. Cd toxicity causes skeletal deformation with severe pain, which is named *Itai-Itai* diseas [56].

18.5 Conclusions

Due to industrial activities, heavy metals are found in every natural media in the environment. Since they accumulate in the environment as a result of continuous input, it is generally possible to quantify them in all aqueous media. Heavy metals exert toxicity on plants, animals and human population even at very low concentration levels. To prevent heavy metal toxicity, some regulations are enacted determining maximum concentration for waterwater discharges to receiving water bodies, drinking water, and agricultural irrigation water. The highest input comes from industrial activities. Therefore, ensuring industrial wastewater treatment to the required extent is an important solution, instead of end-of-pipe actions.

References

1. Sarkar B. Hevy metals in the environment. New York: Marcel Dekker, Inc.; 2002.
2. Khan S, Cao Q, Zheng YM, Huang YZ, Zhu YG. Health risks of heavy metals in contaminated soils and food crops irrigated with wastewater in Beijing, China. Environ Pollut. 2008;152(3):686–92.
3. Zhang MK, Liu ZY, Wang H. Use of single extraction methods to predict bioavailability of heavy metals in polluted soils to rice. Commun Soil Sci Plant Anal. 2010;41(7):820–31.
4. Rai PK. Heavy metal phytoremediation from aquatic ecosystems with special reference to macrophytes. Crit Rev Environ Sci Technol. 2009;39:697–753.
5. Boyd RS. Heavy metal pollutants and chemical ecology: exploring new frontiers. J Chem Ecol. 2010;36:46–58.
6. Vardhan, KH, Kumar PS, Panda, RC. A review on heavy metal pollution, toxicity and remedial measures: current trends and future perspectives. J Mol Liquids. 2019;290:111197.
7. Kumar V, Pandita S, Sidhu GPS, Sharma A, Khanna K, Kaur P, Bali AS, Setia R. Copper bioavailability, uptake, toxicity and tolerance in plants: a comprehensive review. Chemosphere. 2021;262:127810.
8. Rahman Z, Singh VP. The relative impact of toxic heavy metals (THMs) (arsenic(As), cadmium (Cd), chromium (Cr)(VI), mercury (Hg), and lead (Pb)) on the total environment: an overview. Environ Monit Assess. 2019;191:419.
9. WHO, World Health Organization. Guideline for drinking water quality, vol 1. Geneva; 2004.
10. Goyer R. Issue paper on the human health effects of metals, U.S. Environmental Protection Agency Risk Assessment Forum; 2004. https://www.epa.gov/sites/default/files/2014-11/documents/human_health_effects.pdf.
11. Murcott S. Arsenic exposure and health effects. In: Proceedings of 5th international conference on arsenic exposure and health effects, July 14–18, San Diego, California; 2002. pp 451–8.

12. Muhammed Abdul KS, Jayasinghe SS, Chandanaa EPS, Jayasumanac C, De Silvaa PMCS. Arsenic and human health effects: a review. Environ Toxicol Pharmacol. 2015;40:828–46.
13. Smedley PL, Kinniburgh DG. A review of the source, behavior and distribution of arsenic in natural waters. Appl Geochem. 2012;17:517–68.
14. US EPA, U.S. Environmental Protection Agency. Quality criteria for water. Office of Water, Regulation and Standard, Washington, DC 20460, EPA 440/5-86–001; 1986.
15. Flanagan SV, Johnstonb RB, Zhenga Y. Arsenic in tube well water in Bangladesh: health and economic impacts and implications for arsenic mitigation. Bull World Health Organ. 2012;90:839–46.
16. Desbarats J, Koenig CEM, Pal T, Mukherjee PK, Beckie RD. Groundwater flow dynamics and arsenic source characterization in an aquifer system of West Bengal, India. Water Resour Res. 2007;50:4974–5002.
17. Zhang H, Reynolds M. Cadmium exposure in living organisms: a short review. Sci Total Environ. 2019;678:761–7.
18. Song Y, Jinc L, Wang X. Cadmium absorption and transportation pathways in plants. Int J Phytorem. 2017;19(2):133–41.
19. Haider FU, Liqun C, Coulter JA, Cheema SA, Wu J, Zhang R, Wenjun M, Farooq M. Cadmium toxicity in plants: ımpacts and remediation strategies. Ecotoxicol Environ Safety. 2021;211:111887.
20. Sebastian A, Prasad M. Cadmium minimization in rice. A review. In: Agronomy for sustainable development. Springer/EDP Sciences/INRA, vol 34, no 1; 2014. pp 155–73.
21. Barrera-Díaz CE, Lugo-Lugo V, Bilyeu B. A review of chemical, electrochemical and biological methods for aqueous Cr(VI) reduction. J Hazard Mater. 2012;223–224:1–12.
22. Jobby R, Jha P, Yadav AK, Desai N. Biosorption and biotransformation of hexavalent chromium [Cr(VI)]: a comprehensive review. Chemosphere. 2018;207:255–66.
23. Zhitkovich A. Chromium in drinking water: sources, metabolism, and cancer risks. Chem Res Toxicol. 2011;24:1617–29.
24. Gidlow DA. Lead toxicity. Occup Med. 2004;54:76–81.
25. Chaoua S, Boussaa S, El Gharmali A, Boumezzough A. Impact of irrigation with wastewater on accumulation of heavy metals insoil and crops in the region of Marrakech in Morocco. J Saudi Soc Agric Sci. 2019;18:429–36.
26. WHO/FAO. Joint FAO/WHO Food Standard Programme Codex Alimentarius Commission 13th Session. Report of the Thirty-Eight Session of the Codex Committee on Food Hygiene, Houston, United States of America, ALINORM 07/30/13; 2007.
27. Zulkipli SZ, Liew HJ, Ando M, Lim LS, Wang M, Sung YY, Mok WJ. A review of mercury pathological effects on organs specific of fishes. Environ Pollut Bioavail. 2021;33(1):76–87.
28. Murota K, Yoshida M, Ishibashi N, Yamazaki H, Minami T. Direct absorption of methyl mercury by lymph. Biol Trace Elem Res. 2012;145:349–54.
29. Trasande L, Landrigan PJ, Scehechter C. Public health and economic consequences of methyl mercury toxicity to the developing brain. Environ Health Perspect. 2005;113(5):590–6.
30. Jung SA, Chung D, On J, Moon MH, Lee J, Pyo H. Correlation between total mercury and methyl mercury-in whole blood of South Korean. Bull Korean Chem Soc. 2013;34(4):1101–7.
31. Aydin ME, Beduk F, Aydin S, Koyuncu S, Genuit G, Bahadir M. Development of biofilm collectors as passive samplers in sewerage systems—a novel wastewater monitoring method. Environ Sci Pollut Res. 2020;27:8199–209.
32. Zhou Y, Lei J, Zhang Y, Zhu J, Lu Y, Wu X, Fang H. Determining discharge characteristics and limits of heavy metals and metalloids for wastewater treatment plants (WWTPs) in China based on statistical methods. Water. 2018;10:1248.
33. Wojciula A, Boruszko D, Pajewska G. Analysis of heavy metal fraction content in sewage sludge from selected wastewater treatment plants. J Ecol Eng. 2021;22(4):98–105.
34. Kowalik R, Latosinska J, Gawdzik J. Risk analysis of heavy metal accumulation from sewage sludge of selected wastewater treatment plants in Poland. Water. 2021;13:2070.
35. Aydin ME, Aydin S, Beduk F, Tor A, Tekinay A, Kolb M, Bahadir M. Effects of long-term irrigation with untreated municipal wastewater on soil properties and crop quality. Environ Sci Pollut Res. 2015;22:19203–12.
36. Li P, Wang X, Allinson G, Li X, Xiong X. Risk Assessment of heavy metals in soil previously irrigated with industrial wastewater in Shenyang. China, Journal of Hazardous Materials. 2009;161:516–21.
37. Sindern S, Tremöhlen M, Dsikowitzky L, Gronen L, Schwarzbauer J, Hartati Siregar T, Ariyani F, Eko Irianto H. Heavy metals in river and coast sediments of the Jakarta Bay region (Indonesia)-Geogenic versus anthropogenic sources. Mar Pollut Bull. 2016;110:624–33.
38. Pourabadehei M, Mulligan CN. Resuspension of sediment, a new approach for remediation of contaminated sediment. Environ Pollut. 2016;213:63–75.
39. Thevenon F, Graham ND, Chiaradia M, Arpagaus P, Wildi W, Poté J. Local to regional scale industrial heavy metal pollution recorded in sediments of large freshwater lakes in central Europe (lakes Geneva and Lucerne) over the last centuries. Sci Total Environ. 2011;412–413:239–47.
40. Souza PO, Ferreira LR, Pires NRX, Filho PJS, Duarte FA, Pereira CMP, Mesko MF. Algae of economic importance that accumulate cadmium and

lead: a review. Braz J Pharmacognosy. 2012;22 (4):825–37.
41. Mensoor M, Said A. Determination of heavy metals in freshwater fishes of the Tigris River in Baghdad. Fishes. 2018;3:23.
42. Milenkovic B, Stajic JM, Stojic N, Pucarevic M, Strbac S. Evaluation of heavy metals and radionuclides in fish and seafood products. Chemosphere. 2019;229:324–31.
43. Barone G, Giacominelli-Stuffler R, Storelli MM. Comparative study on trace metal accumulation in the liver of two fish species (Torpedinidae): concentration-size relationship. Ecotoxicol Environ Saf. 2013;97:73–7.
44. Unal ME, Wixson BG, Gale N, Pitt JL. Evaluation of toxicity, bioavailability and speciation of lead, zinc and cadmium in mine/mill wastewaters. Chem Speciat Bioavailab. 1998;10(2):37–46.
45. ATSDR. Priority list of hazardous substances. https://www.atsdr.cdc.gov/spl/index.html #modalIdString_-myTable2015; 2015.
46. Neff JM. Ecotoxicology of arsenic in the marine environment. Environ Toxicol Chem. 1997;16 (5):917–27.
47. Foster S, Maher W, Krikowa F. Changes in proportions of arsenic species within an Ecklonia radiata food chain. Environ Chem. 2008;5(3):176–83.
48. Pajany YM, Hurel C, Géret F, Galgani F, Brunet FB, Marmier N, Romeo M. Arsenic in marine sediments from French Mediterranean ports: geochemical partitioning, bioavailability and ecotoxicology. Chemosphere. 2013;90:2730–6.
49. Buka I, Hervouet-Zeiber C. Lead toxicity with a new focus: addressing low-level lead exposure in Canadian children. Paediatr Child Health. 2019;24(4):293.
50. World Health Organization. Lead poisoning and health [Fact sheet]. Media centre. 2017 Aug. http://www.who.int/mediacentre/factsheets/fs379/en/.
51. Beyersmann D, Hartwig A. Carcinogenic metal compounds: recent insight intomolecular and cellular mechanisms. Arch Toxicol. 2008;82:493–512.
52. Kim HJ, Mahboob S, Viayaraghavan P, Al-Ghanim KA, Al-Misned F, Kim YO, Ahmed Z. Determination of toxic effects of lead acetate on different sizes of zebrafish (Danio rerio) in soft and hard water. J King Saud Univ Sci. 2020;32:1390–4.
53. Kershaw J, Hall A. Mercury in cetaceans: exposure, bioaccumulation and toxicity. Sci Total Environ. 2019;694(1):133683.
54. Takeuchi T, Morikawa N, Matsumoto H, Shiraishi Y. A pathological study of Minamata disease in Japan. Acta Neuropathol. 1962;2:40–57.
55. Bernhof RA. Cadmium toxicity and treatment. Sci World J, Hindawi Publishing Corporation. 2013:394652.
56. Wright DA, Welbourn P. Environmental toxicology. In: Cambridge environmental chemistry series. Cambridge University Press; 2002. p 298.

Part VI

Agricultural Water Management

Reuse of Water in Agriculture (Treated Wastewater, Drainage Water)

19

Kemal Güneş, Mehmet Ali Çullu, Mehmet Şimşek, Bülent Topkaya, Mustafa Hakkı Aydoğdu, Mehmet Beşiktaş, and Mehtap Dursun Çelebi

Abstract

It is known that the increasing water demand due to climate change and drought, especially in the Mediterranean basin, will make itself felt more towards the middle of the century. Clearly, this situation will create increasing pressure on the use of treated water and drainage waters, which are alternative water sources in irrigation. Therefore, the establishment of alternative water resource management plans as soon as possible is seen as a necessity, especially for the Mediterranean and MENA regions. In the action plans, it is necessary to highlight issues such as the training of farmers, the creation of irrigation investment plans by considering pressurized irrigation as the conditions allow, and the creation of an irrigation program in accordance with the agricultural plant pattern. Farmers who cannot access water due to lack of sufficient or suitable distribution infrastructure, have already been using wastewater or drainage water as an alternative source for ages. Wastewater was used directly or after passing through certain treatment steps, depending on the treatment technologies that were available during the relevant years. Even today, despite the quite advanced wastewater treatment technologies, wastewater is used mostly directly or after partial treatment in many parts of the world. The reuse of drainage water in irrigation is a common practice in regions where suitable irrigation infrastructure is lacking, or water sources are limited. Direct use of drainage and untreated water in irrigation poses several hazards to humans, the environment and food safety. Such water can be reused in irrigation only after certain precautions are taken in order to protect human health, soil and plants. To this end, preliminary studies must be conducted to investigate under what conditions and on which types of plants treated water will be used, and satisfactory conditions must be met before treatment is performed. Likewise, necessary studies must be conducted before drainage water is reused. Certain regulations set out the required criteria regarding how to use treated wastewater in agriculture. While

K. Güneş (✉) · M. Beşiktaş · M. D. Çelebi
TUBITAK—Marmara Research Center (MRC), Environment and Cleaner Production Institute, GebzeKocaeli, Turkey
e-mail: kemal.gunes@tubitak.gov.tr

M. A. Çullu · M. H. Aydoğdu
Faculty of Agriculture, Department of Soil Science and Plant Nutrition, Harran University, Sanlıurfa, Turkey

M. Şimşek
Faculty of Engineering, Department of Civil Engineering, Sirnak University, Sirnak, Turkey

B. Topkaya
Faculty of Engineering, Department of Environmental Engineering, Antalya University, Antalya, Turkey

M. Bahadir and A. Haarstrick (eds.), *Water and Wastewater Management*,
Water and Wastewater Management, https://doi.org/10.1007/978-3-030-95288-4_19

they may change from country to country, a wholesome approach can be taken as is done in the European Union. The use of treated wastewater and drainage water in agriculture from past to present, problems encountered, necessary infrastructure and legislative gaps are discussed in this study.

Keywords

Treated wastewater reuse · Drainage water reuse · Salinity · Plant tolerance · Nitrogen · Phosphorus · Pesticides

19.1 Introduction

The world's population is increasing rapidly idly and the current world population is around 7.8 billionand is expected to reach 7.8 billion [1] and is expected to reach 9.7 billion in 2050 [2]. This situation requires the sustainable management of existing water and soil resources. However, the expansion of urban areas has steadily eroded the farmlands and threatened food production and security. It is stated that if the pace of urbanization in the world continues in this manner, it will lead to a loss of 1.8–2.4% of global cultivated areas by 2030 [3]. It is an evident paradox that while more croplands are required to meet the increasing need for food caused by the growing population, the total cropland area is decreasing due to the increase in land use for settlement, industry, infrastructure and transportation. A parallel situation exists for water resources. While the demand for water increases due to population growth, pollution, mismanagement and climate change pose a threat to our water supply. This is a serious problem, especially in arid or semi-arid regions. In other words, the increasing population will require additional food and commodities; however, water scarcity is threatening agricultural production. Because of the water shortage, wastewater and irrigation return flow water is being used in some regions of the world. The development of irrigation systems plays an important role both in obtaining more products and against harsh climatic conditions [4, 5].

In arid and semi-arid climate zones worldwide, having a reliable water supply is a crucial to a consistent yield. Throughout the world, irrigation is practiced on 24% of croplands, which produce approximately 40% of the total agricultural commodities [6]. It is estimated that 60% of the world's population may be impacted negatively by water shortages by the year 2025 [7]. Global water consumption at the current level is not sustainable, and it is indicated that 80% of the world's population is under significant threat in terms of hydric security due to climate change, pollution of water resources and the high cost of the water supply, countries' aspirations to produce more for development, degradation of natural resources and the ever-increasing expansion of croplands [8]. The growing demand for irrigation water has led to the use of treated and untreated wastewater for agricultural production in developing countries [9]. As a result of increasing human activities and thus more production in parallel with the growing population in the world, the amount of wastewater tends to increase continuously. In most of the less developed countries, wastewater is released into the environment without treatment and has an adverse effect on human health and the ecosystem. The global freshwater demand has been growing consistently, and it is emphasized that water resources are limited [10]. Because of this increasing demand, approximately 20 million hectares of land in the world are irrigated with untreated or partially treated domestic/industrial wastewater [11–13]. This partially treated or untreated wastewater is mostly used in urban and peri-urban agricultural areas and corresponds to about 11% of the irrigated areas in the world [14].

The use of wastewater in agriculture requires that some conditions be met to preserve both vegetation and human health. Such conditions include not only several treatment parameters but also other irrigation techniques. For example, while the determination of salinity, pathogens, nutrients and heavy metals [15] is prioritized in some studies, pathogens play a more significant role in others. Agriculture is the single largest consumer of freshwater and constitutes approximately 70% of total consumption. As such, the agricultural sector is the one that reuses the most treated wastewater. While it is indicated that the greatest risk related to the reuse of treated water is its contents of salt, nitrogen and pathogens, it is emphasized that the risk from heavy metals, which are the other important contaminants, and of emerging contaminants is lower [16].

19.2 Irrigation Systems Currently Used in Agriculture

There are two ways of supplying water to cultivated plants to help them grow. One of them is rain-fed farming and the other one is irrigation. Irrigation has become even more important today due to the experienced impacts of climate change. However, water resources are becoming increasingly unsuitable with climate change, and a significant proportion of the world's wild and cultivated plants are rain fed. Nevertheless, irrigation is used for agriculture in the remaining regions. There are different methods of irrigation, and the method to be practiced is chosen based on several factors, including the type of plant, climate conditions, the economy, the level of academic background, the sociological structure and the amount of revenue that is expected to be gained from the irrigated product.

Overall, there are 8 main types of irrigation: surface irrigation, localized irrigation, drip irrigation, sprinkler irrigation, centre pivot irrigation, lateral move irrigation, sub-irrigation and manual irrigation [17]. They are presented below.

- **Surface irrigation** refers to systems that deliver water to crops using a gravity-fed flow of water without using any additional power or equipment. This system is divided into the following categories: (i) basin irrigation, (ii) border irrigation, (iii) furrow irrigation and (iv) uncontrolled flooding [18]. The efficiency of this irrigation method ranges between 20 and 95%, depending on the conditions [19]. Among the above-mentioned irrigation methods, the most efficient one is basin irrigation (65–95%), while the least efficient one is the flood irrigation method (20–50%). Figure 19.1 shows surface irrigation applications. In this application, irrigation water use efficiency (IWUE) is recorded as approximately 40%.

- **Localized irrigation** uses a type of drip irrigation technique, which allows water to drip slowly to the roots of plants. It is an irrigation system that saves water, as it loses the minimum amount of water through evaporation. This system can also be named micro, drip or trickle irrigation. The efficiency of this system

Fig. 19.1 Surface irrigation applications (*pictures taken by the author*)

Fig. 19.2 Sprinkler irrigation applications (*pictures taken by the author*)

ranges between 70 and 95%, depending on the conditions [19].

- **Sprinkler irrigation,** which is also known as spray or overhead irrigation, is a method of applying irrigation water in a uniform manner through pressure by using irrigation sprinklers or perforated pipes. The efficiency obtained from field studies using this system is reported to be 75% [20]. An example image of this irrigation system is shown in Fig. 19.2.
- **Centre pivot irrigation** is an irrigation technique in which the equipment rotates around a pivot (circular) and that can be used effectively to irrigate very large fields. The aim of this system is to ensure equal application of water onto the field through the hole diameters of the nozzle, which increase from the centre to the outer parts. They are more expensive than other systems due to the initial investment and power consumption, and they are used to irrigate large areas. This system, which is a type of sprinkler irrigation, can be expected to yield an efficiency of 75%.
- **Lateral move irrigation** refers to an irrigation system in which multiple irrigation nozzles are placed on a line that moves laterally. It is used to irrigate large areas in the shape of a rectangular or square. It differs from centre pivot irrigation in the sense that it moves laterally. Initial investment costs, power costs and other operational conditions are similar to those of the pivot irrigation system.
- **Sub-irrigation (subsurface drip irrigation)** implies a low-pressure system that uses a network of drip tubes buried at certain distances and into a certain depth depending on the type of plants to be watered in order to distribute water to the soil in a uniform manner. As this system prevents significant loss of water through evaporation, its irrigation efficiency is much higher than of other systems. Irrigation efficiencies of overhead irrigation and subsurface drip irrigation in cotton cultivation were compared in a study conducted in Georgia (USA) in two different locations during the years 2004–2005. Subsurface drip irrigation efficiency was found to be higher than overhead irrigation by 23% and 15% [21].
- **Manual irrigation** refers to an irrigation system based completely on human labour. Water is moved plant to plant or applied manually to a certain area. While it is a highly efficient method of irrigation in comparison with methods other than modern irrigation, the efficiency may change depending on the knowledge of the person who performs it, the sufficiency of water resource and the type of plant.

19.3 Irrigation Systems and Socio-Economic Structure Interactions

There is a direct correlation between how advanced irrigation systems are and economic development. This is generally and equally true for both countries and farmers, because the costs of modern irrigation systems are quite high. Such costs vary depending on local conditions, e.g., in Turkey, a pressure irrigation system with a length of 1 km cost approximately USD 1370 to 2400 in 2016. According to data from the year 2000, the average total unit costs of newly built irrigation systems in Sub-Saharan Africa, Non-SSA, Middle East & North Africa, South Asia, Southeast Asia, East Asia and Latin America were 14,455, 6590, 8780, 3393, 9709, 8221 and 4903 USD/ha, respectively [22]. According to a recent study, the 'capital costs' of various irrigation systems have been determined. Considering the year 2014, the capital costs for gravity channel surface irrigation (not pumped) were 6000 USD/ha, pumped pipe and riser surface irrigation 7500 USD/ha, centre pivot irrigation 6500 USD/ha and drip irrigation 10,000 USD/ha [23]. Nonetheless, the maintenance costs of those irrigation systems are higher than those of conventional systems. Annual maintenance costs of centre-pivot, set-sprinkler (sprinkler irrigation), and drip irrigation, which are among pressure irrigation systems, are 35–21, 75–70 and 120–112 €/ha, respectively [24]. Due to the initial investment and operational costs of modern irrigation systems, these effective systems can be included in public investment programs insofar as the economic conditions of countries allow, and they span a long period of time. Not only governments but also different associations or manufacturers themselves can invest in irrigation of fertile croplands that are expected to yield high agricultural revenue. However, irrigation systems are generally built by governments so that they can be more wide-ranging and their management is assigned to private or non-profit associations or NGOs such as irrigation unions. Different tariffs are applied on water use in order to meet some government investment costs. These changes depending on the conditions within countries, and different tariffs can even be applied in the same country, depending on access to water. For example, there is such a case in Portugal [25]. Several studies were conducted in Portugal starting from the 1980s regarding water tariffs and based on the experience obtained from such studies. Water tariffs are divided into three categories today, which are fixed tariffs (14–211 €/ha), water budget charge (0.0093–0.0906 €/m^3) and water markets (formal-informal) [25]. In Turkey, the average irrigation water tariff is approximately 100 USD/ha [26].

Socio-economic underdevelopment hinders the profitable use of irrigation systems. Especially in places where conventional irrigation systems are used, providing local people with training as to how to use profitable irrigation techniques is one of the most important milestones towards management of water resources, as lack of knowledge in such areas that already suffer from low irrigation efficiency would inevitably lead to greater wasting of water resources. Nevertheless, farmers must be provided with more training opportunities regarding other modern irrigation techniques, and it must be ensured that such training opportunities become continuous. In general terms, simultaneously with inclusion of a certain cropland in an investment program for irrigation, a training program must be initiated for farmers in the relevant region. Irrigation trainings must be specific to the types of plants that are currently available and planned to be cultivated in such a region. Accordingly, training must be given by experts who are very well acquainted with local people and their sociological structure. Training must be conducted during seasons when agricultural activities are low so that they can be more engaging for farmers. Trainings should be given not only by irrigation experts, but also with other fields of expertise related to agriculture. In addition, the trainings should be carried out in coordination by a team of experienced persons who know the socio-economic structure of the region to be irrigated.

19.4 The Relationship Between Climate Change and Irrigation Water

Due to climate change, wet areas are becoming even wetter, and arid areas are becoming drier. While 3.6 billion people are facing water scarcity today, it is predicted that this will reach 4.8–5.7 billion in 2050 [27]. In addition, sudden changes in temperature and seasonal differences lead to irregular precipitation regimes. This causes the agricultural sector, which is already under heavy pressure because it must supply food products to the increasing population, to turn fields into croplands and results in higher demand for water and water resources. For this reason, the opening of forest areas and other areas that are not suitable for agriculture, especially in developing countries, causes the need for more water. As a result, it is inevitable that the efficiency of water resources will decrease gradually, and thus it is a vicious circle. Despite many uncertainties, several studies conducted using current models show that climate change will increase the demand for irrigation significantly. According to a study conducted based on the year 2000, it is predicted that irrigated cropland area will increase by 45%, and this will occur mostly in developing countries [28]. For example, in a study conducted to determine the irrigation demand of corn and soybean with projections that span long periods (2020–2039 and 2060–2079) by using SWAT (Soil and Water Assessment Tool) at Kalamazoo River Watershed (Michigan/USA), it was indicated that there would be an increased demand for irrigation from 2020–2039 to 2060–2079, but it was established that there would be a decreased demand for such products [29]. Another study argues that the greater the global warming effect, the higher the irrigation water demand will be, and significant parts of southern and eastern Asia particularly will suffer from water scarcity [30]. In several studies, it is suggested that while climate change will not cause any remarkable change in total irrigation water demand, daily and annual irrigation water demand will increase, or there will be significant changes in plants' needs for irrigation water [31–33].

Nonetheless, it is indicated that developments in irrigation systems and irrigation conveyance infrastructure can compensate for the needed irrigation water demand in parallel with climate change and population growth [34]. Some research shows that reusing treated wastewater in agriculture can help reduce the effects of climate change. It is also reported that reuse of treated wastewater would lead to a decrease in greenhouse gas emissions in the global life-cycle system by 33% [35].

19.5 Reuse of Treated Wastewater in Agriculture

In a study conducted on the irrigation of some plant species in Crete with wastewater from a domestic wastewater treatment plant, it was found that the wastewater used did not meet the criteria specified in the regulation because it was not the water coming out of the tertiary treatment plant. And it is pointed out that if they meet legislative criteria, treated water will meet 4.3% of the total irrigation water demand [36].

According to a study investigating the effects of treated wastewater and treated brackish water on two wheat genotypes in Saudi Arabia, it was stated that the best results were obtained using treated wastewater after examining the pre-harvest parameters in terms of yield components, yield characteristics and protein, micronutrient and heavy metal content, which are the indicators of grain quality [37]. In a study that was conducted in Palestine, The West Bank, regarding difficulties encountered in reuse of wastewater in agriculture, technical, legal, social and economic challenges was evaluated. The most prominent finding was that the community must be informed about the reuse of treated water in legal, social, economic and institutional terms to raise awareness [38]. In a study that addressed Jordan, Syria and Morocco, which are in the Southern and Eastern Mediterranean regions, which suffer the most from water scarcity, it was discussed that climate change will increasingly threaten water resources in the future, but sustainable development models can reverse such effects [39]. However, it is argued that this requires a combination of technical, managerial, economic, social and institutional changes to be implemented in the region in order to encourage structural change. In a survey conducted in Turkey with 500 participants regarding wastewater reuse, 375 participants indicated that they find wastewater reuse acceptable if precautions are taken as required for health [40]. Evidently, if precautions are taken or it is ensured that people are informed to raise awareness and have trust, the reuse of treated water will be more acceptable.

One of the most important criteria for the reuse of treated wastewater is to make sure that its bacteriological content is below a certain level. Relevant values may vary from guideline to guideline. It is indicated that if treated wastewater is to be used for urban, agricultural or industrial purposes, the level of chlorination must be over 4 ppm, while UV can be used as an alternative method of disinfection for other recreational and environmental purposes [41]. There are several studies on different treatment methods for the reuse of treated water in agriculture. For example, in the Czech Republic, water that was obtained from a hybrid wetland system, which produced significantly high removal of Biological Oxygen Demand (BOD_5), Chemical Oxygen Demand (COD), Ammonia (NH_4^+), Total Nitrogen (TN), Total Suspended Solids (TSS) and bacteriological parameters, was used to partially irrigate tomato, potato and lettuce, and it had a fertilizing effect on plants. However, it is emphasized in such studies that reuse of treated water requires due examination for bacteriological content [42]. In a study that was conducted in Qom, Iran regarding reuse of treated water for irrigation, 39 parameters in wastewater were monitored for a year, and the findings were evaluated as per the latest standards on the world irrigation index and Wilcox diagram. Evaluations showed that values were over limits for some parameters (turbidity, total suspended solids, electrical conductivity, sodium, detergents, total coliforms and faecal coliforms, ammonium, residual sodium carbonate) [43].

In recent years, reuse of wastewater has been addressed more rigorously in the Middle East. It is underlined that the reuse of treated wastewater in agriculture is more economic than the reuse of water obtained from desalination systems [44]. Likewise, in Israel, models that are specific to Israel are developed, and the extents of possible environmental and economic outcomes are explained in detail. It is mentioned that if the treated water is used in agriculture, the sea and brackish-water desalination for this purpose will decrease and according to the result of simulating this over a 3-decade period, it will provide an additional welfare of 3.3 billion USD to Israel [45].

19.6 Wastewater Reuse Policy and Standards

Only less than 10% of the entire wastewater collected in the world is treated through any treatment system [46]. Therefore, it is understood that more than 90% of the wastewater collected in the world is either used for direct irrigation or discharged directly to the receiving environment. According to the data of 2002, the average annual water consumption per capita in the world was 617 m^3/yr. This varied between 1630 m^3/yr in North America and 256 m^3/yr in Africa [47]. Considering that the world population is approximately 7.8 billion, the annual amount of wastewater generated in the world is approximately 4.8×10^{12} m^3. Assuming that 10% of this calculated value is treated, it can be said that the amount of wastewater discharged directly to the receiving environment is 4.3×10^{12} m^3. A certain amount of this is used for direct irrigation.

Wastewater that is discharged into receiving environments reaches lakes or seas through streams. However, raw wastewater that joins the stream is diluted depending on stream's flow rate. Nevertheless, they pose a risk because of the contaminants, primarily pathogens that they might contain. In a study conducted in Musi River, where the wastewater of the city of Hyderabad (in India) is discharged, the amount of helminth eggs (*Ascaris*, hookworm, and *Trichuris*) detected at the wastewater discharge point was 133 eggs/L, while the value determined 27.7 km from the discharge point was recorded as 0.1 eggs/L [46]. As it is seen, wastewater can contain measurable pathogens although they run a long way. The risk of such pathogens, which are sensitive to heat, is higher especially during summer months. Such risk is the highest when plants which are edible in raw form are irrigated. Nonetheless, farm laborers or farmers conducting irrigation can also be adversely affected. The amount of agricultural return flow that reaches drainage channels also increases during the summer months, when the intensity of irrigation activities reaches its peak, depending on the method of irrigation. The occupancy of a drainage channel during the intensive irrigation period is given in Fig. 19.3.

When untreated wastewater blends with drainage channels, such water that contains not only pathogens but also organic and inorganic fertilizers threatens the receiving environment and human health. Wastewater, regardless of whether it is treated or not, must meet certain criteria in terms of not only nutrients and pathogens, but also other chemicals that they may contain. Although pathogens are the most important criterion for reuse of treated wastewater in irrigation, it must meet the criteria set for other parameters in order to ensure the health of the soil, plants and the receiving medium. In this regard, the European Union, WHO and USEPA guidelines are given in Table 19.1 as an example. 'Guidelines for interpretations of water quality for irrigation', which was prepared comprehensively by FAO in 1985, is available in Table 19.2. This guideline is still applicable. Yet, this guideline also addresses potential irrigation problems, parameters that affect infiltration, specific ion toxicity, trace elements and other factors, and presents explanations about criteria.

Another issue is knowledge about current land conditions, in other words, soil conditions before application of water or treated wastewater onto a

Fig. 19.3 Peak status of a drainage channel during irrigation period (*pictures taken by the author*)

Table 19.1 EU, WHO and USEPA guidelines regarding reuse of treated wastewater in agriculture

	Reclaimed water quality class	Indicative technology target	Quality criteria					Lit
			E. coli (cfu/100 mL)	BOD_5 (mg/L)	TSS (mg/L)	Turbidity (NTU)	Additional criteria	
EU Guidelines	Class A	Secondary treatment, filtration, and disinfection (advanced water treatment)	$\leq$10 or below dedection limit	$\leq$ 10	$\leq$ 10	$\leq$ 5	*Legionella* spp.: $\leq$ 1000 °Cfu/L, when there is a risk of aerosol formation Intestinal nematodes (helminth eggs): $\leq$1 egg/L, when irrigation of pastures or fodder for livestock is intended	[48]
	Class B	Secondary treatment, and disinfection	$\leq$100	According to Directive 91/271/EEC	According to Directive 91/271/EEC	–		
	Class C	Secondary treatment, and disinfection	$\leq$1,000	According to Directive 91/271/EEC	According to directive 91/271/EEC	–		
	Class D	Secondary treatment, and disinfection	$\leq$10,000	According to Directive 91/271/EEC	According to directive 91/271/EEC	–		

	Exposure scenario	Health based target (DALY* per person per year)	Log_{10} pathogen reduction needed	Number of helminth eggs per liter		Lit
WHO—Health based targets for wastewater use in agriculture	Unrestricted irrigation Lettuce onion	$\leq 10^{-6a}$	6 7	$\leq 1^{b,c}$		[11]
	Restricted irrigation Highly mechanized labour intensive	$\leq 10^{-6a}$	3 4	$\leq 1^{b,c}$		
	Localized (drip) Irrigation High growing crops Low growing crops	$\leq 10^{-6a}$	2 4	No reccommendation [d] $\leq 1^{c}$		

	Agricultural reuse	**Treatment**	**Reclaimed water quality**	**Reclaimed water monitoring**	**Setback distance**	Lit
EPA—Suggested guidelines for water reuse/Agricultural reuse	**Food crops** The use of reclaimed water for surface or spray irrigation of food crops, which are intended for human consumption, consumed raw	• Secondary • Filtration • Disinfection	• pH = 6.0–9.0 • BOD $\leq$ 10 mg/L • NTU $\leq$ 2 • Fecal coliform/100 mL • Not detectable • Cl_2 residual 1 mg/L (min.)	• pH—weekly • BOD—weekly • Turbidity—continuously • Fecal coliform—daily • Cl_2 residual—continuously	50 ft (15 m) to potable water supply wells; increased to 100 ft (30 m) when located in porous media	[49]
	Processed Food Crops The use of reclaimed water for surface irrigation of food crops, which are intended for human consumption, commercially processed **Non-Food Crops** The use of reclaimed water for irrigation of crops, which are not consumed by humans, including fodder, fiber, and seed crops, or to irrigate pasture land, commercial nurseries, and sod farms	• Secondary • Disinfection	• pH = 6.0–9.0 • BOD $\leq$ 30 mg/L • TSS $\leq$ 30 mg/L • Fecal coliform/100 mL • $\leq$200 • Cl_2 residual 1 mg/L (min.)	• pH—weekly • BOD—weekly • TSS—daily • Fecal coliform—daily • Cl_2 residual – continuously	300 ft (90 m) to potable water supply wells 100 ft (30 m) to areas accessible to the public (if spray irrigation)	

Remarks:
*Disability adjusted life year
[a]Rotavirus reduction. The health-based target can be achieved, for unrestricted and localized irrigation, by a 6-7 log unit pathogen reduction (obtained by combination of wastewater treatment and other health protection measures); for restricted irrigation, it is achieved by a 2-3 log unit pathogen reduction.
[b]When children under 15 are exposed, additional health protection measures should be used (e.g. treatment to $\leq$0.1 egg per litre, protective equipment such as gloves or shoes/boots or chemotherapy).
[c]An aritmetic mean should be determined throughtout the irrigation season. The mean value of $\leq$ 1egg per litre should be obtained for at least 90% of samples in order to allow for the occasional value sample (i.e. with >10 eggs per litre). With some wastewater treatment processes (e.g. waste stabilization ponds), the hydraulic retention time can be used as a surrogate to assure compliance with $\leq$1 egg per litre.
[d]No crops to be picked up from the soil

Tab. 19.2 Guidelines for interpretations of water quality for irrigation (Adapted from FAO 1985) [50]

Potential irrigation problem		Units	Degree of restriction on use		
Salinity *(affects crop water availability)*[1]			None	Slightly to Moderate	Severe
	EC_w	dS/m	<0.7	0.7–3.0	>3.0
	TDS	mg/l	<450	450–2000	>2000
Infiltration *(affects infiltration rate of water into the soil; evaluated as Σ EC_w + SAR together)*[2]					
SAR[3]	0–3	**EC^3_w**	>0.7	0.7–0.2	<0.2
	3–6		>1.2	1.2–0.3	<0.3
	6–12		>1.9	1.9–0.5	<0.5
	12–20		>2.9	2.9–1.3	<1.3
	20–40		>5.0	5.0–2.9	<2.9
Specific Ion Toxicity *(affects sensitive crops)*					
	Sodium (Na)[4]				
	surface irrigation	SAR	<3	3–9	>9
	sprinkler irrigation	me/L	<3	> 3	
	Chloride (Cl)[4]				
	surface irrigation	me/L	<4	4–10	>10
	sprinkler irrigation	me/L	<3	>3	
	Boron (B)[5]	mg/L	<0.7	0.7–3.0	>3.0
Miscellaneous Effects *(affects susceptible crops)*					
	Nitrogen (NO_3–N)[6]	mg/L	<5	5–30	>30
	Bicarbonate (HCO_3) (overhead sprinkling only)	me/L	<1.5	1.5–8.5	>8.5
	pH		**Normal range 6.5–8.4**		

Trace elements in irrigation water

Element	Recommended max. conc.[7] (mg/L)	Remarks
Al (aluminium)	5.0	Can cause non-productivity in acid soils (pH < 5.5), alkaline soils (pH > 7.0) precipitate the ion and eliminate its toxicity
As (arsenic)	0.10	Toxicity to plants varies widely, ranging from 12 mg/L for Sudan grass to less than 0.05 mg/L for rice
Be (beryllium)	0.10	Toxicity to plants varies widely, ranging from 5 mg/L for kale to 0.5 mg/L for bush beans
Cd (cadmium)	0.01	Toxic to beans, beets and turnips at concentrations as low as 0.1 mg/L in nutrient solutions. Conservative limits recommended due to its potential for accumulation in plants and soils to concentrations that may be harmful to humans
Co (cobalt)	0.05	Toxic to tomato plants at 0.1 mg/L in nutrient solution. Tends to be inactivated by neutral and alkaline soils
Cr (chromium)	0.10	Not generally recognized as an essential growth element. Conservative limits recommended due to lack of knowledge on its toxicity to plants
Cu (copper)	0.20	Toxic to a number of plants at 0.1 to 1.0 mg/L in nutrient solutions
F (fluoride)	1.0	Inactivated by neutral and alkaline soils

(continued)

Tab. 19.2 (continued)

Trace elements in irrigation water		
Element	Recommended max. conc.[7] (mg/L)	Remarks
Fe (iron)	5.0	Not toxic to plants in aerated soils, but can contribute to soil acidification and loss of availability of essential phosphorus and molybdenum. Overhead sprinkling may result in unsightly deposits on plants, equipment and buildings
Li (lithium)	2.5	Tolerated by most crops up to 5 mg/L; mobile in soil. Toxic to citrus at low concentrations (<0.075 mg/L). Acts similarly to boron
Mn (manganese)	0.20	Toxic to a number of crops at a few-tenths to a few mg/L, but usually only in acid soils
Mo (molybdenum)	0.01	Not toxic to plants at normal concentrations in soil and water. Can be toxic to livestock if forage is grown in soils with high concentrations of available molybdenum
Ni (nickel)	0.20	Toxic to a number of plants at 0.5 mg/L to 1.0 mg/L; reduced toxicity at neutral or alkaline pH
Pb (lead)	5.0	Can inhibit plant cell growth at very high concentrations
Se (selenium)	0.02	Toxic to plants at concentrations as low as 0.025 mg/L and toxic to livestock if forage is grown in soils with relatively high levels of added selenium. An essential element to animals but in very low concentrations
Sn (stannum)		
Ti (titanium)	–	Effectively excluded by plants; specific tolerance unknown
W (tungsten)		
V (vanadium)	0.10	Toxic to many plants at relatively low concentrations
Zn (zinc)	2.0	Toxic to many plants at widely varying concentrations; reduced toxicity at pH > 6.0 and in fine textured or organic soils

[1]EC_w means electrical conductivity, a measure of the water salinity, reported in deciSiemens per meter at 25 °C (dS/m) or in units millimhos per centimeter (mmho/cm)
Both are equivalent. TDS means total dissolved solids, reported in milligrams per liter (mg/L).
[2]SAR is the sodium adsorption ratio; at a given SAR, infiltration rate increases as water salinity increases. SAR is sometimes reported by the symbol RNa.
[3]It indicates the joint effect of EC_w and SAR parameters in irrigation water on infiltration capacity. A high SAR value combined with a low salt content (low EC_w) means that there will be high rates of infiltration problems. However, low EC_w or high SAR can act separately or together to disperse soil aggregates, ultimately reducing the number of large pores in the soil [51].
[4]For surface irrigation, most tree crops and woody plants are sensitive to sodium and chloride; use the values shown. Most annual crops are not sensitive; salinity tolerance tables can be viewed at http://www.fao.org/3/t0234e/T0234E03.htm. For chloride tolerance of selected fruit crops can be viewed at http://www.fao.org/3/T0234E/T0234E05.htm. With overhead sprinkler irrigation and low humidity (<30%), sodium and chloride may be absorbed through the leaves of sensitive crops. For crop sensitivity to absorption can be viewed at http://www.fao.org/3/T0234E/T0234E05.htm.
[5]For boron tolerances, can be viewed at http://www.fao.org/3/T0234E/T0234E05.htm.
[6]NO_3–N means nitrate nitrogen reported in terms of elemental nitrogen (NH_4 -N and Organic-N should be included, when wastewater is being tested).
[7]The recommended maximum concentration is based on a water application rate, which is consistent with good irrigation practices (10,000 m^3/ha*yr).
If the water application rate greatly exceeds this, the maximum concentrations should be adjusted downward accordingly.
No adjustment should be made for application rates less than 10,000 m^3/ha*yr. The values given are for water used on a continuous basis at one site.

Table 19.3 Maximum tolerable soil concentrations of various toxic chemicals based on human health [11]

Chemicals (Element)	Soil concentration (mg/kg)	Organic Pollutants	Soil concentration (mg/kg)
Antimony	36	Aldrin	0.48
Arsenic	8	Benzene	0.14
Barium[a]	302	Cholordane	3
Beryllium[a]	0.2	Cholorobenzene	211
Boron[a]	1.7	Chloroform	0.47
Cadmium	4	2,4-D	0.25
Fluorine	635	DDT	1.54
Lead	84	Dichlorobenzene	15
Mercury	7	Dieldrin	0.17
Molybdenum[a]	0.6	Dioxins TEF	0.00012
Nickel	107	Heptachlor	0.18
Selenium	6	Hexachlorobenzene	1.40
Silver	3	Lindane	12
Thallium[a]	0.3	Methoxychlor	4.27
Vanadium[a]	47	PCBs	0.89
		PAHs (as benzo(a)pyrene)	16
		Pentachlorophenol	14
		Phthalate	13,733
		Pyrene	41
		Styrene	0.68
		2,4,5-T	3.82
		Tetrachloroethylene	0.54
		Toluene	12
		Toxaphene	0.0013
		Trichloroethane	0.68

[a]The computed numerical limits for these elements are within the ranges that are typical for soils

land. In this sense, analyses that are mentioned in Table 19.3 must be conducted beforehand in the area to be irrigated, and the soil's tolerance to the ingredients that may possibly be contained in irrigation water must be evaluated.

19.7 Conclusions

The importance of alternative water resources becomes even more evident, when it is considered that the demand for freshwater is ever increasing, and climate change scenarios show that drought will have greater impacts within the next decades, especially in the Mediterranean Region. As approximately two-thirds of freshwater resources used on the world is consumed by the agricultural sector, effective reuse of treated water, which is one of the alternative water resources, would diminish, though slightly, intense use of limited freshwater resources. Regions which suffer from water scarcity would feel diminishment strongly. Besides, recycling of nutrients contained in treated water will both reduce the waste of our resources and minimize environmental pollution. However, precautions must be taken as required to ensure human and environmental health before wastewater is used for irrigation. Therefore, developing countries must issue their own legislation or adapt a current guideline in line with their domestic conditions in order to regulate reuse of wastewater in irrigation. Furthermore, farmers who take part in irrigation activities must be provided with training opportunities and other incentives at regular intervals, and audits must be conducted accordingly. Moreover, governments must support the relevant authorities, especially regarding domestic wastewater treatment and providing subventions regarding reuse of treated water for irrigating agricultural or recreational fields will contribute to proper and responsible management of wastewater.

Apart from the above-mentioned issues, the rate of freshwater used in agriculture can be reduced, if and only if investments are made in pressurized irrigation systems. Therefore, alternative water resources must be investigated on one hand, while precautions and investments that would prevent wasting of water must be prioritized on the other.

References

1. United Nations Population Fund. World Population Dashboard, vol 16; 2021. p 46. https://www.unfpa.org/data/world-population-dashboard. Accessed 12 Oct 2021.
2. United Nations, Department of economic and social affairs, population division. world population prospects 2019: highlights, ST/ESA/SER.A/423; 2019.
3. Brend'Amour C, Reitsma F, Baiocchi G, Barthel S, Güneralp B, Erb K-H, Haberl H, Creutzig F, Seto KC. Future urban land expansion and implications for global croplands. Proc Natl Acad Sci USA. 2016;114(34):8939–44.
4. Challinor A, Martre P, Asseng S, Thornton P, Ewert F. Making the most of climate impact ensembles. Nat Clim Chang. 2014;4:77–80.
5. Troy TJ, Konar M, Srinivasan V, Thompson S. Moving socio-hydrology forward: a synthesis across studies. Hydrol Earth Syst Sci. 2015;19:3667–79. https://doi.org/10.5194/hess-19-3667-2015.
6. Siebert S, Doll P. Quantifying blue and green virtual water contents in global crop production as well as potential production losses without irrigation. J Hydrol. 2010;384(3–4):198–217.
7. Rijsberman FR. Water scarcity: fact or fiction? Agric Water Manage. 2006;80:5–22.
8. Jacobi PR, Empinotti VL, Schmidt L. Water scarcity and human rights. Ambiente Sociedade. 2016;19 (1):1–5. https://doi.org/10.1590/1809-4422ASOCeditorialV1912016.
9. Drechsel P, Scott CA, Raschid-Sally L, Bahri A, Mara D, Redwood M, Bahri A. Wastewater irrigation and health: assessing and mitigating risk in low-income countries. International Water Management Institute and International Development Research Centre. London, Sterling, VA. UK: Earthscan; 2010. pp 381–94.
10. WWAP (United Nations World Water Assessment Programme). The United Nations World Water Development Report 2017. Wastewater: The Untapped Resource. Paris, UNESCO; 2017. ISBN 978-92-3-100201-4.
11. WHO. Guidelines for the safe use of wastewater, excreta and greywater. Volume 2: wastewater use in agriculture. World Health Organization, Geneva, Switzerland; 2006.
12. Hamilton AJ, Stagnitti F, Xiong X, Kreidil SL, Benke KK, Maher P. Wastewater irrigation: the state of play. Vadose Zone J. 2007;6:823–40.
13. Zhang Y, Shen Y. Wastewater irrigation: past, present, and future. WIREs Water. 2017;6:1–6. https://doi.org/10.1002/wat2.1234.
14. Thebo AL, Drechsel P, Lambin EF. Global assessment of urban and peri-urban agriculture: irrigated and rainfed croplands. Environ Res Lett. 2014;9 (11):114002(9 pp).
15. Norton-Brandão D, Scherrenberg SM, van Lier JB. Reclamation of used urban waters for irrigation purposese. a review of treatment technologies. J Environ Manage. 2013;122:85–98.
16. Chen W, Lu S, Jiao W, Wang M, Chang AC. Reclaimed water: a safe irrigation water source? Environ Dev. 2013;8:74–83.
17. Centers for Disease Control and Prevention, National Center for Emerging and Zoonotic Infectious Diseases (NCEZID), Division of Foodborne, Waterborne, and Environmental Diseases (DFWED), October 11, 2016. https://www.cdc.gov/healthywater/other/agricultural/types.html. Accesed 26 Aug 2021,12:11.
18. Pazouki E. A practical surface irrigation system design based on volume balance model and multi-objective evolutionary optimization algorithms. Agric Water Manage. 2021;248:106755.
19. Gunes K. Agricultural water resources management environmental and financial evaluation –a case study. In: International workshop on "Water perspectives in emerging countries" Water-Energy-Food NEXUS in MENA Region. 2018, Nowember 11–17, Aswan, Egypt. Network partners: Technische Universitat Braunschweig and International Network on Sustainable Water Management in Developing Countries (SWINDON). Funded by: exceed and DAAD.
20. FAO. http://www.fao.org/3/t7202e/t7202e08.htm. Accessed 26 Aug 2021, 17:43.
21. Whitaker JR, Ritchie GL, Bednarz CW, Mills CI. Cotton subsurface drip and overhead irrigation

efficiency, maturity, yield, and quality. Agron J. 2008;100(6):1763–8.
22. Inocencio A, Kikuchi M, Merrey D, Tonosaki M, Maruyama A, de Jong I, Sally H, Penning de Vries F. Lessons form irrigation investment experiences; cost reducing and performance enhancing options for Sub-Saharan Africa. Final report submitted by International Water Management Institute, August 2005; 2005.
23. RMGC, Comparison of irrigation system costs—update, 2018. https://www.gbcma.vic.gov.au/downloads/Farm_Water_Program/2019%20-%20Comparison%20of%20irrigation%20system%20costs.pdf. Accessed 19 Sept 2021, 15:30.
24. Rodrigues GC, Paredes P, Goncalves JM, Alves I, Pereira LS. Comparing sprinkler and drip irrigation systems for full and deficit irrigated maize using multicriteria analysis and simulation modelling: ranking for water saving vs farm economic returns. Agric Water Manage. 2013;126:85–96.
25. Pereiara H, Marques RC. Irrigation water tariffs: lessons for Portugal. Water Policy. 2020;22:887–907.
26. Cakmak EH. Agricultural water pricing: Turkey. OECD; 2010.
27. WWDR, Nature–based solutions for water. The United Nations World Water Development Report. Printed by UNESCO, Paris. ISBN 978-92-3-100264-9; 2018.
28. Fischer G, Tubiello FN, van Velthuizen H, Wiberg DA. Climate change impacts on irrigation water requirements: effects of mitigation, 1990–2080. Technol Forecast Soc Change. 2007;74:1083–107.
29. Woznicki SA, Nejadhashemi AP, Parsinejad M. Climate change and irrigation demand: uncertainty and adaptation. J Hydrol Reg Stud. 2015;3:247–64.
30. Haddeland I, Heinke J, Biemans H, Eisner S, Flörke M, Hanasaki N, Konzmann M, Ludwig F, Masaki Y, Schewe J, Stacke T, Tessler ZD, Wada Y, Wisser D. Global water resources affected by human interventions and climate change. PNAS. 2014;111(9):3251–6.
31. Elgaali E, Garcia LA, Ojima DS. High resolution modeling of the regional impacts of climate change on irrigation water demand. Clim Change. 2007;84:441–61.
32. Shahid S. Impact of climate change on irrigation water demand of dry season Boro Rice In Northwest Bangladesh. Clim Change. 2011;105:433–53.
33. Gorguner M, Kavvas ML. Modeling impacts of future climate change on reservoir storages and irrigation water demands in a Mediterranean Basin. Sci Total Environ. 2020;748:141246.
34. Fader M, Shi S, von Bloh W, Bondeau A, Cramer W. Mediterranean irrigation under climate change: more efficient irrigation needed to compensate for increases in irrigation water requirements. Hydrol Earth Syst Sci. 2016;20:953–73.
35. Miller-Robbie L, Ramaswami A, Amerasinghe P. Wastewater treatment and reuse in urban agriculture: exploring the food, energy, water, and health nexus in Hyderabad, India. Environ Res Lett. 2017;12. https://doi.org/10.1088/1748-9326/aa6bfe.
36. Agrafioti E, Diamadopoulos E. A strategic plan for reuse of treated municipal wastewater for crop irrigation on the island of crete. Agric Water Manag. 2012;105:57–64.
37. Alderfasi AA. Agronomic and economic impacts of reuse secondary treated wastewater in irrigation under arid and semi-arid regions. World J Agric Sci. 2009;5(3):369–74.
38. Mizyed NR. Challenges to treated wastewater reuse in arid and semi-arid areas. Environ Sci Policy. 2013;25:186–95.
39. Varela-Ortega C, Esteve P, Blanco I, Carmona G, Ruiz J; Rabah T. Assessment of socio-economic and climate change effects on water resources and agriculture in Southern and Eastern Mediterranean Countries. MEDPRO Technical Report No. 28/March, 2013. EU-7. Framework Programme. Funded under Socio-ecomnomic Sciences & Humanities.
40. Buyukkamaci N, Alkan HS. Public acceptance potential for reuse applications in Turkey. Resour Conserv Recycl. 2013;80:32–5.
41. Gomez-Lopez MD, Bayo J, Garcia-Cascales MS, Angosto JM. Decision support in disinfection technologies for treated wastewater reuse. J Clean Prod. 2009;17:1504–11.
42. Seres M, Innemanova P, Hnatkova T, Rozkosny M, Stefanakis A, Semerad J, Caithaml T. Evaluation of hybrid constructed wetland performance and reuse of treated wastewater in agricultural irrigation. Water. 2021;13:1165.
43. Rahimi MH, Kalantari N, Sharifidoost M, Kazemi M. Quality assessment of treated wastewater to be reused in agriculture. Global J Environ Sci Manage. 2018;4(2):17–230.
44. Jasim SY, Saththasvisam J, Loganathan K, Ogunbiyi OO, Sarp S. Reuse of treated sewage effluent (TSE) in Qatar. J Water Process Eng. 2016;11:174–82.
45. Reznik A, Feinerman E, Finkelshtain I, Fisher F, Huber-Lee A, Joyce B, Kan I. Economic implications of agricultural reuse of treated wastewater in Israel: a statewide long-term perspective. Ecol Econ. 2017;135:222–33.
46. Thebo AL. Wastewater reuse in irrigated agriculture: global perspectives on water quantity, quality, and exposure to health risks, 2016. A dissertation submitted in partial satisfaction of the requirements of the degree of Doctor of Philosophy in Engineering—Civil and Environmental Engineering in the Graduate Division of the University of California, Berkeley.
47. URL: https://stats.areppim.com/archives/print_water20080823.pdf. Accessed 30 Aug 2021.
48. URL, Minimum quality requirements for water reuse in agricultural irrigation and aquifer recharge. Towards a water reuse regulatory instrument at EU level. https://doi.org/10.2760/804116; 2017. https://publications.jrc.ec.europa.eu/repository/handle/JRC109291. Accessed 19 Aug 2021.

49. EPA, USEPA (US Environmental Protection Agency). Guidelines for water reuse; US Environmental Protection Agency: Anchorage, AK, USA; 2012.
50. (FAO) Food and Agriculture Organization of the United Nations. Water quality for agriculture—FAO irrigation and drainage paper, 29 Rev. 1. Reprinted 1989, 1994: M-56 ISBN 92-5-102263-1, Rome, Italy; 1985. http://www.fao.org/3/t0234e/t0234e00.htm. Accessed 9 Sept 2021, 15:20.
51. Grattan SR. Irrigation water salinity and crop production. University of California, Agriculture and National Resources – National Resources Conservation Service, 2002, Publication 8066, fwqp reference sheet 9.10. https://doi.org/10.3733/ucanr.8066.

Irrigation Management by Using Digital Technologies

20

Eyüp Selim Köksal, Emre Tunca, and Sakine Çetin Taner

Abstract

Population growth and changes in people's consumption habits cause an increase in the need for agricultural production. All over the world, generally, most of the water resources under control are used for irrigation in agriculture. Especially in arid and semi-arid regions, plant production without irrigation is often not economically sustainable. Although most of the agricultural needs are provided from irrigated agricultural areas, the amount of crops produced per unit of water is relatively low due to the low irrigation efficiency. Pressurized and modern irrigation systems are supported and encouraged by governments in many countries. However, the widespread use of these systems cannot provide the expected increase in irrigation efficiency, because irrigation management is more critical than irrigation infrastructure in irrigation efficiency. During the 1990's, to improve water use efficiency in agriculture, training farmers and using computer-based decision support systems about irrigation were essential activities. Recently, irrigation management systems based on digital technologies have come to the fore. These digital-based systems have made it possible for farmers to use recent scientific approaches related to soil, plant, atmosphere, and water relationships. These systems primarily work based on data obtained from sensors used to monitor soil, vegetation, and meteorological parameters, and spectral and thermal images acquired by satellite and unmanned aerial vehicles (UAV). Many computer software and applications for portable devices have been developed to convert these data into information to be used in irrigation management. By integrating these systems with equipment that will allow the irrigation systems to be operated automatically, producers can have the opportunity to perform much more precise and/or variable rate irrigation. These digital systems can be used in individual farms, large irrigation networks, and water resources management in water basins. This review aims to evaluate the current research studies and developed systems for the use of digital technologies in irrigation management.

Keywords

Agriculture · Irrigation management · Digital technologies · Remote sensing · Unmanned aerial vehicle

E. S. Köksal (✉) · E. Tunca · S. Ç. Taner
Faculty of Agriculture, Department of Agricultural Structures and Irrigation, Ondouz Mayıs University, Samsun, Turkey
e-mail: eselim@omu.edu.tr

M. Bahadir and A. Haarstrick (eds.), *Water and Wastewater Management*,
Water and Wastewater Management, https://doi.org/10.1007/978-3-030-95288-4_20

Abbreviations

ALEXI	Atmosphere-Land Exchange Inverse
BLUE	Blue region of electromagnetic spectrum
CWSI	Crop Water Stress Index
ETa	Actual evapotranspiration
ETc	The evapotranspiration of a selected crop
EVI	Enhanced Vegetation Index
FDR	Frequency domain reflectometers
G	Soil heat flux
GREEN	Green region of electromagnetic spectrum
H	Sensible heat flux
Kc	Plant coefficient
Kcb	Basal plant coefficient
K_y	Water yield response factor
LE	Latent heat flux
LSWI	Land Surface Water Index
LWP	Leaf water potential
MAD	Moisture Allowable Deficit
METRIC	Mapping ET at high resolution and with Internalized Calibration
NDVI	Normalized Difference Vegetation Index
NDWI	Normalized Difference Water Index
NIR	Near Infrared region of electromagnetic spectrum
NMM	Neutron moisture meter
RED	Red region of electromagnetic spectrum
RED-EDGE	Red-Edge region of electromagnetic spectrum
Rn	Net radiation
SAVI	Soil Adjusted Vegetation Index
SEBAL	Surface Energy Balance Algorithm for Land
SEBS	Surface Energy Balance System
SR	Simple Ratio
SWIR	Short wave infrared region of electromagnetic spectrum
SWP	Stem water potential
TAW	Total Available Water
TDR	Time Domain Reflectometer
TSEB	Two Source Energy Balance
Ts–Ta	The difference between surface temperature and atmosphere temperature
UAV	Un-manned aerial vehicle
VRI	Variable rate irrigation
WDI	Water Deficit Index
WUE	Water use efficiency
ε_a	Dielectric permittivity

20.1 Introduction

Agriculture does not only meet the nutritional needs, which is the basis of human life, but also to fulfill the raw material needs of many sectors. Due to the global climate change and drought experienced in recent years, the importance of the agricultural sector is gradually increasing. According to studies on climate change, water constraints resulting from population growth will be severe, especially in arid and semi-arid areas [1]. Although the available natural resources in the world are constant, the human population is increasing. The need for water is gradually increasing due to the increasing population and the corresponding increasing food demand. However, available water resources are also decreasing. The amount of water used in agriculture, especially for irrigation and food production, constitutes one of the greatest pressures on the available water resources in the world [2, 3].

The agricultural sector uses approximately 70% of water resources under control, especially for irrigation [3–7]. For this reason, optimum use of water resources for agricultural production has become one of the most fundamental problems worldwide. Only about 17% of the world's cultivated land is irrigated. However, 40% of the total food is produced in this irrigated land [5]. According to the researches, approximately 30–70% of water used in agricultural areas can be saved with more efficient irrigation practices. With the proper irrigation management practices, the yield obtained from the unit agricultural area can be increased by approximately 20–90% [8].

In arid and semi-arid areas, agriculture relies heavily on irrigation of plots with water oriented from rivers. However, with global climate change, drought, and changes in land-use intensity, it has been observed that the flow rates of many rivers have decreased [9]. At the same time, the increase in industrial, domestic, environmental, and agricultural water use causes an increase in pollution and a decrease in the amount of available water resources [10]. For this reason, in the agriculture sector, new approaches developed for saving water are adopted instead of traditional irrigation management for sustainable water use. In recent years, fast, easy to obtain, and reliable data have become necessary to determine effective irrigation management strategies. Today, field-level sensors, automatic meteorology stations, satellite systems, unmanned aerial vehicle systems (UAV), irrigation automation-control systems, variable rate irrigation (VRI) equipments, and mobile applications can be counted as the primary systems that can be used in irrigation management.

Irrigation water is one of the most important inputs used in plant production and significantly affects plant nutrition use efficiency [11, 12]. The crop water requirement must be accurately determined and considered to obtain crop yield at a high level. This amount should be fully met by precipitation and/or irrigation. In conditions of insufficient water supply, irrigation should be planned to achieve the highest possible yield. While the crop water requirement varies according to the phenological stage and climatic factors, the irrigation water requirement in a certain period can be determined according to the crop water need and the effective precipitation. Thus, there is a need to decide when and how much irrigation water should be applied, either the water supply is sufficient or scarce [13].

Water and food scarcity can be considered a worldwide problem [14]. Irrigation is widely perceived as the cause of water scarcity in many parts of the world [15]. Increasing water scarcity does not only limit irrigated areas but also reduces agricultural production due to limited irrigation [15]. Water scarcity limits agricultural production, complicates socio-economic conditions in rural areas and threatens food security. The critical role of irrigation in agricultural production will also remain in the future [16].

Smart irrigation applications based on digital technology have gained importance in recent years to effectively manage water use in agriculture [3, 17]. Agricultural meteorology stations, software, and applications that can make irrigation programming based on the data measured by these systems are the oldest and most up-to-date systems in this field. Automated and advanced irrigation technologies based on soil moisture sensors play an essential role in precision agriculture. Such techniques include soil moisture sensors embedded in the soil and a control module connected to an irrigation timer [18–20]. In addition to these techniques, systems based on satellite or UAV images, including multi-spectral and thermal band images, are developed. [3, 6, 21].

In this literature survey, meteorological systems, soil water content measurement systems, and remote sensing systems, which are the main digital possibilities that can be used in irrigation water management, were evaluated. In this context, the primary literature on the subject and current studies have been brought together and evaluated from a scientific perspective.

20.2 Soil, Plant, Atmosphere and Water Relationship

The mechanical, impedance, water, aeration, and temperature properties of the soil are of great importance in plant cultivation, and the most important one of them is water [22–24]. Since water is the most critical environmental factor limiting plant growth, examining the relationship between soil, plant, and water is very important in terms of adequacy and continuity in plant production [24].

There are many studies on the physical, chemical, and biological aspects of the plant, water, and soil relationships [24–26]. Studies in various environments that differ in climate, soil type, and vegetation indicate strong interactions between these characteristics. It is known that the water content in the crop rooting environment controls the temporal and spatial dynamics of the crop water uptake mechanisms [27–29]. According to these studies, plants show different characteristics in terms of accessing water at different depths in the soil. It is well known that plant roots grow and use deeper soil water under drought conditions [30–32].

Transpiration is defined as the water that plants give to the atmosphere through their vegetative parts [24, 33]. As a result of transpiration, water potential decreases in the plant organs that lose water. Water moves from a higher potential to a lower potential within the plant structure [34]. This mechanism also exists between soil or growing medium and plant roots. Thus, water in the soil enters the plant through the roots, and as a result of transpiration, water gets out from the plant through the stomata in the leaves [35]. Accordingly, the presence of water at a suitable potential (close to -0.33 atm) in the crop rooting zone is critical for the plant to grow and develop without suffering from a lack of water. In cases where this water cannot be met by precipitation, it should be provided with irrigation.

Appropriate irrigation management and irrigation scheduling are crucial for optimum water use in irrigated crops [36]. The response of agricultural crops to water deficiency has gained a scientific identity with Stewert's approach [37]. According to this, proportional water deficiency in crops has a linear relationship with proportional yield reduction (Eq. 20.1). This relationship is represented by the concept of a water yield response factor (Ky). Ky value of 1.0 indicates yield reduction at an equal rate of water deficiency. If the Ky value is lower than 1.0, it means drought resistance, and if it is higher than 1.0, it means water sensitivity (Fig. 20.1). As the Ky values of agricultural crops differ, the response of a crop to water deficiency may also vary between phenological periods [38, 39].

$$1 - \left(\frac{Y_a}{Y_m}\right) = Ky\left(\frac{ET_a}{ET_m}\right) \quad (20.1)$$

A wide variety of approaches have been proposed for accurate and precise irrigation scheduling. These approaches are generally based on meteorological parameters, soil water content, and crop monitoring-based systems [36, 40]. It is known that the water potential in plant cells and tissues is a direct indicator of crop water stress [36, 41]. According to researchers, it has been determined that many parameters that can be defined by plant sampling, direct measurement, or remote sensing have a high correlation with leaf water potential (LWP) [42–44].

20.3 Meteorological Parameters and Evapotranspiration

Decrease of water in the crop rooting zone occurs by transpiration from the plant and evaporation from the surface that is called crop Evapotranspiration (ETc). For this reason, how much irrigation water crops and cultivated lands need is often calculated based on ETc estimations and precipitation measurements. There are many equations developed for ETc estimation purpose, such as Blaney Criddle [45, 46], Hargreavest [47, 48], Makkink [49], Jensen Haise [50], and Penman Monteith [51]. Today, systems for irrigation management based on meteorological data use various variants of such models.

In many of these approaches, the first step is to estimate the reference evapotranspiration (ETo) based on meteorological data. ETo calculation equations are mostly developed using the specifications of the grass plant. The second step involves multiplying ETo with a proportional crop coefficient (Kc), which represents the difference in water consumption of grass and given crop, to calculate ETc (Eq. 20.2). Thus, the ETc of a selected crop can be estimated daily, weekly, monthly, or seasonally for all or a part of its

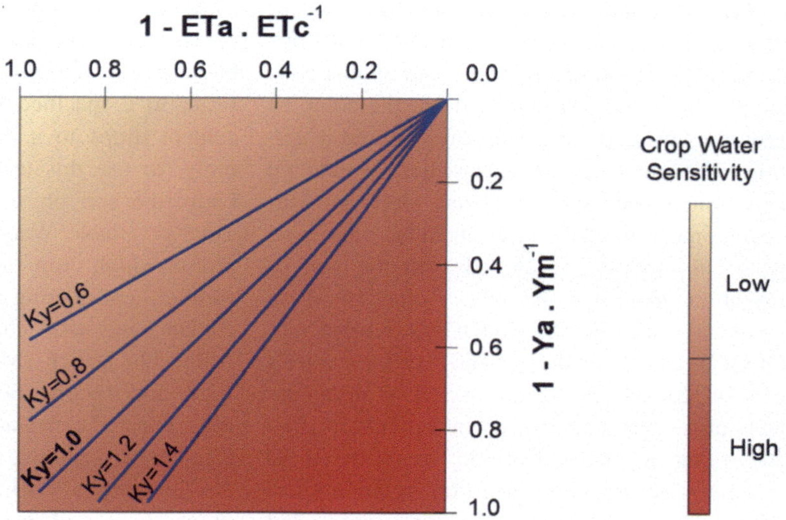

Fig. 20.1 Relationship between relative deficiency in crop evapotranspiration and relative yield decrease represented by water yield response factors (Ky) to show crop water sensitivity

growing season [45, 51–53]. In Fig. 20.2, the daily ETo values calculated by the Standardized Penman Monteith method using the 30-years average daily meteorological data of Ankara province and the changes in the daily ETc values computed using the Kc data of the sugar beet over time are given.

$$ET_c = ET_o \ Kc \tag{20.2}$$

Estimated ETc values using meteorological data of the past long years are used for purposes such as hydrological calculations, planning of water resources, design of water structures and irrigation systems, planning of pre-season plant patterns, and general irrigation scheduling. It can be used for real-time irrigation scheduling with ETc values calculated based on daily meteorological data. Calculation of ETo and ETc, determination of Kc values and their correct use, preparation of irrigation schedules by considering soil physical properties, and other factors require significant technical knowledge and computer skills.

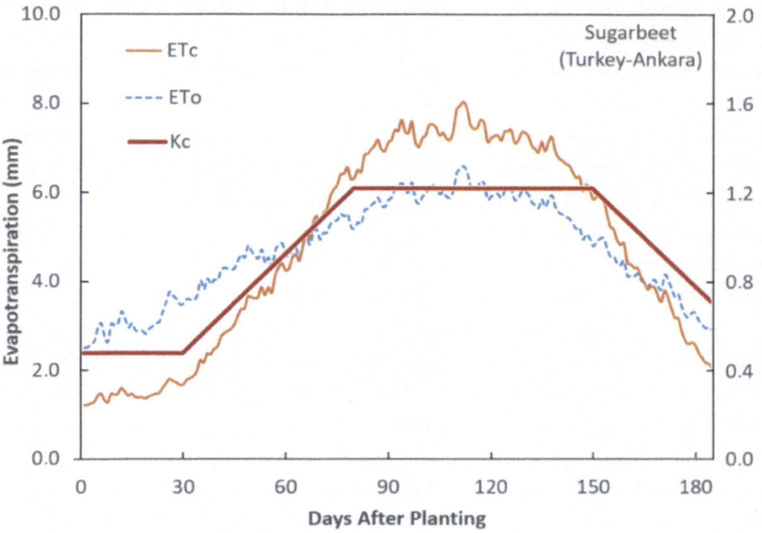

Fig. 20.2 Reference evapotranspiration (ETo), crop coefficient (Kc), and crop evapotranspiration (ETc) for sugar beet estimated with 30 years average daily meteorological data of Ankara, Turkey [60]

Computer models are generally designed to mimic the behavior of a system [54]. With this point of view, various software and applications have been developed that facilitate the complex processes of crop water consumption and irrigation scheduling. These approaches, which are also integrated into various crop growth-development models, are included in more detail in software developed with the aim of irrigation management only. Some of these software's are IRSIS, BUDGED, AquaCrop, CROPWAT, and SUET [55–60]. This software's can be considered as an important computer technology application used in agriculture [61]. According to their scientific approach, these software's are primarily based on meteorological data such as precipitation, temperature, wind speed, relative humidity, solar radiation, and local atmospheric pressure. In addition to estimating ETo, ETc, irrigation time, and irrigation amount, these software's are also used to predict the impact of climate change on agriculture, to predict crop yields, to support field management decisions, and to implement various management strategies at low cost [62, 63].

20.4 Soil Water Content Measurement

Irrigation scheduling based on monitoring or estimating soil water content is the most widely used technique globally. In assessing soil water content, the ETc parameter, which represents both crop characteristics and changes in meteorological factors, is often used [64, 65]. Systems consisting of various sensors and software have been developed to determine how the soil water content changes during the irrigation season. While soil water content data was mostly used for scientific purposes until 2015, producers have also used it for irrigation management in fields where high-value crops are grown in recent years [66]. This trend has brought soil water content monitoring systems to an important position among the emerging agricultural digital technologies and accelerated the developments in this field.

Systems developed to monitor the soil water content, and irrigation management are based on the principle of applying an amount of irrigation water to a specified level when the soil water content drops to a critical level [67]. Accordingly, to use this technique, it is necessary to know the soil physical properties to calculate Total Available Water (TAW) capacity of soil profile which crop rooting will be occurred. In addition, considering the irrigation method, soil, climate, and crop characteristics, the Moisture Allowable Deficit (MAD) level as the ratio of TAW and the irrigation strategy should be decided. Thus, the irrigation schedule can be prepared by calculating the irrigation interval and the amount of irrigation water. Figure 20.3 shows how soil water content changes with respect to TAW and MAD under optimum irrigation program conditions. As seen here, the decrease in soil water content is due to the effect of ETc, while the increase is due to irrigation. To prepare a precision irrigation schedule, there is a need to accurately measure or estimate the soil water content at sufficient soil depths and appropriate areas during the irrigation season, and at required time intervals.

The most reliable method for determining the soil water content is based on taking soil samples and determining the sample's wet and dry weight [68]. Either a volume-based or weight-based approach can be adopted in determining soil water content based on sampling. In calculations related to irrigation, volumetric soil water content information is required. Although sampling-based soil water content determination methods have various limitations in terms of irrigation management, it is the most valid method for the calibration of many new-generation digital soil water content determination and irrigation management systems [69, 70].

The fact that soil water content determination by sampling is a reliable method has enabled it to be used in many studies [71–73]. However, the method has significant limitations, such as; require intensive labor and time, and low spatial representation and destoration of land in repeated measurements [74, 75]. Therefore, irrigation management is not possible using this method

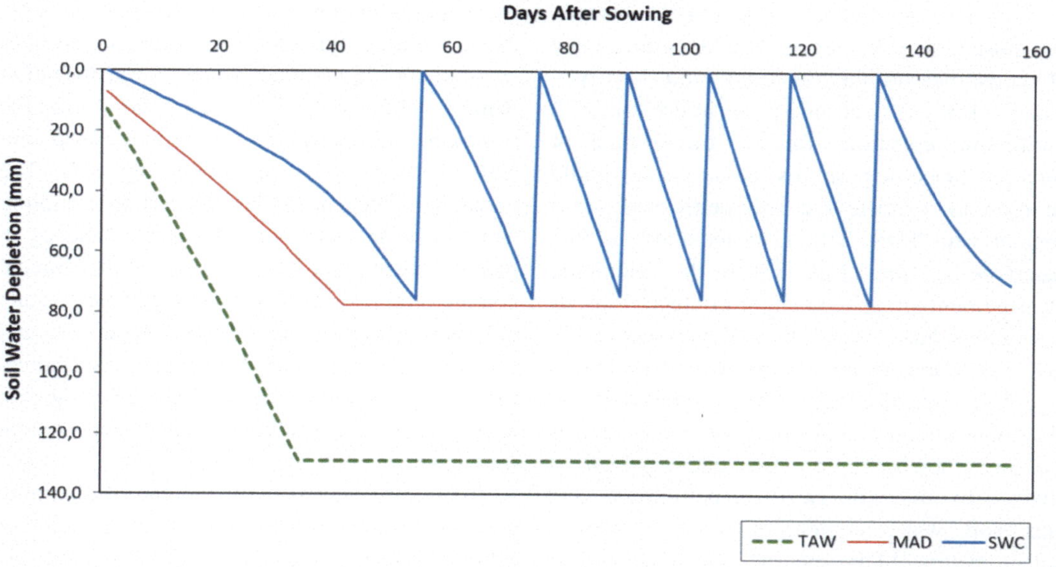

Fig. 20.3 Soil Water Content (SWC) variation under optimum irrigation water management conditions with respect to Total Available Water (TAW) capacity of soil and Moisture Allowable Deficit (MAD) level defined according to crop specifications, irrigation method and soil physical structure

under farm conditions. For this reason, the use of digital technologies in the measurement of soil water content is becoming increasingly common. The sensors used in these technologies generally measure the change of a parameter associated with water in the soil. These methods are included in the literature as surrogate soil water measurement techniques. Surrogate volumetric soil water content is mainly measured using the neutron moisture meter method and electromagnetic methods (time domain reflectometer, frequency domain reflectometer).

Neutron moisture meter method (NMM) depends on the slowing down of neutrons from a source that emits fast neutron (mean energy of 5 meV) by soil water and measuring the number of slowed neutrons (~0.025 eV at 27 °C) with special meters [76]. For this purpose, hollow pipes are placed at the point to be measured. The radioactive material, which usually consists of a mixture of americium and beryllium, is suspended at the depth at which moisture can be measured. If the device is well calibrated to the measuring soil, highly reliable data can be obtained with this method. NMM can be affected by soil hydrogen, chloride, boron, and soil density in general. Due to the use of radioactive sources in the system, the use of NMM is subject to legal regulations and restrictions regarding training and monitoring on issues such as transportation, storage, safety, and use. However, it is the most preferred and most reliable method in researches on irrigation and crop water consumption. In a study, neutron probe and capacitance probes were compared. As a result of the study, it was stated that more successful results were obtained in soil moisture monitoring with the neutron probe, but it was stated that the spatial resolution of the soil moisture determined by this technique is very low [77]. Evet et al. [78] investigated the sensitivity of various soil moisture sensors in determining soil moisture content and the effect of these determined values on ETa estimation and water use efficiency (WUE). As a result of the study, the determination of soil moisture by NMM and direct soil sampling was considered more reliable than the other techniques. There are numerous studies involving the monitoring of soil moisture with NMM, [79–81].

Time Domain Reflectometer (TDR) systems, the most widely known of the Electromagnetic Methods, measure the travel time of a short-rise-time (~ 150 ps) electronic pulse in the soil [82]. Calibration equations for the travel time or apparent dielectric permittivity (εa) values should be used to estimate the volumetric soil water content with TDR. The TDR method was first described by Topp et al. [83] for the volumetric measurement of soil moisture content. Various researchers have used this method successfully [84, 85]. Capacitance sensors work based on a repetitive sinusoidal waveform produced by an electronic circuit called oscillator. Such devices measure the oscillation frequency, which is affected by soil bulk electrical permittivity and thus soil water content. Frequency domain reflectometers (FDR) measure the frequency of an electronic pulse reflected from the ends of the probe. Systems operating with the FDR principle detect the change in signal propagation velocity in the sensor. This change is mainly due to the change of the dielectric constant, which is affected by the soil water content. Similar to the other systems, calibrating the capacitance and FDR systems is a necessary pre-process. It should be noted that, generally, electromagnetic systems are affected by salinity, temperature, and some other soil properties [18, 81]. However, uninterrupted measurement, continuous transmission of data to irrigation programming software, and the user's ability to continuously monitor the change in soil water content and irrigation need in the measurement area are the most outstanding aspects of electromagnetic sensors [86, 87].

20.5 Remote Sensing and Crop Monitoring

The moisture level in the soil is not always an accurate indicator of crop water requirement because factors that affect the crop and soil, such as plant disease or salinity, prevent the crop from benefiting from available water. Methods based on soil water content measurement in large areas have limitations, such as the need for calibration, point representation capability, animal damage, damage during agricultural operations, and high costs. In the light of all these views, irrigation management can also be done by monitoring the symptoms caused by water stress in the crop with methods based on crop monitoring [88]. The parameters related to the destructive sampling and contact measurement of water deficiency in plants include plant-leaf water potential, stomatal conductivity-resistance, photosynthesis rate, chlorophyll content, CO_2 exchange levels, and sap flow rate (amount). However, these indicators have important limitations in irrigation management as similar to soil water content monitoring.

Remote sensing is the technique of obtaining and evaluating information about natural or artificial objects with measuring instruments placed at certain distances from the target. The soil and plants reflect sunlight, and spectral reflectance characteristics provide information about crop growth levels, the effects of plant diseases and pests, and other factors that affect crop production. Remote sensing techniques, including the spectral reflectance values at various wavelengths of the electromagnetic spectrum and the surface canopy temperature, provide essential information about the water levels of the crops. These data can be obtained easily and at a low cost in large areas. For this reason, the use of remotely sensed data in irrigation management is increasingly coming to the fore. These methods allow the evaluation of the factors that limit the crop water use and irrigation management in larger areas in a shorter time and with high sensitivity levels.

20.5.1 Spectral Reflectance and Vegetation Indexes

Spectral reflectance can be defined as the ratio of the solar radiation reflected from an object to the solar radiation to which it is exposed in a particular region of the electromagnetic spectrum. Spectral vegetation indexes can be calculated by mathematical combinations of spectral reflectance values of different spectral bands. They can be used to monitor the seasonal, inter-annual, and long-term changes of vegetation's structural, phonological, and biophysical properties [89]. Enhanced Vegetation Index (EVI), Normalized Difference Vegetation Index (NDVI), Simple Ratio (SR), Soil Adjusted Vegetation Index (SAVI), Normalized Difference Water Index (NDWI) and Land Surface Water Index (LSWI) are spectral indexes used extensively in the literature [90–96]. The equations used in the calculation of these spectral indexes are given in Table 20.1. The data obtained from the Near Infrared (NIR) region is primarily used to calculate spectral indexes. In addition, RED-EDGE, RED, GREEN, and BLUE band data are also widely used to determine vegetation properties. In addition, the sensitivity of the short wave infrared (SWIR) region to water has enabled the data of this region to be used in water-related indexes. It is also recommended to use various correction coefficients (L, G, C1, C2) for some indexes.

Spectral indexes are successful in revealing the growing level of the vegetation [97]. This amount of vegetation can be associated with physiological events such as transpiration and photosynthesis [98]. Zhao et al. [99] determined a high correlation between NDVI calculated from high-resolution UAV images and stem water potential (SWP). Zúñiga et al. [100] evaluated the relationship of WUE with spectral vegetation indexes in vineyards irrigated by sub-surface drip irrigation and surface irrigation. Vegetation indexes calculated from remotely sensed data have been associated with crop coefficients (Kc) in recent years. Studies have shown significant statistical relationships between Kc and vegetation indexes [101, 102]. These relationships were found in various crops [103–105]. Neale et al. [106] used SAVI to estimate Kc data, while Tasumi et al. [107] used NDVI. Bausch and Neale [108] revealed the relationship between maize's NDVI calculated from handheld radiometers and basal crop coefficient (Kcb). Duchemin et al. [109] determined the relationship between NDVI and Kcb calculated from the high-resolution Quickbird satellite image. Also, ETa prediction maps can be created by associating the evapotranspiration fraction (ETrF), calculated with the surface energy balance, with the spectral vegetation indexes. Mokhtari et al. [110] correlated ETrF calculated with Landsat 8 satellite images and METRIC model with various indexes calculated from high-resolution multi-spectral UAV images, and high-resolution ETa maps were obtained. NDVI, SAVI, EVI, NDWI, and LSWI were used in the study. According to the results obtained from the researches, spectral vegetation indexes to be obtained from multi-spectral images taken by satellite or UAV systems can be used in irrigation management.

Table 20.1 Equations of some well known spectral indeces

Spectral index	Formulation	References	Equation number
Normalized difference vegetation index (NDVI)	$\frac{(NIR-RED)}{(NIR+RED)}$	[91]	3
Soil adjusted vegetation index (SAVI)	$(1+L)\times\frac{(NIR-RED)}{(L+NIR+RED)}$	[92]	4
Band ratio (RATIO)	$\frac{NIR}{RED}$	[93]	5
Enhanced vegetation index (EVI)	$G\times\frac{(NIR-RED)}{NIR+(C1\times RED-C2\times BLUE)+L}$	[94]	6
Normalized difference water index (NDWI)	$\frac{NIR-SWIR}{NIR+SWIR}$	[95]	7
Land surface water index (LSWI)	$\frac{NIR-SWIR}{NIR+SWIR}$	[96]	8

20.5.2 Canopy-Surface Temperature and Water Stress Indexes

Plants take CO_2 from their stomata for photosynthesis and lose a significant amount of water from their bodies in the same way [111]. When this evaporated water is not replaced through the roots, the amount of water in the plant decreases, the stomata in the leaves are closed proportionally, and the vegetation temperature increases at the same rate [36, 66, 112]. Water stress causes the vegetation temperature to approach and exceed the air temperature [66, 113]. According to Jackson et al. [114], vegetation temperature measured radiometrically is a significant indicator of water stress compared with a reference temperature (air temperature). The results of a study, conducted by Wiegand and Namken [115] on cotton crops, showed that a decrease in the relative water content of the leaf between 83 and 59% causes an increase in leaf temperature by 3.6 °C, and an increase in solar radiation by 350–1000 J/m^2 causes an increase in leaf temperature by 9.0–10.0 °C.

In the beginning, plant thermocouples were used for canopy temperature measurements [116]. The instruments developed to determine the surface temperature by remote sensing are based on the thermal infrared measurement and surface emissivity [36, 117, 118]. In a study conducted by Fucs and Tanner [119] to determine the emissivity, it was defined that the emissivity values of alfalfa and grass plants varied between 0.97 and 0.98, and the leaf emissivity values of beans and tobacco were 0.96 and 0.97, respectively. One of the first studies based on remotely sensed canopy temperature was carried out by Ehrler [120]. Jackson et al. [121] developed an indicator called Stress Degree Day (SDD). Subsequently, various water stress indexes have been developed such as; Crop Water Stress Index (CWSI) [122], Water Deficit Index (WDI) [123], Temperature Vegetation Index (TVI) [124], Temperature Vegetation Dryness Index (TVDI) [125] and Vegetation Temperature Condition Index (VTCI) [126]. CWSI and WDI are the two most commonly used indexes in the literature. In Fig. 20.4, basic graphics developed for CWSI and WDI are given. CWSI and WDI calculation formulations prepared according to these basic graphics are given in Eqs. 20.9 and 20.10, respectively. Parameters A, B and C in the CWSI equation are in Tc-Ta. Similarly, the variables D, E and F in the WDI equation are in Ts–Ta.

There are many studies in which the effects of different irrigation strategies on canopy-surface temperature are monitored, and various indexes are calculated by using these data and the meteorological data together [36, 129–132]. Today, studies continue to apply these indexes to images

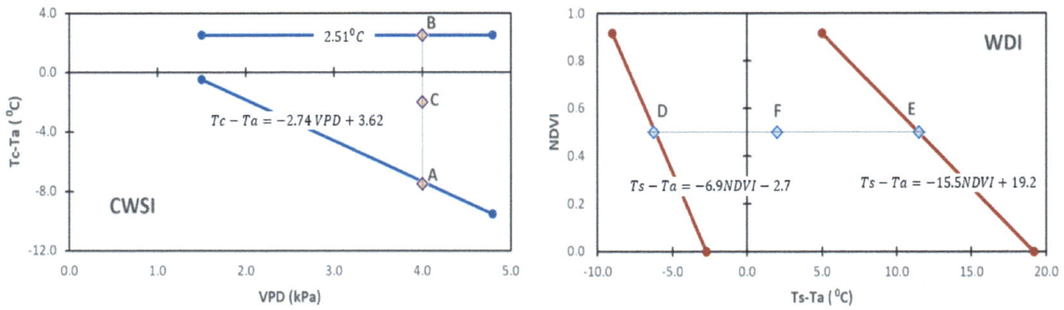

Fig. 20.4 Baseline graphs developed for Crop Water Stress Index (CWSI) and Water Deficit Index (WDI) calculations [127, 128]

obtained from satellite and UAV systems and the monitoring of water use and drought effects in agricultural lands [133–135].

$$CWSI = \frac{A - C}{A - B} \quad (20.9)$$

$$WDI = \frac{D - F}{D - E} \quad (20.10)$$

20.5.3 Energy Balance and Evapotranspiration Mapping

In regions where crop water demand is higher than natural precipitation, knowledge of ETc and/or actual evapotranspiration (ETa) is crucial for water resources management [136]. Today, lysimetry [137], eddy covariance [138], bowen ratio energy balance [139], and soil water budget methods [140] are commonly used to determine ETa. However, these methods are often expensive, complex, and more suitable for research purposes [141]. Therefore, ETc is commonly estimated using the Kc approach, which details were given above [142]. However, unsuitable conditions of some meteorological stations used in the Kc approach, the deterioration of the sensors' calibration over time, the completion and correction of incomplete and erroneous data can be counted among the critical shortcomings of this method. Moreover, the ETc calculated with this approach represents standard conditions, and actual conditions of ETa often differ from theoretical standard conditions. Kelley and Pardyjak [143] noted that this approach is often not suitable for use in real-time irrigation management. Therefore, it is necessary to use remote sensing techniques that reveal spatial and temporal variation of ETa, for more effective management of water resources.

Remote sensing platforms can be examined under three groups: satellite systems, aerial vehicles, and field-level systems. According to Matese et al. [144], all these remote sensing platforms have advantages and disadvantages in application, technology, and economy. In general, it is not easy to classify ETa estimation methods that depend on remotely sensed data. However, Courault et al. [145] divided these methods into four different categories; empirical methods, surface energy balance-based methods, deterministic methods, and methods that depend on vegetation indexes. Among these, the most frequently used evapotranspiration approaches depend on energy balance.

The energy balance allows the estimation of ETa from a surface [146]. The surface energy balance takes into account the parameters of net radiation (Rn), soil heat flux (G), sensible heat flux (H), and latent heat flux (LE) (Eq. 20.11). In calculating H, besides the aerodynamic resistance, the difference in surface temperature and atmosphere temperature (Ts-Ta) is a crucial variable [147]. The use of remotely sensed data in determining the surface energy balance components Rn and Ts-Ta makes it possible to use the temperature and absorbed solar radiation values related to the surface in ETa from the surface [148]. The evaporation calculated in this way represents the actual properties of the surface where evaporation occurs.

$$R_n = LE + H + G \quad (20.11)$$

Numerous studies have been conducted on remotely sensed surface temperature and spectral vegetation indexes in surface energy balance. Brown and Rosenberg [149] conducted one of the first experimental studies on this subject. The study reported by Hatfield et al. [148] shows that different crops are grown in weighing lysimeters established in many regions of the USA, and the ETa values measured in the lysimeters have a significant correlation with the ETa values estimated using the surface energy balance. Jackson et al. [150] include a fundamental approach for converting instantaneous ETa values calculated by remotely sensed data, to daily ETa values. This approach later inspired new models developed in the 1990s and 2000s, such as Surface Energy Balance Algorithm for Land (SEBAL) [146, 151], Mapping ET at High Resolution and with Internalized Calibration (METRIC) [152, 153], Surface Energy Balance System (SEBS) [154], Atmosphere-Land Exchange Inverse

(ALEXI) [155], and Two Source Energy Balance (TSEB) [156].

Real-time ETa estimation and irrigation scheduling were made using a combination of ETa estimated with METRIC and soil water budget [157]. Trezza et al. [158] applied the METRIC model to MODIS satellite images and compared the monthly, and annual ETa estimates obtained with ETa values calculated from Landsat satellite images. Regional-scale ETa estimation was performed by Elhaddad and Garcia [159] using the ReSeT model. In recent years, studies on high-resolution ETa maps have intensified with the effective use of UAV systems in agriculture. For example, Ortega-Farías et al. [160] successfully predicted the energy balance components in olive trees using the METRIC model.

Hoffmann et al. [161] estimated ETa in barley using the TSEB model. French et al. [162] applied the METRIC and TSEB models to the images acquired by a manned aerial vehicle system. Mokhtari et al. [110] created high-resolution ETa maps by fusing Landsat 8 satellite images with UAV images. Although UAVs images provide an advantage with their high spatial resolution, several new problems have emerged in ETa mapping by using UAVs images. Aboutalebi et al. [163] determined that the tree shadows in the UAV images significantly affect the NDVI and ETa values. Nassar et al. [164] discussed the effects of spatial resolution of UAVs images on the TSEB model. Nassar et al. [165] used and evaluated different models to convert instantaneous ETa values calculated with the TSEB model to daily ETa.

The second most frequently used method after the energy balance approach is ETa estimation based on spectral vegetation indexes. In this method, Kc maps are created by using the statistical relationships between Kc and spectral vegetation indexes. These Kc maps are converted to ETa maps based on the ETo value of the corresponding day. Johnson and Trout (2012) used NDVI obtained from satellite images to monitor the ETa of crops in California benefiting from the NDVI - fc and fc - Kc relationships in their study. Singh and Irmak [166] developed Kc regression models for various crops (corn, soybean, sorghum, and alfalfa) based on NDVI obtained from Landsat satellite images.

20.6 Dicussion

The digital systems used for irrigation management are depending on meteorological parameters, soil water content and/or remote sensing. These data can also be used together with a holistic system approach, as summarized in Fig. 20.5.

Although meteorology stations are established and operated by governments around the world, there are also stations established by the private sector. These stations, which are also used for early warning in various agricultural issues, can be used effectively in irrigation water management. This information can be used in irrigation management by calculating water consumed by a crop under standard conditions with the data measured at the stations. In the use of meteorology stations, there are some disadvantages, such as the unsuitability of the station ground and its surroundings, the calibration deficiencies of the sensors, the meteorological differences between the measurement and calculation points, the differences between the ETc and ETa, and high investment and operating costs. However, the estimated ETc values based on meteorological data have an important place in the use of digital technologies in irrigation water management.

The measurement of the water content of the root zone in the soil constitutes an essential factor that can be used directly for irrigation management. The soil water content data has been the most important basis of many irrigation management techniques developed to date. Volumetric soil water content values can be determined reliably by taking soil samples. However, there are difficulties in using this method even in research studies, and it is almost impossible to use it in irrigation management of large areas. Measuring a parameter affected by the amount of water in the soil and calibrating this parameter according to the change in soil

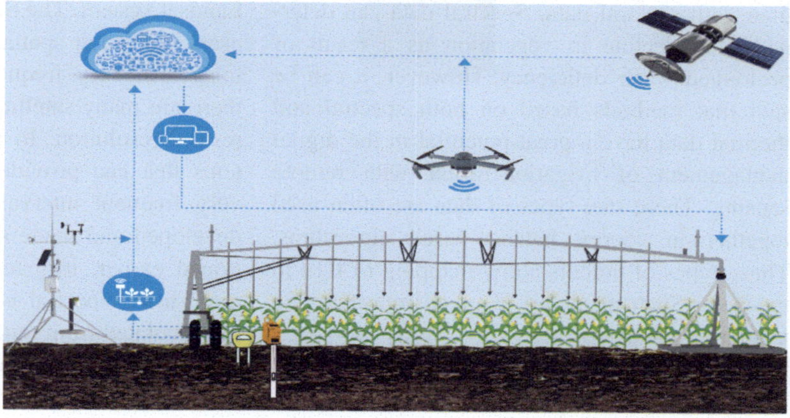

Fig. 20.5 A general concept of a digital irrigation management system

water content has made these measurements more practical and usable. Although NMM is the most used technique in irrigation management research, the necessity of moving the device to each measurement point and using the sensors separately at each measurement depth prevents this system from being used in the scope of digitalization. The risk of adversely affecting human health because it contains radioactive material and its legal obligations and restrictions can be counted as other negative aspects of the NMM system.

For this reason, although soil water content measurement is still valid in both research and application activities related to irrigation management, the NMM system is only used in some researches. Still, in practice, it is preferred to use TDR, FDR and capacitance sensors working after electromagnetic principles. The continuous data measurement of such sensors can send this data to the digital cloud with various data transfer equipment. Soil water content values measured in this way are converted into information by being processed with software and presented to the user via a personal computer, tablet, and mobile phone. In addition, the continuous measurement of soil water content data, which changes in line with the actual precipitation, irrigation, and crop water consumption, reveals the possibility of calibration with machine learning instead of calibration with site-specific sampling. Thus, such systems enable digital irrigation management based on soil water content. However, the sensitivity of electromagnetic-based systems to salt, temperature, and some soil properties can significantly reduce their sensitivity to soil water content, which indicates the possibility of giving results including some errors. Most of the time, these errors could be of lower effect on field level water use efficiency than mistakes that farmers make in the field due to deficit-irrigation or over-irrigation applications. However, there are still important limitations in using these systems in scientific research, in which crop water consumption and related findings will be reported.

Instead of monitoring the soil water content, monitoring the symptoms created by the water level in the crop with remote sensing techniques is among the important and innovative subjects that have been discussed in research since the 1960s. Until the end of the 2010s, these studies were mostly carried out in the light of data obtained from handheld radiometers and satellite systems. The results showed that spectral signatures, spectral vegetation indexes, surface or canopy temperature values provide important information in terms of irrigation management. Indexes based on thermal data have a high correlation with ETa. It has been determined that there is an indirect interaction between spectral indexes and ETa. Because the spectral indexes can reveal the vegetation level of the plants, where the amount of irrigation plays a significant role, in other words, water deficiency in crops can be determined faster with thermal data rather than with spectral data. Spectral data can determine the decline in vegetation as a result of prolonged water deficiency. However, it can be said that methods based on both spectral and thermal data have a great potential in the digital management of irrigation water with remote sensing. These two types of data are often used together in energy balance-based algorithms. These types of models allow mapping of ETa in the field. Such maps are of great importance in irrigation management.

The use of remotely sensed data in monitoring crop production has led to the development of practical handheld devices, new satellite systems, and aircraft mountable cameras. Today, especially UAV and satellite systems come to the fore in this field. Multi-spectral and thermal cameras large enough to be carried by UAV sytems have been developed. Studies show that these images obtained by UAV systems, and the indicators to be calculated based on this data, can be used successfully in irrigation management. However, some limitations of UAV systems prevent their use in large areas. Among these constraints are the legal regulations regarding the flights of UAV systems, the total viewing area, which varies according to the whole flight time of the UAV systems, flight speeds, and flight altitudes. Increasing the flight altitude to enlarge the imaging field results in a decrease in spatial resolution and the need for atmospheric correction. In such a case, satellite systems may be preferred. Today, the Landsat system provides the main satellite images that can be used in irrigation management. Landsat captures both multi-spectral and thermal images. However, the spatial resolution of Landsat thermal images makes it impossible to use this system on lands smaller than 10 ha. In addition, in large scaled agrciultral lands, significant data loss occurs at the land borders. Another disadvantage of the Landsat system is its temporal resolution (16 days). Despite this, the fact that it offers both multi-spectral and thermal images carries the Landsat system to a very important point in irrigation management. The highest terrestrial resolution images that allow ETa mapping with energy balance based models are provided by the Landsat system. The Sentinel system provides its users with high spatial resolution multi-spectral images at very frequent intervals. In addition, there are many satellite systems with higher terrestrial resolution. In recent years, satellite systems that can provide multi-spectral images at very frequent intervals (daily) have also been developed and made available images. As mentioned earlier, irrigated farmland can be monitored using spectral vegetation indexes that can be calculated from multi-spectral data. ETa maps can be created by calculating the crop coefficient Kc images using these indexes. It should be noted that the accuracy of these ETa maps will be lower than energy balance-based ETa maps.

Recent developments reported and discussed in the literature survey above predict that remote sensing systems have an excellent potential for irrigation management in agriculture. Technological developments in unmanned aerial vehicle and satellite systems in the future, and machine learning-based artificial intelligence applications will be used in processing and presentation of the big data gained. However, there is a need for further scientific research to determine the source of the deficiencies determined by remote sensing for crop production areas. Irrigation, crop diseases, nutrient deficiency, physical–chemical properties of the soil, and extreme meteorological events can cause the poor growth of crops, and remotely sensed data could not reveal which of them is most responsible for this.

20.7 Conclusions

In addition to population growth, changes in consumption and waste habits in people's daily lives play a significant role in increasing demand for agricultural products worldwide. On the other hand, despite the waste in rich societies, there are various difficulties in accessing agricultural products in poor communities. In regions where agricultural production is limited, prices of even essential food products increase with transportation costs and supply–demand balance. People have aimed to increase agricultural production in unit areas from past to present. Irrigation management in crop production is one of the determining factors on yield. For this reason, most of the controlled water resources around the world are used for irrigation purposes. However, in many regions, where irrigated agriculture is common, water use efficiency is very low. Thus, scientists and engineers have tried to develop various irrigation methods, systems that monitor water use, and irrigation scheduling techniques for more efficient use of water in agriculture. In today's world, where digital possibilities are at the highest level, the use of data measured by various sensors in irrigation water management and software that transforms these data into information within the framework of a system approach has increased rapidly.

As a result, increasing the efficiency of irrigation water management is an important issue all over the world. With the development of electronic and software systems, the use of data and information in this field is becoming increasingly common. Irrigation systems can be operated based on digital data in both point scale and map format, and irrigation systems suitable for automation can be managed with such digital systems, according to fixed irrigation or variable-rate irrigation approaches in large areas. It can be stated that digital opportunities will have a great place in the management of water resources through irrigation water management in the future.

References

1. Anderson MC, Allen RG, Morse A, Kustas WP. Use of Landsat thermal imagery in monitoring evapotranspiration and managing water resources. Remote Sens Environ. 2012;122:50–65.
2. UNESCO, United Nations Educational, Scientific and Cultural Organization. Securing the food supply. Paris: UNESCO; 2001.
3. Al-Naji A, Fakhri AB, Gharghan SK, Chahl J. Soil color analysis based on a RGB camera and an artificial neural network towards smart irrigation: a pilot study. Heliyon. 2021;7(1):e06078.
4. FAO. Water for sustainable food and agriculture a report produced for the G20 Presidency of Germany food and agriculture organization of the United Nations—FAO, Rome; 2017.
5. FAO. Water and agriculture an issues note produced for the G20 Presidency of the Kingdom of Saudi Arabia—FAO, Rome; 2021.
6. Quebrajo L, Perez-Ruiz M, Pérez-Urrestarazu L, Martínez G, Egea G. Linking thermal imaging and soil remote sensing to enhance irrigation management of sugar beet. Biosys Eng. 2018;165:77–87.
7. Campos N, Rocha AR, Gondim R, Coelho da Silva TL, Gomes DG. Gomes smart & green: an internet-of-things framework for smart irrigation. Sensors. 2020;20:190.
8. Saccon P. Water for agriculture, irrigation management. Appl Soil Ecol. 2018;123:793–6.
9. Zhang Q, Singh VP, Sun P, Chen X, Zhang Z, Li J. Precipitation and streamflow changes in China: changing patterns, causes and implications. J Hydrol. 2011;410:204–16.
10. Vörömarty CJ, Green P, Salisbury J, Lammers RB. Global water resources: vulnerability from climate change and population growth. Science. 2000;289:284–8.
11. Akmal M, Janssens MJJ. Productivity and light use efficiency of perennial ryegrass with contrasting water and nitrogen supplies. Field Crops Res. 2004;88:143–55.
12. Gonzalez-Dugo B, Durand JL, Gastal F, Picon-Cochard C. Shortterm response of the nitrogen nutrition status of tall fescue and Italian ryegrass swards under water deficit. Aust J Agric Res. 2005;56:1269–76.
13. Haverkort A. Handbook of precision agriculture. Princ Appl Euphytica. 2007;156(1):269–70.
14. Jury WA, Vaux H. The role of science in solving the world's emerging water problems. Proc Natl Acad Sci. 2005;102(44):15715–20.
15. Rosegrant MW, Ringler C, Zhu T. Water for agriculture: Maintaining food security under growing scarcity. Annu Rev Environ Resour. 2009;34(1):205–22.
16. Zhu T, Ringler C, Rosegrant MW. Viewing agricultural water management through a systems analysis lens. Water Resour Res. 2019;55(3):1778–91.
17. Nikolidakis SA, Kandris D, Vergados DD, Douligeris C. Energy efficient automated control of irrigation in agriculture by using wireless sensor networks. Comput Electron Agric. 2015;113:154–63.
18. Ferrarezi RS, Dove SK, van Iersel MW. An automated system for monitoring soil moisture and controlling irrigation using low-cost open-source microcontrollers. HortTechnology. 2015;25(1):110–8.
19. Jain P, Kumar P, Palwalia D. Irrigation management system with micro-controller application. In: 2017 1st international conference on electronics,

materials engineering and nano-technology (IEMENTech). IEEE; 2017. pp 1–6.
20. Taneja K, Bhatia S. Automatic irrigation system using Arduino UNO. In: 2017 international conference on intelligent computing and control systems (ICICCS). IEEE; 2017. pp 132–5.
21. Tunca E, Köksal ES, Çetin S, Ekiz NM, Balde H. Yield and leaf area index estimations for sunflower plants using unmanned aerial vehicle images. Environ Monit Assess. 2018;190(11):1–12.
22. Shaw BT. Soil physical conditions and plant growth. New York: Academic Press; 1952.
23. Kirkham D. Soil physics and soil fertility. Bulletin des Recherches Agronomiques de Gembloux Faculté des Sciences Agronomiques de l'État (new series). 1973;8(2):60–88.
24. Kirkham MB. Principles of soil and plant water relations. Academic Press; 2014.
25. Philip JR. Plant water relations: some physical aspects. Annu Rev Plant Physiol. 1966;17:245–68.
26. Asbjornsen H, Goldsmith GR, Alvarado-Barrientos MS, Rebel K, van Osch FP, Rietkerk M, Chen J, Gotsch S, Tobón C, Geissert DR, Gómez-Tagle A, Vache K, Dawson TE. Ecological advances and applications in plant–water relations research: a review. J Plant Ecol. 2011;4:3–22.
27. Jackson RB, Sperry JS, Dawson TE. Root water uptake and transport: using physiological processes in global predictions. Trends Plant Sci. 2000;5:482–8.
28. Schwendenmann L, Pendall E, Sanchez-Bragado R, Kunert N, Hölscher D. Tree water uptake in a tropical plantation varying in tree diversity: interspecific differences, seasonal shifts and complementarity. Ecohydrology. 2015;8:1–12.
29. Dai Y, Zheng X, Tang L, Li Y. Stable oxygen isotopes reveal distinct water use patterns of two Haloxylon species in the Gurbantonggut Desert. Plant Soil. 2015;389:73–87.
30. Penna D, Oliviero O, Assendelft R, Zuecco G, van Meerveld HJ, Anfodillo T, Carraro V, Borga M, Dalla FG. Tracing the water sources of trees and streams: isotopic analysis in a small pre-alpine catchment. Proc Environ Sci. 2013;19:106–12.
31. White JC, Smith WK. Seasonal variation in water sources of the riparian tree species Acer negundo and Betula nigra, southern Appalachian foothills, USA. Botany. 2015;93:519–28.
32. Barbeta A, Mejía-Chang M, Ogaya R, Voltas J, Dawson TE, Peñuelas J. The combined effects of a long-term experimental drought and an extreme drought on the use of plantwater sources in a Mediterranean forest. Glob Chang Biol. 2015;21:1213–25.
33. Sterling TM. Transpiration: water movement through plants. J Nat Resour Life Sci Educ. 2005;34(1):123–123.
34. Scharwies JD, Dinneny JR. Water transport, perception, and response in plants. J Plant Res. 2019;132(3):311–24.
35. Lucas WJ, Groover A, Lichtenberger R, et al. The plant vascular system: evolution, development and functions. J Integr Plant Biol. 2013;55:294–388.
36. Khorsand A, Rezaverdinejad V, Asgarzadeh H, Majnooni-Heris A, Rahimi A, Besharat S, Sadraddini AA. Linking plant and soil indices for water stress management in black gram. Sci Rep. 2021;11(1):1–19.
37. Stewart JI, Cuenca RH, Pruitt WO, Hagan RM, Tosso J. Determination and utilization of water production functions for principal California crops. W-67 CA Contributing Project Report. University of California, Davis, USA; 1977.
38. Steduto P, Raes D, Hsiao TC, Fereres E, Heng L, Izzi G, Hoogeveen J. AquaCrop: a new model for crop prediction under water deficit conditions. FAO, Rome. 2009;33(80).
39. Steduto P, Hsiao TC, Fereres E, Raes D. Crop yield response to water. Irrigation and drainage paper no. 66, FAO, Rome; 2012.
40. Jones HG. Irrigation scheduling: advantages and pitfalls of plant-based methods. J Exp Bot. 2004;55(407):2427–36.
41. Sharatt BS, Reicosky DC, Idso SB, Baker DG. Relationships between leaf water potential, canopy temperature, and evapotranspiration in irrigated and nonirrigated alfalfa. Agron J. 1983;75(6):891–4.
42. Poblete T, Ortega-Farías S, Moreno MA, Bardeen M. Artificial neural network to predict vine water status spatial variability using multispectral information obtained from an unmanned aerial vehicle (UAV). Sensors. 2017;17(11):2488.
43. Maciel DA, Silva VA, Alves HMR, Volpato MML, Barbosa JPRAD, Souza VCOD, Oliveira dos Santos J. Leaf water potential of coffee estimated by landsat-8 images. Plos one. 2020;15(3):e0230013.
44. Tosin R, Pôças I, Novo H, Teixeira J, Fontes N, Graça A, Cunha M. Assessing predawn leaf water potential based on hyperspectral data and pigment's concentration of Vitis vinifera L. in the Douro Wine Region. Sci Horticul. 2021;278:109860.
45. Doorenbos J, Pruitt WO. Guidelines for predicting crop water requirements. FAO irrigation and drainage paper 24, 2nd ed. Rome, Italy; 1977.
46. Allen RG, Pruitt WO. Rational use of the FAO Blaney-Criddle formula. J Irrig Drain Eng. 1986;112(2):139–55.
47. Hargreaves GH. Climate and irrigation requirements for Brazil. Logan: Utah State University; 1976. p. 44.
48. Hargreaves GH, Samani ZA. Estimating potential evapotranspiration. J Irrig Drain Engr ASCE. 1982;108(IR3):223–30.
49. Makkink GH. Ekzameno de la formula de Penman, Netherlands. J Agric Sci Wageningen. 1957;5:290–305.
50. Jensen ME, Haise HR. Estimating evapotranspiration from solar radiation. J Irrig Drain Div. 1963;89(4):15–41.

51. Allen RG, Pereira LS, Raes D, Smith M. Crop evapotranspiration: guidelines for computing crop water requirements. Rome: FAO, 300 p (Irrigation and Drainage Paper, 56); 1998.
52. Pereira AR, Pruitt WO. Adaptation of the Thornthwaite scheme for estimating daily reference evapotranspiration. Agric Water Manag. 2004;66(3):251–7.
53. Allen RG, Walter IA, Elliot R, Howell T, Itenfisu D, Jensen M. The ASCE standardized reference evapotranspiration equation, environment and water resources Institute of the Am. Soc. Civil Eng. (ASCE), Task Committee on Standardization of Reference Evapotranspiration, Final Rep., ASCE, Reston, VA, USA; 2005.
54. Abedinpour M, Sarangi A, Rajput TBS, Singh M, Pathak H, Ahmad T. Performance evaluation of AquaCrop model for maize crop in a semi-arid environment. Agric Water Manag. 2012;110:55–66.
55. Raes D, Geerts S, Kipkorir E, Wellens J, Sahli A. Simulation of yield decline as a result of water stress with a robust soil water balance model. Agric Water Manag. 2006;81:335–57.
56. Raes D, Lemmens H, Van Aelst P, Vanden Bulcke M, Smith M. IRSIS—irrigation scheduling information system, vol 1. Manual. K.U.Leuven, Dep. Land Management, Reference Manual 3; 1988.
57. FAO. Cropwat 8.0 for windows user guide. Rome, Italy; 2009.
58. Steduto P, Hsiao TC, Raes D, Fereres E. AquaCrop—the FAO crop model to simulate yield response to water: I. concepts and underlying principles. Agron J. 2009;101.
59. Katerji N, Campi P, Mastrorilli M. Productivity, evapotranspiration, and water use efficiency of corn and tomato crops simulated by AquaCrop under contrasting water stress conditions in the Mediterranean region. Agr Water Manage. 2013;130(4):14–26.
60. Anonymous; 2021. https://tagemsuet.tarimorman.gov.tr/pages/login.
61. Guo D, Zhao R, Xing X, Ma X. Global sensitivity and uncertainty analysis of the AquaCrop model for maize under different irrigation and fertilizer management conditions. Arch Agron Soil Sci. 2020;66(8):1115–33.
62. Kassie BT, Asseanonyng S, Porter CH, Royce FS. Performance of DSSAT-Nwheat across a wide range of current and future growing conditions. Eur J Agron. 2016;81:27–36.
63. Ran H, Kang SZ, Hu XT, Li FS, Du TS, Tong L, Li S, Ding RS, Zhou ZJ, Parsons D. Newly developed water productivity and harvest index models for maize in an arid region. Field Crop Res. 2019;234:73–86.
64. Cai W, Cowan T, Briggs P, Raupach M. Rising temperature depletes soil moisture and exacerbates severe drought conditions across southeast Australia. Geophys Res Lett. 2009;36:L21709.
65. Keshta N, Elshorbagy A, Carey S. Impacts of climate change on soil moisture and evapotranspiration in reconstructed watersheds in northern Alberta, Canada. Hydrol Process. 2012;26:1321–31.
66. Carroll DA, Hansen NC, Hopkins BG, DeJonge KC. Leaf temperature of maize and crop water stress index with variable irrigation and nitrogen supply. Irrig Sci. 2017;35(6):549–60.
67. Hignett and Evett S. Direct and surrogate measures of soil water content. In: Field estimation of soil water content, a practical guide to methods, instrumentation and sensor technology, training course series no. 30, International Atomic Energy Agency, Vienna; 2008.
68. Evett S. Gravimetric and volumetric direct measurements of soil water content. In: Field estimation of soil water content, a practical guide to methods, instrumentation and sensor technology, training course series no. 30, International Atomic Energy Agency, Vienna; 2008.
69. Dobriyal P, Qureshi A, Badola R, Hussain SA. A review of the methods available for estimating soil moisture and its implications for water resource management. J Hydrol. 2012;458:110–7.
70. Birendra K, Chau HW, Mohssen M, Cameron K, Safa M, McIndoe I, Rutter H, Dark A, Lee M, Pandey VP. Assessment of spatial and temporal variability in soil moisture using multi-length TDR probes to calibrate Aquaflex sensors. Irrig Sci. 2021:1–11.
71. Demir AO, Göksoy AT, Büyükcangaz H, Turan ZM, Köksal ES. Deficit irrigation of sunflower (Helianthus annuus L.) in a sub-humid climate. Irrig Sci. 2006;24(4):279–89.
72. Aydinsakir K, Dinc N, Buyuktas D, Kocaturk M, Ozkan CF, Karaca C. Water productivity of soybeans under regulated surface and subsurface drip irrigation conditions. Irrig Sci. 2021:1–15.
73. Wang D, Li G, Mo Y, Zhang D, Xu X, Wilkerson CJ, Hoogenboom G. Evaluation of subsurface, mulched and non-mulched surface drip irrigation for maize production and economic benefits in northeast China. Irrig Sci. 2021;39(2):159–71.
74. Dane JH, Topp C. Methods of soil analysis. 1st Edn., Soil Science Society of America, Madison, WI, ISBN-10: 089118841X; 2002. p 866.
75. Sharma PK, Kumar D, Srivastava HS, Patel P. Assessment of different methods for soil moisture estimation: a review. J Remote Sens GIS. 2018;9(1):57–73.
76. Evett S. Neutron moisture meters, field estimation of soil water content. In: Field estimation of soil water content, a practical guide to methods, instrumentation and sensor technology, training course series no. 30, International Atomic Energy Agency, Vienna; 2008.
77. Fityus S, Wells T, Huang W. Water content measurement in expansive soils using the neutron probe. Geotech Test J. 2011;34(3):255–64.

78. Evett SR, Schwartz RC, Casanova JJ, Heng LK. Soil water sensing for water balance, ET and WUE. Agric Water Manag. 2012;104:1–9.
79. Barker JB, Heeren DM, Neale CM, Rudnick DR. Evaluation of variable rate irrigation using a remote-sensing-based model. Agric Water Manag. 2018;203:63–74.
80. Alfonso C, Barbieri P, Hernandez MD, Lewczuk N, Martinez JP, Echarte MM, Echarte L. Water productivity in soybean following a cover crop in a humid environment. Agric Water Manage. 2020;232:106045.
81. Moldero D, López-Bernal Á, Testi L, Lorite IJ, Fereres E, Orgaz F. Long-term almond yield response to deficit irrigation. Irrig Sci. 2021:1–12.
82. Evett S, Heng LK. Conventional time domain reflectometry systems. In: Field estimation of soil water content, a practical guide to methods, instrumentation and sensor technology, training course series no. 30, International Atomic Energy Agency, Vienna; 2008.
83. Topp GC, Davis J, Annan AP. Electromagnetic determination of soil water content: measurements in coaxial transmission lines. Water Resour Res. 1980;16(3):574–82.
84. Amente G, Baker JM, Reece CF. Estimation of soil solution electrical conductivity from bulk soil electrical conductivity in sandy soils. Soil Sci Soc Am J. 2000;64(6):1931–9.
85. Yu X, Drnevich VP. Soil water content and dry density by time domain reflectometry. J Geotech Geoenviron Eng. 2004;130(9):922–34.
86. Evett S, Cepuder P. Capacitance sensors for use in access tubes. In: Field estimation of soil water content, a practical guide to methods, instrumentation and sensor technology, training course series no. 30, International Atomic Energy Agency, Vienna; 2008.
87. Evett S, Laurent JP. TRIME® FM3 moisture meter and T3 access tube probe. In: Field estimation of soil water content, a practical guide to methods, instrumentation and sensor technology, training course series no. 30, International Atomic Energy Agency, Vienna; 2008.
88. Shackel KA. plant-based approach to defcit irrigation in trees and vines. HortScience. 2011;46 (2):173–7.
89. Tittebrand A, Spank U, Bernhofer C. Comparison of satellite-and ground-based NDVI above different land-use types. Theoret Appl Climatol. 2009;98 (1):171–86.
90. Pearson RL, Miller LD. Remote spectral measurements as a method for determining plant cover (Doctoral dissertation, Colorado State University. Libraries). 1972.
91. Rouse JW, Haas RH, Schell JA, Deering DW. Monitoring vegetation systems in the Great Plains with ERTS. In: NASA Goddard Space Flight Center 3d ERTS-1 Symp., vol 1, Sect. A; 1973. pp 309–17.
92. Huete AR. A soil-adjusted vegetation index (SAVI). Remote Sens Environ. 1988;25(3):295–309.
93. Tucker CJ. Red and photographic infrared linear combinations for monitoring vegetation. Remote Sens Environ. 1979;8(2):127–50.
94. Liu HQ, Huete AR. A feedback based modification of the NDVI to minimize canopy background and atmospheric noise. IEEE Trans Geosci Remote Sens. 1995;33:457–65.
95. Gao BC. NDWI—a normalized difference water index for remote sensing of vegetation liquid water from space. Remote Sens Environ. 1996;58:257–66.
96. Chandrasekar K, Sai MVRS, Roy PS, Dwevedi RS. Land Surface Water Index (LSWI) response to rainfall and NDVI using the MODIS Vegetation Index product. Int J Remote Sens. 2010;31 (15):3987–4005.
97. Marino S, Basso B, Leone A, Alvino A. Agronomic traits and vegetation indices of two onion hybrids. Sci Hortic. 2013;155:56–64.
98. Alvino A, Marino S. Remote sensing for irrigation of horticultural crops. Horticulturae. 2017;3(2):40.
99. Zhao H, Xu Z, Zhao J. Development and application of agricultural drought index based on CWSI and drought event rarity. Trans Chin Soc Agric Eng. 2017;33(9):116–25.
100. Zúñiga CE, Khot LR, Jacoby P, Sankaran S. Remote sensing based water-use efficiency evaluation in subsurface irrigated wine grape vines. In: Autonomous air and ground sensing systems for agricultural optimization and phenotyping. International Society for Optics and Photonics; 2016. p 98660O.
101. Neale CM, Bausch WC, Heermann DF. Development of reflectance-based crop coefficients for corn. T Asae. 1990;32(6):1891–900.
102. Hunsaker DJ, Pinter PJ, Kimball BA. Wheat basal crop coefficients determined by normalized difference vegetation index. Irrig Sci. 2005;24(1):1–14.
103. Serrano L, González-Flor C, Gorchs G. Assessment of grape yield and composition using the reflectance based Water Index in Mediterranean rainfed vineyards. Remote Sens Environ. 2012;118:249–58.
104. Er-Raki S, Rodriguez J, Garatuza-Payan J, Watts C, Chehbouni A. Determination of crop evapotranspiration of table grapes in a semi-arid region of Northwest Mexico using multi-spectral vegetation index. Agric Water Manag. 2013;122:12–9.
105. Consoli S, Vanella D. Comparisons of satellite-based models for estimating evapotranspiration fluxes. J Hydrol. 2014;513:475–89.
106. Neale C, Ahmed R, Moran M, Pinter P, Qi J, Clarke T. Estimating seasonal cotton evapotranspiration using canopy reflectance. In: International evapotranspiration irrigation scheduling conference; 1996.
107. Tasumi M, Trezza R, Allen RG, Wright JL. Operational aspects of satellite-based energy balance models for irrigated crops in the semi-arid US. Irrig Drain Syst. 2005;19(3–4):355–76.

108. Bausch WC, Neale CM. Crop coefficients derived from reflected canopy radiation: a concept. T Asae. 1987;30(3):703–0709.
109. Duchemin B, Hadria R, Erraki S, Boulet G, Maisongrande P, Chehbouni A, Escadafal R, Ezzahar J, Hoedjes J, Kharrou M. Monitoring wheat phenology and irrigation in Central Morocco: On the use of relationships between evapotranspiration, crops coefficients, leaf area index and remotely-sensed vegetation indices. Agric Water Manag. 2006;79(1):1–27.
110. Mokhtari A, Ahmadi A, Daccache A, Drechsler K. Actual evapotranspiration from UAV images: a multi-sensor data fusion approach. Remote Sens-Basel. 2021;13(12):2315.
111. Poirier-Pocovi M, Volder A, Bailey BN. Modeling of reference temperatures for calculating crop water stress indices from infrared thermography. Agric Water Manage. 2020;233:106070.
112. Gutiérrez S, Diago MP, Fernández-Novales J, Tardaguila J. Vineyard water status assessment using on-the-go thermal imaging and machine learning. PLoS One. 2018;13(2):e0192037.
113. Walker GK, Hatfield JL. test of stress-degree-day concept using multiple planting dates of red kidney beans. Agron J. 1979;71:967–71.
114. Jackson RD, Pinter PJ Jr, Reginato RJ, Idso SB. Detection and evaluation of plant stress for crop management decisions. IEEE Trans Geosci Remote Sens. 1986;24(1):99–106.
115. Wiegand CL, Namken LN. Influences of plant moisture stress, solar radiation, and air temperature on cotton leaf temperature. Agron J. 1966;58:582–6.
116. Blad BL, Rosenberg NJ. Measurement of crop temperature by leaf thermocouple, infrared thermometry and remotely sensed thermal imagery. Agron J. 1975;65:635–41.
117. Hatfield JL. Measuring plant stress with an infrared thermometer. J Hortic Sci. 1990;25:1535–8.
118. Irmak S, Haman DZ, Bastug R. Determination of crop water stress index for irrigation timing and yield estimation of corn. Agron J. 2000;92(6):1221–7.
119. Fucs M, Tanner CB. Infrared thermometry of vegetation. Agron J. 1966;58:597–601.
120. Ehrler WL. Cotton leaf temperatures as related to soil water depletion and meteorological factors 1. Agron J. 1973;65(3):404–9.
121. Jackson RD, Idso SB, Reginato RJ. Remote sensing of crop canopy temperatures for scheduling irrigations and estimating yields. In: Proceedings of symposium on remote sensing of natural resources, Utah State University, Logan. UT; 1977.
122. Idso SB, Jackson RD, Pinter PJ Jr, Reginato RJ, Hatfeld JL. Normalizing the stress-degree-day parameter for environmental variability. Agric Meteorol. 1981;24:45–55.
123. Moran M, Clarke T, Inoue Y, Vidal A. Estimating crop water deficit using the relation between surface-air temperature and spectral vegetation index. Remote Sens Environ. 1994;49(3):246–63.
124. Prihodko L, Goward SN. Estimation of air temperature from remotely sensed surface observations. Remote Sens Environ. 1997;60(3):335–46.
125. Sandholt I, Rasmussen K, Andersen J. A simple interpretation of the surface temperature/vegetation index space for assessment of surface moisture status. Remote Sens Environ. 2002;79(2–3):213–24.
126. Wang C, Qi S, Niu Z, Wang J. Evaluating soil moisture status in China using the temperature–vegetation dryness index (TVDI). Can J Remote Sens. 2004;30(5):671–9.
127. Köksal E, İlbeyi A, Üstün H, Özcan H. Yeşil fasulye sulama suyu yönetiminde örtü sıcaklığı ve spektral yansıma oranı değerlerinin kullanım olanakları. Tagem Yayın No: Tagem-Bb-Topraksu-. 2007;29:26.
128. Köksal ES. Irrigation water management with water deficit index calculated based on oblique viewed surface temperature. Irrig Sci. 2008;2008(27):41–56. https://doi.org/10.1007/s00271-008-0120-5.
129. Taghvaeian S, Chávez JL, Hansen NC. Infrared thermometry to estimate crop water stress index and water use of irrigated maize in northeastern Colorado. Remote Sens. 2012;4:3619–37.
130. Taghvaeian S, Chávez JL, Bausch WC, DeJonge KC, Trout TJ. Minimizing instrumentation requirement for estimating crop water stress index and transpiration in maize. Irrig Sci. 2014;32:53–65.
131. Kullberg EG, DeJonge KC, Chávez JL. Evaluation of thermal remote sensing indices to estimate crop evapotranspiration coefficients. Agric Water Manag. 2017;179:64–73.
132. Kirnak H, Irik H, Unlukara A. Potential use of crop water stress index (CWSI) in irrigation scheduling of drip-irrigated seed pumpkin plants with different irrigation levels. Sci Horticul. 2019;256:108608.
133. Son NT, Chen CF, Chen CR, Chang LY, Minh VQ. Monitoring agricultural drought in the Lower Mekong Basin using MODIS NDVI and land surface temperature data. Int J Appl Earth Obs Geoinf. 2012;18:417–27.
134. Zhao T, Stark B, Chen Y, Ray A, Doll D. More reliable crop water stress quantification using small unmanned aerial systems (suas). IFAC-PapersOnLine. 2016;49(16):409–14.
135. Wu H, Xiong D, Liu B, Zhang S, Yuan Y, Fang Y, … Dahal NM. Spatio-temporal analysis of drought variability using CWSI in the Koshi River Basin (KRB). Int J Environ Res Public Health. 2019;16(17):3100.
136. Gowda PH, Chavez JL, Colaizzi PD, Evett SR, Howell TA, Tolk JA. ET mapping for agricultural

water management: present status and challenges. Irrig Sci. 2008;26(3):223–37.
137. Ding R, Kang S, Li F, Zhang Y, Tong L, Sun Q. Evaluating eddy covariance method by large-scale weighing lysimeter in a maize field of northwest China. Agric Water Manag. 2010;98(1):87–95.
138. Foken T, Aubinet M, Leuning R. The eddy covariance method. In: Eddy covariance. Springer; 2012. pp 1–19.
139. Irmak S, Skaggs KE, Chatterjee S. A review of the Bowen ratio surface energy balance method for quantifying evapotranspiration and other energy fluxes. T Asabe. 2014;57(6):1657–74.
140. Xue B-L, Wang L, Li X, Yang K, Chen D, Sun L. Evaluation of evapotranspiration estimates for two river basins on the Tibetan Plateau by a water balance method. J Hydrol. 2013;492:290–7.
141. Alam MS, Lamb DW, Rahman MM. A refined method for rapidly determining the relationship between canopy NDVI and the pasture evapotranspiration coefficient. Comput Electron Agric. 2018;147:12–7.
142. Genaidy M. Estimating of evapotranspiration using artificial neural network. MISR J Agric Eng. 2020;37(1):81–94.
143. Kelley J, Pardyjak ER. Using neural networks to estimate site-specific crop evapotranspiration with low-cost sensors. Agronomy. 2019;9(2):108.
144. Matese A, Toscano P, Di Gennaro SF, Genesio L, Vaccari FP, Primicerio J, Belli C, Zaldei A, Bianconi R, Gioli B. Intercomparison of UAV, aircraft and satellite remote sensing platforms for precision viticulture. Remote Sens-Basel. 2015;7(3):2971–90.
145. Courault D, Seguin B, Olioso A. Review on estimation of evapotranspiration from remote sensing data: from empirical to numerical modeling approaches. Irrig Drain Syst. 2005;19(3–4):223–49.
146. Bastiaanssen WGM, Pelgrum H, Wang J, Ma Y, Moreno JF, Roerink GJ, van der Wal T. A remote sensing surface energy balance algorithm for land (SEBAL) 1.Formulation. J Hydrol. 1998;212–213:213–29.
147. Monteith JL, Unsworth MH. Principles of environmental physics, 2nd edn; 1990. 291 p.
148. Hatfield JL, Reginato RJ, Idso SB. Evaluation of canopy temperature-evapotranspiration models over various crops. Agric Forest Meteorol. 1984;32:41–53.
149. Brown KW, Rosenberg NJ. A resistance model to predict evapotranspiration and its application to a sugar beet field. Agron J. 1973;65(3):341–7.
150. Jackson RD, Hatield JL, Reginato RJ, Idso SB, Pinter PJ Jr. Estimation of daily evapotranspiration from one time-of-day measurements. Agric Water Manage. 1983;7:51–362.
151. Bastiaanssen WGM, Pelgrum H, Wang J, Ma Y, Moreno JF, Roerink GJ, van der Wal T. A remote sensing surface energy balance algorithm for land (SEBAL) 2.Validation. J Hydrol. 1998;212–213:213–29.
152. Allen RG, Tasumi M, Morse A, Trezza R, Wright JL, Bastiaanssen W, Kramber W, Lorite I, Robison CW. Satellite-based energy balance for mapping evapotranspiration with internalized calibration (METRIC)—Applications. J Irrig Drain Eng. 2007;133(4):395–406.
153. Allen RG, Tasumi M, Trezza R. Satellite-based energy balance for mapping evapotranspiration with internalized calibration (METRIC)—Model. J Irrig Drain Eng. 2007;133(4):380–94.
154. Su Z. The Surface Energy Balance System (SEBS) for estimation of turbulent heat fluxes. Hydrol Earth Syst Sc. 2002;6(1):85–100.
155. Anderson MC, Norman JM, Mecikalski JR, Otkin JA, Kustas WP. A climatological study of evapotranspiration and moisture stress across the continental United States based on thermal remote sensing: 1. Model formulation. J Geophys Res Atmos. 2007;112(D10).
156. Norman JM, Kustas WP, Humes KS. Source approach for estimating soil and vegetation energy fluxes in observations of directional radiometric surface temperature. Agric For Meteorol. 1995;77(3–4):263–93.
157. Santos C, Lorite I, Tasumi M, Allen R, Fereres E. Integrating satellite-based evapotranspiration with simulation models for irrigation management at the scheme level. Irrig Sci. 2008;26(3):277–88.
158. Trezza R, Allen RG, Tasumi M. Estimation of actual evapotranspiration along the Middle Rio Grande of New Mexico using MODIS and landsat imagery with the METRIC model. Remote Sens-Basel. 2013;5(10):5397–423.
159. Elhaddad A, Garcia LA. Using a surface energy balance model (ReSET-Raster) to estimate seasonal crop water use for large agricultural areas: Case study of the Palo Verde irrigation district. J Irrig Drain Eng. 2014;140(10):05014006.
160. Ortega-Farías S, Ortega-Salazar S, Poblete T, Kilic A, Allen R, Poblete-Echeverría C, Ahumada-Orellana L, Zuñiga M, Sepúlveda D. Estimation of energy balance components over a drip-irrigated olive orchard using thermal and multispectral cameras placed on a helicopter-based unmanned aerial vehicle (UAV). Remote Sens-Basel. 2016;8(8):638.
161. Hoffmann H, Nieto H, Jensen R, Guzinski R, Zarco-Tejada P, Friborg T. Estimating evapotranspiration with thermal UAV data and two source energy balance models. Hydrol Earth Syst Sci Discus. 2015;12(8).
162. French AN, Hunsaker DJ, Thorp KR. Remote sensing of evapotranspiration over cotton using the TSEB and METRIC energy balance models. Remote Sens Environ. 2015;158:281–94.

163. Aboutalebi M, Torres-Rua AF, Kustas WP, Nieto H, Coopmans C, McKee M. Assessment of different methods for shadow detection in high-resolution optical imagery and evaluation of shadow impact on calculation of NDVI, and evapotranspiration. Irrig Sci. 2019;37(3):407–29.
164. Nassar A, Torres-Rua A, Kustas W, Nieto H, McKee M, Hipps L, Stevens D, Alfieri J, Prueger J, Alsina MM. Influence of model grid size on the estimation of surface fluxes using the two source energy balance model and suas imagery in Vineyards. Remote Sens-Basel. 2020;12(3):342.
165. Nassar A, Torres-Rua A, Kustas W, Alfieri J, Hipps L, Prueger J, Nieto H, Alsina MM, White W, McKee L. Assessing daily evapotranspiration methodologies from one-time-of-day sUAS and EC Information in the GRAPEX project. Remote Sens-Basel. 2021;13(15):2887.
166. Singh RK, Irmak A. Estimation of crop coefficients using satellite remote sensing. J Irrig Drain Eng. 2009;135(5):597–608.

21 Fruit Production in Brazil's Desert and Sustainability Aspects of Irrigated Family Farming Along the Lower-Middle Sao Francisco River: A Case Study

Heinrich Hagel, Daniela Gomez Rincon, and Reiner Doluschitz

Abstract

Population growth and changing dietary habits increase the worldwide food demand. Climate change effects and land degradation intensify the pressure on arable land. Resource efficient agricultural production is crucial to ensure food and nutrition security for the global population. Similar to many semiarid regions worldwide, water for irrigation is the limiting factor of agricultural production in Brazil's semiarid Northeast. Constant water supply favored the implementation of irrigation schemes along the São Francisco River since the 1960s to ensure food security, to decrease poverty, and to reduce rural exodus. However, increasing water demand due to expansion of irrigated agriculture and a water diversion project (Transposição), periods of limited water supply caused by droughts occurring in the high and upper part of the river catchment, and conflicts of interests between several stakeholder groups require highly efficient water use. We analyzed three existing agricultural production systems—extensive smallholder farming, small-scale irrigated agriculture, and middle-scale export oriented fruit and vegetable production—in regards to productivity, economic feasibility, and water use efficiency. Within this study, we conducted 60 expert interviews and 193 farm household interviews, and reviewed secondary data to perform an in-depth analysis of crop and livestock production and to identify the most promising approaches to increase water use efficiency along the lower-middle São Francisco River. Results show high vulnerability of irrigated agriculture towards climate change and market effects. Inefficient water use persisted in large areas of the study region. However, given adequate infrastructure and market access, irrigated agriculture generated adequate farm income. Education, agricultural extension, market access, and incentives to economize water consumption were identified as crucial for sustainable water use in the study area. The study area partly serves as an example to other semiarid regions to establish irrigated crop production to serve as nutritious food source and to provide adequate income for the local population.

Keywords

Caatinga · Systematic literature review · Econometrics · Water use efficiency

H. Hagel (✉) · R. Doluschitz
Food Security Center, Institute of Farm Management, University of Hohenheim, Stuttgart, Germany
e-mail: heinrich.hagel@uni-hohenheim.de

D. G. Rincon
Institute of Farm Management, University of Hohenheim, Stuttgart, Germany

M. Bahadir and A. Haarstrick (eds.), *Water and Wastewater Management*,
Water and Wastewater Management, https://doi.org/10.1007/978-3-030-95288-4_21

21.1 Introduction

In 2019, approximately 2 billion people around the world were affected by moderate or extreme food insecurity and lacked regular access to safe, nutritious, and sufficient food. Nearly 750 million people, which equals almost 10 percent of the global population, were extremely food insecure—all even before the Covid-19 pandemic [1]. Population growth and dietary changes increase the demand for food, whereas land degradation and negative climate change threaten global food production [2, 3]. In hand with unfavorable economic conditions, these trends are driving factors threatening not only the achievement of the Sustainable Development Goal (SDG) "Zero Hunger", but even increasing the number of people affected by hunger until 2030 [1, 4]. Increasing demand for food on the one side, and negatively affected production conditions on the other side, intensify the pressure on available arable land to feed the global population [4, 5]. At the same time, climate change negatively affects water availability in many regions around the world [6] and desertification risks will increase in areas all over the world [7]. Conflicts between different water users already exist worldwide and are projected to increase with ongoing climate change [8–10]. Given these developments, resource efficient irrigated agriculture is crucial to ensure sufficient production of healthy and nutritious food for the population in water scarce regions [8, 11].

The situation of irrigated family farming in Brazil's semi-arid northeast is a suitable representative example for this problem setting. Written records on severe droughts in this region date back to the country's colonization in the early seventeenth century [12]. Since the 1960s, the Brazilian government strengthened its efforts to promote irrigated family farming to ensure food security and to provide income opportunities in one of the country's most vulnerable regions, and therewith to reduce migration to bigger cities [13]. During that time, Brazil's military administration fostered the construction of large dams and reservoirs for hydropower to meet the growing electricity demand [13, 14]. Constant water availability and warm temperatures throughout the year enabled the implementation of irrigation schemes in the lower-middle São Francisco River basin. Production of mainly fruits and vegetables creates high agricultural value added and provides sufficient and nutritious food to the local population [16–18]. Irrigated production systems include extensive subsistence farming, fruit and vegetable production on small scales for local markets, but also advanced export-oriented horticulture on larger sales [16, 18, 19]. Until the 1980s, Brazilian policies concentrated on supporting modern irrigated agriculture. Although they considered and promoted small-scale family farming since the 1990s, modern high value fruit and vegetable production systems have better access to irrigation water, and production and commercialization infrastructure [21].

In the last decade, irrigated areas increased constantly [22], and the water diversion project *Transposição do rio São Francisco* even transfers water to dry regions in the north of the basin [23]. However, recent research forecasts decreased water availability due to negative climate change effects in this region [24]. This development will strongly affect irrigated agriculture, being the main water consumer in the country [25]. Given the planned expansion of irrigated areas and negative climate change effects in the region, highly efficient water management is necessary [26].

In this study, we evaluate the existing production systems in the lower-middle São Francisco river basin concerning (a) their income potentials for family farms, and (b) their water use efficiency. The combination of primary data and a systematic review of recent literature allows the assessment of achievements, challenges, and opportunities of irrigated family farming in Northeast Brazil. This study adds to literature combining most recent findings in the study area and may serve as an example for other semiarid and arid regions around the world.

21.2 Materials and Methods

The study concentrates on the lower-middle São Francisco River basin between the cities Petrolina and Petrolândia in Pernambuco state (Fig. 21.1). Temperatures are relatively constant at around 25 °C throughout the year [27]. Since 1935, annual precipitation ranged from 50 to 930 mm in Petrolândia (mean 366 mm) and from 127 to 1060 mm in Petrolina [28]. Irregularly occurring severe droughts occur mainly due to the El Niño and the La Niña phenomena [27, 28].

Traditional land use comprises substantial livestock and partially irrigated subsistence crop production, mainly along the river margins or on the edges of ephemeral reservoirs [15, 29]. With the construction of large dams for hydropower generation, irrigated family agriculture within irrigation schemes increased strongly since the early 1990s [13]. In the Petrolina region, large irrigation systems were planned to promote social and economic growth, using the most recent production technologies and concentrating on high-value perennial fruits for exportation [16]. Around 50% of the irrigable areas are accessible to private investors or enterprises, and the remaining 50% was set aside for smaller family farms [13]. This approach established the cities Petrolina in Pernambuco and Juazeiro on the Bahian side of the São Francisco as regional economic centers. In the Itaparica region, smaller irrigation units were established to compensate local farmers for flooded land within the Itaparica reservoir construction. Production in this region orientates rather on local markets and creates less economic revenue [16, 19].

To assess agricultural production systems, we conducted a random sample of 193 farm household interviews, of which

- 141 took place in the irrigation schemes Apolônio Sales, Barreiras, and Icó-Mandantes nearby the city Petrolândia,
- 21 in the Manga de Baixo irrigation scheme at the Western border of the Itaparica region,
- 30 with independently irrigating subsistence farmers around the cities Petrolândia and Itacuruba in the Itaparica region,
- and 22 with export oriented family farmers in the Nilo Coelho irrigation scheme in the Petrolina region (see [30]).

A team of one researcher and three former agricultural advisors conducted the interviews

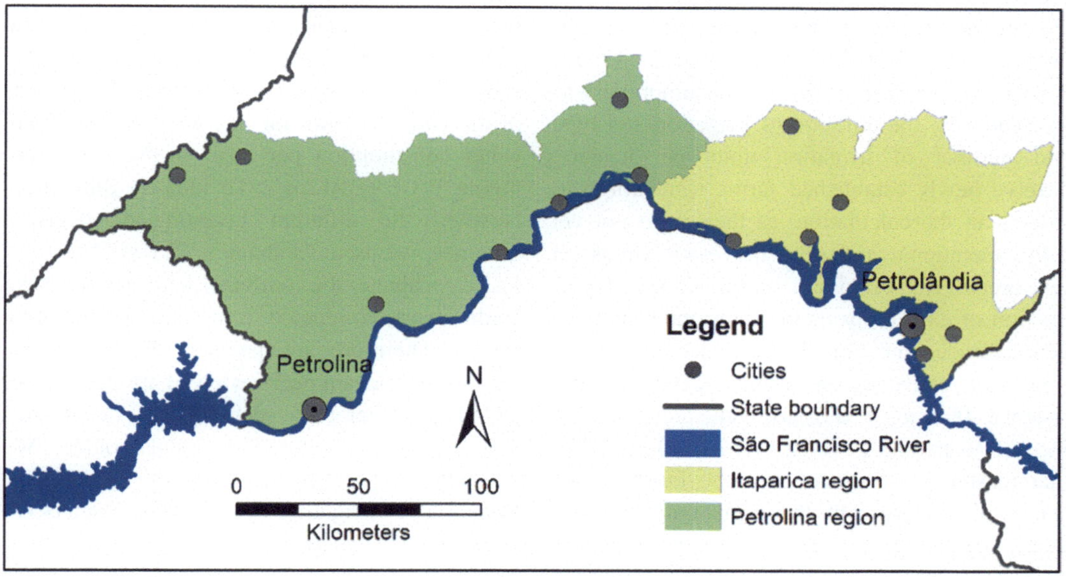

Fig. 21.1 The study region at the lower-middle São Francisco River [30]

from February until June 2013. In many cases, we collected data asking for farmers' most regularly used units (e.g., "number of harvested boxes per area") to obtain accurate findings and converted received data into computable units (e.g., "tons per hectare") after the interviews. We crosschecked collected data on viability together with the team of local agricultural advisors, but also with available reports on agricultural production from the irrigation schemes' operator *Companhia de Desenvolvimento dos Vales do São Francisco e do Parnaíba* (CODEVASF) and free accessible production data of the *Brazilian Corporation of Agricultural Research* (EMBRAPA) [31].

Detailed socio-economic analysis on the farm income within the irrigation schemes of the Itaparica region had been published [18]. In this study, we concentrate on farm productivity to evaluate assessed production systems regarding their water use efficiency (WUE). We used the concept of the ratio of output per water volume to calculate WUE [11]. We hereby distinguished by agricultural water productivity (yield achieved per water volume consumed) and economic WUE (CM per water volume consumed) [32]. We calculated the contribution margin (CM = sales revenues less variable costs) per hectare for each planted crop over a year to estimate the overall farm income. The currency for this calculation was Brazilian Reais (BRL) (2013: 1 BRL $\approx$ 0.5 USD). Water consumption was calculated by the duration of irrigation and type and number of irrigation units per hectare. Twelve newly established farms were not considered in the calculation, as they were not yet fully operational and only harvested yields on few areas [18]. Previous studies identified production of cash crops as the only opportunity to generate adequate farm income in crop production in the study region, whereas extensive subsistence farming could not generate sufficient income to escape poverty [18–20, 30, 33]. For that reason, we focused on these farms in the statistical analyses. In the sample of independently irrigating farmers, there were only six cash crop producers with modern irrigation infrastructure, so significant statistical analysis considering this group was not possible. We also removed the twelve newly established farms in the Barreiras irrigation scheme, as their fruit plantations had not yet provided yields [18]. Hence, we used production data of 107 established farms of irrigation schemes in the Itaparica region and 22 farms in the Petrolina region for the following analyses.

Contrary to the other two farm groups, farmers in the Nilo Coelho irrigation scheme nearby Petrolina paid water prices, which consisted of a fixed price per area for provision and maintenance of irrigation infrastructure, and a usage-bound water price. For that reason, farmers irrigated based on Software calculations for each crop's necessities on each lot to maximize their utility from consumed water [34]. We used these calculations based on official data from Brazilian authorities and invoices for water prices received from interviewed farmers to estimate their water consumption per crop. According to interviewed farmers and the former agricultural advisors of the team, we assumed irrigation efficiency of 70% for furrow irrigation, 80% for sprinkler, and 95% for micro-sprinkler and drip irrigation [34, 35]. We finally compared water consumption and WUE with official recommendations of the Brazilian National Water and Sanitation Agency (ANA) [36]. We conducted descriptive and statistical data analyses using the software IBM SPSS Statistics 27 for Windows 10. Levene's test showed heterogeneity of variances of the variables farm income per area (BRL/ha), water consumption per area (m^3/ha), and economic WUE (BRL/m^3). To identify differences between the different investigated irrigation schemes, we used Tamhane's T2 test [37].

To evaluate the findings with most recent studies, we conducted a systematic literature review. Therefore, we specified research criteria to identify relevant peer-reviewed articles and so-called gray literature, such as data, documents, and reports from international stakeholders. We used the keywords "agriculture" AND "São Francisco", "irrigation" AND "São Francisco", "Petrolina", "Itaparica", and "Petrolândia". Following these criteria, we screened the databases AgEcon Search, Agris FAO, CAB Abstracts, and

Scopus for recently published literature in the period from 2016 to 2021 in English and Portuguese language. In addition, we crosschecked the findings using Google Scholar. We screened over 3,000 documents and identified 38 relevant peer reviewed research papers, which we analyzed using content analysis [38].

Following the primary data analysis, literature review, and recent reports on irrigation potential and climate trends in the study region, we finally evaluated the potential of irrigated family farming as a sustainable system to ensure food and nutrition security and create value added in the study region.

21.3 Results and Discussion

In this section, we will first present the results of the farm household survey and describe and analyze the main productivity and WUE parameters of the typical farm types identified in the study region. Secondly, we will assess different levels of WUE, and thirdly, we will evaluate the findings with the results of the systematic review.

21.3.1 Agricultural Production Systems

Agricultural production systems differed between the three interviewed farm types [30]. In the following, we will present the main farm characteristics distinguished by farm type (Table 21.1). In the illustration, we concentrated on coconut, banana, and mango cultivation, as these crops were present in all farm types. Subsistence farming and annual crops did not exist in the Nilo Coelho irrigation scheme and did not provide adequate farm income in the other cases.

(I) The interviewed farms in the irrigation schemes in the Itaparica region differed by mean farm size and cultivated crops, but had similar access to irrigation water and infrastructure. Interviewed farmers in the irrigation schemes in the Itaparica region cultivated perennial fruits, mainly coconut, banana, and mango, and the annuals *Cucurbitaceae* and beans. Farmers in the irrigation scheme Apolônio Sales concentrated rather on the perennials banana and coconut, whereas annual crop production dominated in the irrigation schemes Barreiras and Icó-Mandantes. Few farmers held livestock by preference and as financial investments. A severe drought during the study period led to high producer prices, especially in the cases of beans, tomatoes, and onions. Three farmers of the Icó-Mandantes irrigation scheme benefited extremely from this, and they generated significantly higher profits than the other farmers of that sample [18]. Production in the irrigation scheme Manga de Baixo had nearly stopped due to lack of suitable soils for irrigation. Few farmers cultivated beans, coriander, and onions on small areas (max. 1.5 ha) using drip and sprinkler irrigation. To adapt to unprofitable crop production, 14 of the 21 interviewed farmers held livestock as an income alternative. As crop production in this irrigation scheme did not play a central role regarding value added and water consumption, we excluded this part of the sample from further calculations [30].

(II) Independently irrigating farms outside the official irrigation schemes were the most diverse farm group. Only six of the 30 interviewed farmers cultivated perennial cash crops, such as coconut, mango, and papaya close to the riverbank. Those farmers benefited from constant water availability of the São Francisco River, had reliable irrigation equipment, and possessed advanced knowledge on irrigated fruit production. The other 24 farmers grew mainly the annual crops beans, maize, onions, and forage grass. Production relied on rainfall, and irrigation was only used complementary during the driest periods of the cultivation period. Irrigation water was provided via small diesel pumps directly from the river. The income situation of small subsistence farmers was less stable, as lack of irrigation water caused frequent harvest losses. Within this sample, livestock presented a relevant income alternative, whereas interviewed farmers only sold animals as and when required. During the severe drought in the study period,

Table 21.1 Main characteristics of the interviewed farm households

	Itaparica system*	Independent irrigators	Nilo Coelho
Irrigable area (ha)	4.55 ± 2.32	7.25 ± 6.11	8.83 ± 3.18
Drylands (ha)	0.55 ± 5.93	5.70 ± 5.66	6.93 ± 12.82
Livestock (TLU)**	4.02 ± 8.80	5.89 ± 9.97	1.29 ± 2.89
Main crops	Beans, Banana, Cucurbitaceae, Coconut, (Mango)	Beans, Coconut, Cassava, Mango	Acerola, Coconut, Guava, Grape, Mango
Mean yields (kg/ha)			
Banana	9,394 ± 1,702	10,000	12,000 ± 1,633
Coconut	23,335 ± 6,544	11,667 ± 3,605	39,110 ± 4,891
Mango	16,625 ± 5,118	12,000 ± 2,645	23,867 ± 9,775
Mean producer prices (R$/kg)***			
Banana	0.998 ± 0.071	1.6	1.250 ± 0.289
Coconut	0.254 ± 0.049	0.219 ± 0.011	0.367 ± 0.137
Mango	0.515 ± 0.108	0.467 ± 0.076	0.833 ± 0.287
Main irrigation systems	Sprinkler	Furrow, sprinkler	Micro-sprinkler
Water demand (m^3/ha)			
Banana	13,010 (17,281) ****	13,010 (17,281) ****	(16,677) ****
Coconut	9,750 (15,768) ****	9,750 (17,281) ****	(15,263) ****
Mango	19,500 (12,204) ****	19,500 (17,281) ****	(12,961) ****

* Apolônio Sales, Barreiras, and Icó-Mandantes irrigation schemes
** The Tropical Livestock Unit (TLU) of a cow is 0.7, the TLU of a small ruminant is 0.1 [39]
*** (2013: 1 BRL ≈ 0.5 USD)
**** Calculation based on [34, 35]

livestock also played a more important role in water consumption, as irrigable areas were used for fodder production.

(III) All interviewed farmers in the Nilo Coelho irrigation scheme nearby the city Petrolina, were highly specialized high-value fruit producers. Annual crop production was not present in the sample. Knowledge on efficient production and availability of capital favored the utilization of more advanced production methods, such as soil analyses, site-adapted fertilization, and irrigation based on actual necessities on each irrigable area. These technologies, among others, led to higher yields than in the other farm types (Table 21.1). Producer prices were higher due to improved market access and the presence of commercialization infrastructure.

Water consumption in banana and coconut plantations in the Itaparica region was lower than the official recommendations by [35], and as in the Petrolina region. This may be a result of underestimated or understated information in the interviews. However, in the case of the high-value crop mango, water consumption exceeded the official recommendations. This higher water consumption is in line with personal observations in the field and may rather reflect the actual water consumption.

21.3.2 Water Use Efficiency (WUE)

In this section, we assess the relative profitability and WUE of agricultural production of the interviewed farms per cultivated area. Total farm

Table 21.2 Profits, water consumption, and economic water use efficiency of the investigated farm households

	Profit by area (BRL/ha)	Water consumption (m^3/ha)	Water Use Efficiency (BRL/m^3)
Apolônio Sales (n = 34)	3,374 ± 1,871	9,699 ± 2,070	0.3636 ± 0.2163
Barreiras (n = 27)	3,188 ± 2,179	10,745 ± 2,694	0.3217 ± 0.2184
Icó-Mandantes (n = 46)	3,646 ± 3,840	7,192 ± 3,914	0.6600 ± 0.6154
Nilo Coelho (n = 22)	14,642 ± 9,198	13,587 ± 1,081	1.0821 ± 0.6789

income in Apolônio Sales differed significantly between the one in Barreiras and Icó-Mandantes irrigation schemes. This was mainly due to larger farm sizes and crop choice, as analyzed in an earlier study [18]. Mean profit by area appeared significantly higher in the Nilo Coelho irrigation scheme, whereas it did not differ strongly between the irrigation schemes of the Itaparica region. Profitability, water consumption, and water use efficiency were highest in the Nilo Coelho irrigation scheme. High water use efficiency resulted from significantly higher profits, whereas higher profits resulted from both, higher yields and higher producer prices in the Petrolina region (Table 21.2).

Mean profit by area was similar within the irrigation schemes of the Itaparica region. However, due to larger irrigable areas, overall farm income was significantly higher in the Apolônio Sales irrigation scheme. High profits in the Barreiras and Icó-Mandantes irrigation schemes were due to drought-induced high producer prices of annual crops during the study period. Three farmers in the Icó-Mandantes irrigation scheme benefited disproportionately from high drought-induced producer prices during the study period. This reflects in the high standard deviations of profit and WUE in that group and may explain the high WUE.

Statistical analysis, as shown in Table 21.3, approves the findings of the descriptive statistics, with significantly higher WUE in the Nilo Coelho irrigation and in the Petrolina region compared with the ones in the Apolônio Sales and Barreiras irrigation schemes ($P < 0.01$). Higher WUE in the Petrolina irrigation scheme was due to significantly higher profits generated by both, higher yield levels and higher producer

Table 21.3 Mean differences of economic water use efficiency between the irrigation schemes (Tamhane-T2)

(I) Irr Scheme	(J) Irr Scheme	Mean Diff. (I − J)	Std. Error	P Value
Apolônio Sales	Barreiras	0.04186	0.05606	0.975
	Icó-Mandantes	−0.29641*	0.09803	0.022
	Nilo Coelho	−0.71858**	0.14942	0.000
Barreiras	Apolônio Sales	−0.04186	0.05606	0.975
	Icó-Mandantes	−0.33827**	0.10000	0.007
	Nilo Coelho	−0.76044**	0.15072	0.000
Icó-Mandantes	Apolônio Sales	0.29641*	0.09803	0.022
	Barreiras	0.33827**	0.10000	0.007
	Nilo Coelho	−0.42217	0.17084	0.104
Nilo Coelho	Apolônio Sales	0.71858**	0.14942	0.000
	Barreiras	0.76044**	0.15072	0.000
	Icó-Mandantes	0.42217	0.17084	0.104

* and ** indicate significance at the 0.05 level and the 0.01 level, respectively

prices. Although descriptive statistics indicate higher WUE in the Petrolina irrigation scheme compared with the one in Icó-Mandantes, there was no statistical evidence. However, given the low P value, this might be due to the particular situation with three wealthier farmers in the sample of Icó-Mandantes as mentioned earlier.

21.3.3 Systematic Literature Review and Discussion

We identified 38 recently published peer-reviewed scientific articles, to assess current developments regarding agricultural production and water availability in the study region. After screening all selected publications, we classified them into six primary topics (Table 21.4). Whenever several topics overlapped in one publication, we classified the publication to the most suitable topic. Most publications on agricultural production dealt with the Petrolina region, whereas climate and water management related publications targeted mostly the overall São Francisco river catchment.

Irrigation levels in mango production, as identified in the farm household survey, were in line with reference [40] and official recommendations [35]. In the Petrolina region, mango production is valuated as efficient and competitive on the global level [45]. Lower irrigation levels in the Petrolina region compared to the ones in the Itaparica region may be due to the absence of water prices in the Itaparica region. In addition, there exist positive effects of deficit irrigation, which has higher economic WUE in the study region due to reduced water consumption [41]. Similar findings exist in sugarcane and grape production [42–44]. Highest economic WUE was reported for mango and grape production, which dominate in the Petrolina region [45]. However, these crops require higher inputs in form of labor, knowledge, and capital.

During the interviews in the Petrolina region, farmers classified water costs as justified and fair. This is in line with calculations of the water's economic value, which exceeded water provision costs and water prices in the region [71]. Water saving potentials were identified in replacing micro sprinkler systems with drip irrigation [41], whereas local consultants classified both systems as similarly efficient.

Family farmers in Brazil's semiarid northeast are generally disadvantaged against the ones in larger irrigation schemes. Lack of knowledge, infrastructure, and production technologies are mainly due to lack of political focus on these production systems since the 1960s [21]. This applies also in the comparison between the Itaparica and Petrolina regions [30]. However, WUE in both regions was higher than in newly established crop production system along the diversion project (Transposição) in the north of the study region [47].

General recommendations towards more sustainable irrigated agriculture include education towards production methods to reduce negative health effects on farmers themselves by pesticide application [59], site-adapted crop and soil management [56, 57, 60], tools for adequate water allocation [71], and access to modern production technologies, capital, and markets [58].

Table 21.4 Overview on reviewed literature

Topic	Number of publications	References
Irrigated agriculture	10	[39–48]
Environmental impacts of agricultural production	9	[49–57]
Agricultural production conditions	6	[18, 21, 58–61]
Rural sociology	2	[62, 63]
Climate change impacts	5	[23, 64–67]
Water resource management	7	[68–73]

In 2017, irrigation accounted for 77% of the overall water outtake from the São Francisco River [74]. Climate change and related extreme weather events, and reduced precipitation together with increasing demand for electric energy are likely to affect water availability negatively in the river catchment. Extreme weather events, such as the severe drought from 2012–2015, may occur more frequently that heavily affect all water consumers [65–67, 76]. Whereas the diversion project (Transposição) did not significantly influence the hydrodynamics, expansion of agricultural areas will negatively affect the water availability along the lower São Francisco [23, 64]. Besides positive effects of irrigated agriculture, such as the provision of livelihood, food security, and even improved microclimate, negative environmental effects have to be considered [18, 49–54]. Good governance and inter- and transdisciplinary system thinking in the water management of the river catchment is crucial to sustainably benefit from the scarce resource water in Brazil's semiarid northeast [70–72].

21.4 Conclusions

During the severe droughts along the river catchment from 2011 to 2015 governmental poverty alleviation programs, efficient irrigation technologies, and income diversification, were successful adaptation strategies [75].

Given the high vulnerability of Brazil's semiarid northeastern region to negative climate effects, its high susceptibility to desertification and already occurred extreme droughts, several additional measures seem necessary to increase resilience and sustainability of irrigated agriculture in the study region [76–78]. Subsistence farmers, who have been disadvantaged against those in irrigation schemes, require intensive support from authorities and extension. Negative environmental effects of furrow irrigation, such as low WUE and erosion, exceed its benefits and threaten the farmers' livelihoods [79].

Still, local challenges appear similar to those in semiarid regions on the global level [80, 81]. Global competitiveness of some production systems in the study region and successful adaptation measures serve as a model for semiarid regions worldwide to provide livelihoods and availability of nutritious food to the rural population.

Acknowledgements Primary data collection was funded by the Federal Ministry of Education and Research (BMBF) within the project "INNOVATE" (01LL0904C). The authors thank all farmers who participated in the survey for their patience and willingness to share their private information.

References

1. FAO, Ifad, UNICEF, WFP and WHO The State of Food Security and Nutrition in the World. Transforming food systems for affordable healthy diets. Rome, Italy. 2020;2020:320.
2. United Nations World Population Prospects 2019: Data Booklet; Statistical Papers—United Nations (Ser. A), Population and Vital Statistics Report; UN; 2019. ISBN 978-92-1-004247-5.
3. Béné C, Prager SD, Achicanoy HAE, Toro PA, Lamotte L, Cedrez CB, Mapes BR. Understanding food systems drivers: a critical review of the literature. Glob Food Sec. 2019;23:149–59. https://doi.org/10.1016/j.gfs.2019.04.009.
4. Mc Carthy U, Uysal I, Badia-Melis R, Mercier S, O'Donnell C, Ktenioudaki A. Global food security—issues, challenges and technological solutions. Trends Food Sci Technol. 2018;77:11–20. https://doi.org/10.1016/j.tifs.2018.05.002.
5. Ehrlich PR, Ehrlich AH, Daily GC. Food security, population and environment. Popul Dev Rev. 1993;19:1–32. https://doi.org/10.2307/2938383.
6. Hulme M, Kelly M. Exploring the links between desertification and climate change. Environ Sci Policy Sustain Dev. 1993;35:4–45. https://doi.org/10.1080/00139157.1993.9929106.
7. Huang J, Zhang G, Zhang Y, Guan X, Wei Y, Guo R. Global desertification vulnerability to climate change and human activities. Land Degrad Dev. 2020;31:1380–91. https://doi.org/10.1002/ldr.3556.
8. Prosekov AY, Ivanova SA. Food security: the challenge of the present. Geoforum. 2018;91:73–7. https://doi.org/10.1016/j.geoforum.2018.02.030.
9. Gleick PH, Cooley H, Morikawa M, Morrison J, Cohen MJ. The World's water 2008–2009: the Biennial report on freshwater resources. Island Press; 2009. ISBN 978-1-59726-505-8.
10. Levy BS, Sidel VW. Water rights and water fights: preventing and resolving conflicts before they boil over. Am J Public Health. 2011;101:778–80. https://doi.org/10.2105/AJPH.2010.194670.

11. Howell TA. Enhancing water use efficiency in irrigated agriculture. Agron J. 2001;93:281–9. https://doi.org/10.2134/agronj2001.932281x.
12. Hall AL. Drought and irrigation in North-East Brazil.; Cambridge Univ. Press: Cambridge, UK; 1978. ISBN 0-521-21811-X.
13. Damiani O. Beyond market failures: irrigation, the state, and non-traditional agricultures in Northeast Brazil. Thesis, Massachusetts Institute of Technology; 1999.
14. Costa MMD, Cohen C, Schaeffer R. Social features of energy production and use in Brazil: goals for a sustainable energy future. Nat Res Forum. 2007;31:11–20. https://doi.org/10.1111/j.1477-8947.2007.00134.x.
15. Costa A. da Sustainable Dam Development in Brazil: Between global norms and local practices. German Development Institute/Deutsches Institut für Entwicklungspolitik (DIE); 2010.
16. Andrade MC. de O A terra e o homem no Nordeste: contribuição ao estudo da questão agrária no Nordeste; Cortez Editora; 2005. ISBN 978-85-249-1115-6.
17. Ferreira Irmão J, Hagel H, Hoffmann C, Amazonas AP, Flávio A. Macroeconomic aspects of the micro-regions São Francisco and Itaparica. In: Gunkel G, da Silva JA, do Sobral MC (eds) Editora Universitária, Universidade Federal de Pernambuco (UFPE): Recife; 2013. pp 245–64.
18. Hagel H, Hoffmann C, Irmão JF, Doluschitz R. Socio-economic aspects of irrigation agriculture as livelihood for rural families in Brazil's Semi-Arid Northeast. J Agric Rural Dev Tropics Subtropics (JARTS). 2019;120:157–69. https://doi.org/10.17170/kobra-20191127814.
19. Sietz D, Untied B, Walkenhorst O, Lüdeke MKB, Mertins G, Petschel-Held G, Schellnhuber HJ. Smallholder agriculture in Northeast Brazil: assessing heterogeneous human-environmental dynamics. Reg Environ Change. 2006;6:132–46. https://doi.org/10.1007/s10113-005-0010-9.
20. Untied B. Bewässerungslandwirtschaft Als Strategie Zur Kleinbäuerlichen Existenzsicherung in Nordost-Brasilien? Handlungsspielräume von Kleinbauern Am Mittellauf Des São Francisco. Marburg: Universität Marburg; 2005.
21. Brandão EAF, Rist S. The Agrarian Space of the Brazilian Semi-Arid Region: The dichotomies between the space of irrigated agriculture and the space of traditional agriculture. Stud Agric Econ. 2020;122:140–52. https://doi.org/10.7896/j.2082.
22. IBGE, Instituto Brasileiro de Geografia e Estatística Censo Agropecuário,. Agricultural Censos. Rio de Janeiro: Instituto Brasileiro de Geografia e Estatística; 2017. p. 2017.
23. dos Ferrarini ASF, de Ferreira Filho JBS, Cuadra SV, de Victoria DC. Water demand prospects for irrigation in the São Francisco River: Brazilian public policy. Water Policy. 2020;22:449–67. https://doi.org/10.2166/wp.2020.215.
24. Vieira RMSP, Tomasella J, Alvalá RCS, Sestini MF, Affonso AG, Rodriguez DA, Barbosa AA, Cunha APMA, Valles GF, Crepani E, et al. Identifying areas susceptible to desertification in the Brazilian Northeast. Solid Earth. 2015;6:347–60. https://doi.org/10.5194/se-6-347-2015.
25. Agência Nacional de Águas e Saneamento Básico (ANA) Conjuntura Dos Recursos Hídricos No Brasil 2019: Informe Anual; Brasilia, DF; 2019.
26. Krol M, Jaeger A, Bronstert A, Güntner A. Integrated modelling of climate, water, soil, agricultural and socio-economic processes: a general introduction of the methodology and some exemplary results from the Semi-Arid North-East of Brazil. J Hydrol. 2006;328:417–31. https://doi.org/10.1016/j.jhydrol.2005.12.021.
27. da Parahyba RBV, da Silva FHBB, de Araújo Filho JC, Lopes PRC, da Silva DF, Lima PC. de Diagnóstico Agroambiental do Município de Petrolândia—Estado de Pernambuco; Circular Técnica; Embrapa Solos: Rio de Janeiro; 2004. p 25.
28. Agência Pernambucana de Águas e Clima (APAC) Monitoramento Pluviométrico. Data of weather stations 5 and 49 from 1935 to 2020; 2021.
29. Hastenrath S, Heller L. Dynamics of climatic hazards in Northeast Brazil. Q J R Meteorol Soc. 1977;103:77–92. https://doi.org/10.1002/qj.49710343505.
30. Hagel H. Socio-economic benefits and limitations of irrigated family farming in Brazil's Semi-Arid Region. Dissertation, University of Hohenheim: Stuttgart-Hohenheim; 2017.
31. Empresa Brasileira de Pesquisa Agropecuária (EMBRAPA) Sistemas de Produção Embrapa.
32. Sharma BR, Molden DJ, Cook SE. Water use efficiency in agriculture: measurement, current situation and trends. In: Drechsel P, Heffer P, Magen H, Mikkelsen R, Wichelns D. (eds) Managing water and fertilizer for sustainable agricultural intensification. International Fertilizer Industry Association (IFA), International Water Management Institute (IWMI), International Plant Nutrition Institute (IPNI), International Potash Institute (IPI): Paris, France; 2015. pp 39–64.
33. Hagel H, Hoffmann C, Doluschitz R. Mathematical programming models to increase land and water use efficiency in Semi-Arid NE-Brazil. Int J Food Syst Dyn. 2014;5:173–81. https://doi.org/10.18461/ijfsd.v5i4.542.
34. Mantovani EC, Zinato CE, Simão FR. Manejo de irrigação e fertirrigação na cultura da goiabeira. In: Rozane DE, Couto FAA, editors. Cultura da goiabeira: tecnologia e mercado. Viçosa: Universidade Federal de Viçosa; 2003. p. 243–302.
35. Agência Nacional de Águas e Saneamento Básico (ANA) Atlas Irrigação. Uso Da Água Na Agricultura

Irrigada. Coeficientes Técnicos Para Agricultura Irrigada. https://app.powerbi.com/view?r=eyJrIjoiYWM0NDMzNmYtNTYxZC00ZThjLWIyYjctM2NlMDVjZTQxOWI3IiwidCI6ImUwYmI0MDEyLTgxMGItNDY5YS04YjRkLTY2N2ZjZDFiYWY4OCJ9.
36. Agência Nacional de Águas e Saneamento Básico (ANA) Atlas Irrigação. Uso Da Água Na Agricultura Irrigada; Brasilia, DF; 2021.
37. Tamhane AC. Multiple comparisons in model I one-way anova with unequal variances. Commun Stat Theory Methods. 1977;6:15–32. https://doi.org/10.1080/03610927708827466.
38. Mayring P. Qualitative Inhaltsanalyse. In: Mey G, Mruck K (eds) Handbuch Qualitative Forschung in der Psychologie. VS Verlag für Sozialwissenschaften: Wiesbaden; 2010. pp 601–13. ISBN 978-3-531-92052-8.
39. Chilonda P, Otte J. Indicators to monitor trends in livestock production at national, regional and international levels. Livestock Research for Rural Development; 2006:18.
40. de Simões WL, Andrade VPM, do Carmo Mouco MA, de Sousa JSC, de Lima JRF. Production and quality of Kent Mango (Mangifera Indica L.) under different irrigation systems in the Semi-Arid Northeastern Brazil. Revista em Agronegocio e Meio Ambiente. 2021;14. https://doi.org/10.17765/2176-9168.2021V14N2E7832.
41. de Simões WL, do Carmo Mouco MA, de Andrade VPM, Bezerra PP, Coelho EF. Fruit yield and quality of palmer mango trees under different irrigation systems. Comun Sci. 2020;11. https://doi.org/10.14295/CS.V11I0.3254.
42. Cardozo NP, de Oliveira Bordonal R, La Scala N. Sustainable intensification of sugarcane production under irrigation systems, considering climate interactions and agricultural efficiency. J Clean Prod. 2018;204:861–71. https://doi.org/10.1016/j.jclepro.2018.09.004.
43. Simões WL, Calgaro M, Guimarães MJM, Oliveira ARD, Pinheiro MPMA. Sugarcane crops with controlled water deficit in the Sub-Middle São Francisco Valley, Brazil. Revista Caatinga. 2018;31:963–71. https://doi.org/10.1590/1983-21252018v31n419rc.
44. De Andrade VPM, Da Silva Dias M, Da Silva JAB, De Sousa JSC, Simões WL. Yield and quality of Italia' grapes submitted to irrigation and fertilization control at the San Francisco Valley, Brazil. Comun Sci. 2016;7:175–82. https://doi.org/10.14295/CS.v7i2.1762.
45. Silva FB, Pereira SB, Martinez MA, da Silva DD, Vieira NPA. Water needs and equivalence relations for different irrigated crops in the São Francisco Basin. Revista Ciencia Agronomica. 2018;49:409–19. https://doi.org/10.5935/1806-6690.20180046.
46. Müller Carneiro J, Dias AF, Barros VS, Giongo V, Folegatti Matsuura MIS, Brito de Figueirêdo MC. Carbon and water footprints of Brazilian mango produced in the Semiarid Region. Int J Life Cycle Assess. 2019;24:735–52. https://doi.org/10.1007/s11367-018-1527-8.
47. De Moraes MGA, Carneiro ACG, Da Silva MPR. Technical coefficients of direct use of water in monetary terms for river basin districts in areas of irrigation for agriculture and urban water supply: the case of one of the recipients of the North Basin of the North Axis of Sao Francisco transboundary project. Engenharia Sanitaria e Ambiental. 2016;21:469–77. https://doi.org/10.1590/S1413-41522016116473.
48. Cavalcante ES, de Lacerda CF, Costa RNT, Gheyi HR, Pinho LL, Bezerra FMS, de Oliveira AC, Canjá JF. Supplemental irrigation using brackish water on maize in Tropical Semi-Arid Regions of Brazil: yield and economic analysis. Sci Agricola. 2021;78. https://doi.org/10.1590/1678-992X-2020-0151.
49. Cabral Júnior JB, Silva CMS, de Almeida HA, Bezerra BG, Spyrides MHC. Detecting linear trend of reference evapotranspiration in irrigated farming areas in Brazil's semiarid region. Theo Appl Climatol. 2019;138:215–25. https://doi.org/10.1007/s00704-019-02816-w.
50. Schulz C, Koch R, Cierjacks A, Kleinschmit B. Land change and loss of landscape diversity at the Caatinga Phytogeographical domain—analysis of pattern-process relationships with MODIS land cover products (2001–2012). J Arid Environ. 2017;136:54–74. https://doi.org/10.1016/j.jaridenv.2016.10.004.
51. Salazar AA, Arellano EC, Muñoz-sáez A, Miranda MD, da Silva FO, Zielonka NB, Crowther LP, Silva-ferreira V, Oliveira-reboucas P, Dicks LV. Restoration and conservation of priority areas of Caatinga's semi-arid forest remnants can support connectivity within an agricultural landscape. Land. 2021;10. https://doi.org/10.3390/land10060550.
52. Schulz K, Guschal M, Kowarik I, Silva de Almeida-Cortez J, Valadares de Sá Barreto Sampaio E, Cierjacks A. Grazing reduces plant species diversity of Caatinga dry forests in Northeastern Brazil. Appl Veg Sci. 2019;22:348–59. https://doi.org/10.1111/avsc.12434.
53. Selge F, Matta E, Hinkelmann R, Gunkel G. Nutrient load concept-reservoir vs. bay impacts: a case study from a semi-arid watershed. Water Sci Technol. 2016;74:1671–9. https://doi.org/10.2166/wst.2016.342.
54. Gunkel G, Selge F, Keitel J, Lima D, Calado S, Sobral M, Rodriguez M, Matta E, Hinkelmann R, Casper P, et al. Water Management and aquatic ecosystem services of a tropical reservoir (Itaparica, São Francisco, Brazil). Reg Environ Change. 2018;18:1913–25. https://doi.org/10.1007/s10113-018-1324-8.
55. da Silva JLA, do Araújo MSB, de Sampaio EVSB, Ludke JV, Primo DC. Management of sludge from fish ponds at the edge of the Itaparica Reservoir (Brazil): an alternative to improve agricultural production. Reg Environ Change. 2018;18:1999–2004.
56. Dos Santos LR, Lima AMN, Rodrigues MS, Cunha JC, Dos Santos LPA, Soares EMB, Da

Silva AVL, De Souza IM. Does the irrigated mango cultivation in the semiarid change the physical and chemical attributes of the soil? Comun Sci. 2019;10:402–14. https://doi.org/10.14295/cs.v10i3.2966.
57. dos Santos LR, Nascimento Lima AM, Cunha JC, Rodrigues MS, Barros Soares EM, dos Santos LPA; Lopes da Silva AV, Ferreira Fontes MP. Does irrigated mango cultivation alter organic carbon stocks under fragile soils in semiarid climate? Sci Horticul. 2019;255:121–7. https://doi.org/10.1016/j.scienta.2019.05.015.
58. Goulart DF, de Almeida RP, Resende KC, da Costa FAM, Bezerra JRC. O Desafio Da Estruturação Da Cadeia Produtiva Do Amendoim No Semiárido Do Nordeste. Organizações Rurais e Agroindustriais/Rural and Agro-Industrial Organizations. 2017;19:47–59. https://doi.org/10.22004/ag.econ.265530.
59. Corcino CO, De Andrade Teles RB, Da Silva Almeida JRG, Da Silva Lirani L, Araújo CRM, De Assis Gonsalves A, De Azevedo Maia GL. Evaluation of the effect of pesticide use on the health of rural workers in irrigated fruit farming. Ciencia e Saude Coletiva. 2019;24:3117–28. https://doi.org/10.1590/1413-81232018248.14422017.
60. Cardoso JAF, Lima AMN, Cunha TJF, Rodrigues MS, Hernani LC, Cunha JC, Do Amaral AJ, De Oliveira Neto MB. Changing in chemical and physical attributes of a sandy soil under irrigated mango cultivation in Semiarid Region. Comun Sci. 2017;8:404–13. https://doi.org/10.14295/CS.v8i3.2544.
61. Cierjacks A, Pommeranz M, Schulz K, Almeida-Cortez J. Is crop yield related to weed species diversity and biomass in coconut and banana fields of Northeastern Brazil? Agr Ecosyst Environ. 2016;220:175–83.
62. Da Paz MDM, Vieira DD. Agribusiness in Familial Agriculture of Mangoes in the Maniçoba District, Juazeiro, Brazil. Revista em Agronegocio e Meio Ambiente. 2017;10:33–49. https://doi.org/10.17765/2176-9168.2017v10nEd.esp.p33-49.
63. Gominho KC, Carneiro HF. Old Petrolândia: memories of a city lost in the semiarid of Pernambuco. Desenvolvimento e Meio Ambiente. 2020;55:262–79. https://doi.org/10.5380/dma.v55i0.73278.
64. Matta E, Koch H, Selge F, Simshäuser MN, Rossiter K, da Silva GMN, Gunkel G, Hinkelmann R. Modeling the impacts of climate extremes and multiple water uses to support water management in the Icó-Mandantes Bay, Northeast Brazil. J Water Clim Change. 2019;10:893–906. https://doi.org/10.2166/wcc.2018.254.
65. Da Silva MVM, Da Silva Silveira C, Da Costa JMF, Martins ESPR, Vasconcelos FDC. Projection of climate change and consumptive demands projections impacts on hydropower generation in the Sao Francisco River Basin, Brazil. Water (Switzerland); 2021. p 13. https://doi.org/10.3390/w13030332.
66. Lucas MC, Kublik N, Rodrigues DBB, Meira Neto AA, Almagro A, Melo DCD, Zipper SC, Oliveira PTS. Significant baseflow reduction in the Sao Francisco River Basin. Water (Switzerland). 2021:13. https://doi.org/10.3390/w13010002.
67. Marengo JA, Tomasella J, Nobre CA. Climate change and water resources. In: Galli CS, Abe DS (eds) Waters of Brazil: strategic analysis; 2016. pp 171–86.
68. Alcoforado de Moraes MMG, Biewald A, Carneiro ACG, Souza da Silva GN, Popp A, Lotze-Campen H. The impact of global change on economic values of water for public irrigation schemes at the São Francisco River Basin in Brazil. Reg Environ Change. 2018;18:1943–55. https://doi.org/10.1007/s10113-018-1291-0.
69. Alencar KM, Moreira MC, Silva DDD. Cost of charging for water use in the Brazilian Cerrado Hydrographic Basin. Revista Ambiente e Agua. 2018;13. https://doi.org/10.4136/ambi-agua.2238.
70. Siegmund-Schultze M, Köppel J, Sobral MC. Unraveling the water and land nexus through inter- and transdisciplinary research: sustainable land management in a semi-arid watershed in Brazil's Northeast. Reg Environ Change. 2018;18:2005–17. https://doi.org/10.1007/s10113-018-1302-1.
71. Souza da Silva GN, de Moraes MMGA. Economic water management decisions: trade-offs between conflicting objectives in the Sub-Middle Region of the São Francisco Watershed. Reg Environ Change. 2018;18:1957–67. https://doi.org/10.1007/s10113-018-1319-5.
72. Rodorff V, Siegmund-Schultze M, Guschal M, Hölzl S, Köppel J. Good governance: a framework for implementing sustainable land management, applied to an agricultural case in Northeast-Brazil. Sustainability (Switzerland). 2019;11. https://doi.org/10.3390/su11164303.
73. Marques ÉT, Gunkel G, Sobral MC. Management of tropical river basins and reservoirs under water stress: experiences from Northeast Brazil. Environments MDPI. 2019;6. https://doi.org/10.3390/environments6060062.
74. Agência Nacional de Águas (ANA) Manual de Usos Consuntivos Da Água No Brasil; Brasilia, DF; 2019. p 75.
75. Herwehe L, Scott CA. Drought adaptation and development: small-scale irrigated agriculture in Northeast Brazil. Climate Dev. 2018;10:337–46. https://doi.org/10.1080/17565529.2017.1301862.
76. Guimarães DP, Landau EC, de Sousa DL. Agricultura irrigada e estiagem na Bacia do Rio São Francisco. In: Proceedings of the Simpósio regional de geoprocessamento e sensoriamento remoto; Embrapa: Aracaju, Brasil, vol 7; 2014. pp 17–21.
77. Marengo JA, Torres RR, Alves LM. Drought in Northeast Brazil—past, present, and future. Theoret Appl Climatol. 2017;129:1189–200. https://doi.org/10.1007/s00704-016-1840-8.

78. do Santos A, Oliveira Lopes PM, da Silva MV, Maniçoba da Rosa Ferraz Jardim A, Barbosa de Albuquerque Moura G, Siqueira Tavares Fernandes G, de Oliveira Silva DA, Bezerra da Silva JL, de Moraes Rodrigues JA. Araújo Silva E et al. Causes and consequences of seasonal changes in the water flow of the São Francisco River in the Semiarid of Brazil. Environ Sustain Indic. 2020;8:100084. https://doi.org/10.1016/j.indic.2020.100084.
79. Araujo DF, Costa RN, Mateos L. Pros and cons of furrow irrigation on smallholdings in Northeast Brazil. Agric Water Manag. 2019;221:25–33. https://doi.org/10.1016/j.agwat.2019.04.029.
80. Wheeler SA, Zuo A, Bjornlund H, Mdemu MV, Rooyen A, van Munguambe P. An overview of extension use in irrigated agriculture and case studies in South-Eastern Africa. Int J Water Resour Dev. 2017;33:755–69. https://doi.org/10.1080/07900627.2016.1225570.
81. Koech R, Langat P. Improving irrigation water use efficiency: a review of advances, challenges and opportunities in the Australian context. Water. 2018;10:1771. https://doi.org/10.3390/w10121771.

The manufacturer's authorised representative in the EU is Springer Nature Customer Service Centre GmbH, Europaplatz 3, 69115 Heidelberg, Germany. If you have any concerns regarding our products, please contact ProductSafety@springernature.com

Printed and bound by CPI Group (UK) Ltd, Croydon, CR0 4YY
30/09/2026
02225496-0018